Tropical Tuber Starches

Tropical Tuber Starches

Structural and Functional Characteristics

S.N. Moorthy

Formerly Principal Scientist, CTCRI, India

M.S. Sajeev

Principal Scientist, CTCRI, India

R.P.K. Ambrose

Assistant Professor, Center for Particulate Products and Processes, Purdue University, USA

and

R.J. Anish

Research Scholar, Department of Biochemistry, Kerala University, Trivandrum, India

CABI

CABI is a trading name of CAB International

CABI
Nosworthy Way
Wallingford
Oxfordshire OX10 8DE
UK

Tel: +44 (0)1491 832111
Fax: +44 (0)1491 833508
E-mail: info@cabi.org
Website: www.cabi.org

CABI
745 Atlantic Avenue
8th Floor
Boston, MA 02111
USA

T: +1 (617)682-9015
E-mail: cabi-nao@cabi.org

British Library Cataloguing-in-Publication Data

A catalogue record for this book is available from the British Library.

Library of Congress Cataloging-in-Publication Data

Names: Moorthy, S. N., author. | Sajeev, M. S., author. | Ambrose, R. P., author. | Anish, R. J., author.
Title: Tropical tuber starches : structural and functional characteristics / S. N. Moorthy, M. S. Sajeev, R. P. K. Ambrose, and R. J. Anish.
Description: Boston, MA : CABI Publishing, 2021. | Includes bibliographical references and index.
Identifiers: LCCN 2019980351 (print) | LCCN 2019016380 (ebook) | ISBN 9781786394811 (hardback) | ISBN 9781786394835 (epub) | ISBN 9781786394828 (pdf) | ISBN 9781786394828 (pdf) | ISBN 9781786394835 (epub) | ISBN 9781786394811 (hardback)
Subjects: LCSH: Tuber crops--Research--Tropics. | Starch--Synthesis--Research--Tropics.
Classification: LCC SB210.T73 M66 2021 (ebook) | LCC SB210.T73 (print) | DDC 633.6/8--dc23
LC record available at https://lccn.loc.gov/2019980351

ISBN: 978 1 78639 481 1 (hardback)
978 1 78639 482 8 (ePDF)
978 1 78639 483 5 (ePub)

Commissioning Editors: Rachael Russell and Rebecca Stubbs
Editorial Assistant: Tabitha Jay
Production Editor: Shankari Wilford

Typeset by Exeter Premedia Services Pvt Ltd, Chennai, India
Printed and bound in the UK by Severn, Gloucester

Contents

1

Introduction – Tropical Tuber Crops and Their Importance

Food security and climate change are among the top themes included in the present-day development agenda (Nayar, 2014). Tuber crops are the second most important group of crop plants after cereals (Table 1.1), but a wide gap exists between them.

In fact, the importance of cassava can be gauged from a comment by Rosling who referred to it as the 'Cinderella of the poor' because its rapid spread in Africa led originally to an improvement in agricultural productivity and prevented potential famine in some areas (Rosling, 1987). While cereals contribute 49% energy towards global food security, the share of tuber crops is very meagre (5.4%) (Table 1.2).

Presently, tropical tuber crops (TTC) are being consumed as staple foods, as secondary staple foods, and as vegetables, mainly in the tropics. The main TTCs are listed below: (Kay, 1987).

1. Arracacha, Peruvian carrot, Peruvian parsnip

Botanical name	*Arracacia xanthorrhiza*
Family	Umbelliferae
Other names	Apio, arrecate, Batata baroa
Botany	Stout semi-caulescent herb
Origin and distribution	Origin in Peru, Ecuador, Brazil, Mexico
Growth period	Secondary tubers mature in 10–14 months after planting
Primary product	Tubers with delicate flavour, crisp texture, white creamy yellow or purple flesh
Yield	3–18 t/ha depending upon growing conditions

2. Cassava, Manihot, Manioc, Tapioca

Botanical name	*Manihot esculenta*
Family	Euphorbiaceae
Other names	Aipi, Brazilian arrowroot
Botany	Perennial shrub with latex in all its parts which produces enlarged tuberous roots.
Origin and distribution	West Africa, Brazil, Indonesia, Thailand, Peru, Mexico
Growth period	9–24 months
Primary product	Dark fleshy, cylindrical tubers
Yield	30–40 t/ha

Table 1.1. Food balance sheets of tuber crops, 1961 and 2009 (FAO, 2009).

Particulars	All tuber crops	All cereals	Potato	Cassava	Sweet potato	Yam	Roots, other
Production (Mt)	445.2/720.4	779.0/2251.6	270.4/331.4	62.4/219.8	97.1/101.8	8.0/47.0	6.6/20.3
Feed (000 Mt)	99.3/163.8	288.4/746.2	70.9/37.7	16.3/79.0	10.4/39.8	0.9/6.6	0.8/0.7
Processing (000 Mt)	79.8/62.2	31.6/95.2	24.2/25.3	6.0/22.5	5.3/6.6	0.9/5.2	0.8/2.7
Food (000 Mt)	407.5/406.9	590.5/976.7	108.9/219.3	37.6/95.0	80.3/55.4	46/25.6	4.7/13.6
Food supply (kg capita^{-1} year^{-1})	77.4/61.1	128.0/146.7	35.7/32.6	12.3/14.3	26.3/8.3	1.5/3.8	1.5/2.0
Food supply (kg capita^{-1} day^{-1})	176/136	1078/1292	65/61	32/37	71/22	4/11	4/6

The first figure in each case is for 1961 and the second is for 2009.

Table 1.2. Contribution of tuber crops to food and nutrition, 2009. (Source: FAO, 2009; 'contributions' calculated by the author.)

Item	Food supply (kg capita^{-1} year^{-1})	Energy (kcals capita^{-1} day^{-1})	Contribution (%)
Tuber crops	61.1	136	5.4
Cassava	14.3	37	1.9
Sweet potatoes	8.3	22	1.5
Yams	3.8	11	0.3
Roots (yautia, tannia)	2.0	6	trace
Cereals	146.7	1,292	49.0
Rice	52.9	532	20.5

Others items' contribution: oils = 12.5%; meat and eggs = 9.5%; sugars = 9.0%; fruits = 3.0%.

3. Chinese water chestnut, matai, waternut

Botanical name	*Eleocharis dulcis*
Family	Cyperaceae
Other names	Apulid, buslig, potok
Botany	Variable, annual, stout, tufted, aquatic, sedge plant
Origin and distribution	Grows wild in India and south-east Asia
Growth period	Maturity in 7–8 months
Primary product	Corms
Yield and starch content	20–40 t/ha

4. Sweet potato

Botanical name	*Ipomoea batatas*
Family	Convolvulaceae
Other names	Apichu, Skurar-kanda
Botany	Herbaceous perennial vine
Origin and distribution	South and central America
Growth period	3–8 months
Primary product	Starchy tuber
Yield	8–30 t/ha

5. Taro

Botanical name	*Colocasia esculenta*
Family	Araceae
Other names	Abalong, arvi
Botany	Small herbaceous plant
Origin and distribution	Most parts of the tropics; very important in the Pacific regions
Growth period	8–10 months
Primary product	Corms and cormels
Yield and starch content	5–10 t/ha; 12–20%

6. Tannia

Botanical name	*Xanthosoma sagittifolium*
Family	Araceae
Other names	Badoo, Chou
Botany	Large herbaceous plant
Origin and distribution	Tropical central and south America
Growth period	9–12 months
Primary product	Cormels
Yield and starch content	20–35 t/ha; 20%

7. Elephant foot yam

Botanical name	*Amorphophallus paeoniifolius*
Family	Araceae
Other names	Anto, Telinga potato
Botany	Large perennial herb
Origin and distribution	Tropical Asia and Africa
Growth period	8–12 months
Primary product	Corms
Yield and starch content	20 t/ha; approx. 20%

8. Yams

Botanical name	*Dioscorea* sp.
Family	Dioscoreaceae
Other names	Cara, ighnam
Botany	Subterranean tubers or rhizomes

Botanical name	*Dioscorea* sp.
Origin and distribution	West Africa, Caribbean and tropical America
Growth period	8–12 months
Primary product	Tubers
Yield	12.5–25 t/ha

9. African yam bean

Botanical name	*Sphenostylis stenocarpa*
Family	Leguminosae
Other names	Akitereku, yam pea
Botany	Herbaceous climbing vine, reaching 1.5–2 m in height.
Origin and distribution	Ethiopia, Africa
Growth period	5–10 months
Primary product	Tubers
Yield and starch content	3461–3872 kg/ha; 25%

10. East Indian arrowroot

Botanical name	*Tacca leontopetaloides*
Family	Taccaceae
Other names	Kabitsa, makmok
Botany	Perennial herb with a tuberous rhizome producing large and small tubers
Origin and distribution	South-east Asia
Growth period	8–10 months
Primary product	Tuberous rhizomes
Starch content	20–30%

11. Giant taro

Botanical name	*Alocasia macrorrhizos*
Family	Araceae
Other names	Alavu, Boro
Botany	Tall succulent herbaceous plant; grows 4.5 m in height
Origin and distribution	Asia and South America
Growth period	12–18 months
Primary product	Fleshy aerial stem
Starch content	17–25%

12. Coleus

Botanical name	*Plectranthus rotundifolius, Solenostemon rotundifolius*
Family	Lamiaceae
Other names	Chinese potato, Hausa potato
Botany	Small herbaceous annual, grows 15–30 cm in height
Origin and distribution	West Africa, south Africa, Madagascar, south and south-east Asia
Growth period	6 months
Primary product	Adventitious roots modified to edible tubers
Yield and starch content	7–15 t/ha; 20–24%

13. Kudzu

Botanical name	*Pueraria lobata*
Family	Leguminoseae
Other names	Arrowroot vine, magnagna
Botany	Small perennial twining herb or shrub with elongated tuberous roots
Origin and distribution	China, Japan, east Asia
Growth period	2 years
Primary product	Root tubers
Yield and starch content	5–7 t/ha; 20%

14. Lotus root

Botanical name	*Nelumbo nucifera*
Family	Nymphaeceae
Other names	Ambuj, kamal
Botany	Perennial aquatic herb
Origin and distribution	South and south-east Asia, Africa and Australia
Growth period	6–9 months
Primary product	White globulous rhizomes
Yield and starch content	5 t/ha; 18%

15. Oca

Botanical name	*Oxalis tuberose*
Family	Oxalidaceae
Other names	Apioblanco, quiba
Botany	Small compact annual tuberous herb; grows 20–30 cm in height
Origin and distribution	Andes and south Colombia
Growth period	8 months
Primary product	Rhizomatous tubers
Yield and starch content	4–5 t/ha; 12%

16. Queensland arrowroot (*Canna* sp.)

Botanical name	*Canna indica*
Family	Cannaceae
Other names	Achera, purple arrowroot
Botany	Perennial, herbaceous monocotyledon
Origin and distribution	Andean region of South America
Growth period	6–8 months
Primary product	Rhizomes
Yield and starch content	25–50 t/ha; 24–30%

17. Shoti

Botanical name	*Curcuma zedoaria*
Family	Zingiberaceae
Other names	Indian arrowroot
Botany	Perennial with fleshy, branching rhizomes
Origin and distribution	India, south-east Asia
Growth period	10 months
Primary product	Rhizomes
Yield and starch content	8–12 t/ha; 12–15%

18. Swamp taro

Botanical name	*Cyrtosperma merkusii*
Family	Araceae
Other names	Paluka, puraka
Botany	Giant herbaceous perennial, grows 3–4 m in height, with very large leaves
Origin and distribution	Indonesia, Philippines
Growth period	2–3 years
Primary product	Corms
Yield and starch content	7–10 t/ha; 28–30%

19. Winged bean

Botanical name	*Psophocarpus tetragonolobus*
Family	Fabaceae
Other names	Goa bean, dragon bean
Botany	Leguminous, climbing perennial
Origin and distribution	Asia, Africa
Growth period	5–8 months
Primary product	Root tubers
Yield and starch content	2–6 t/ha; 20%

20. Buffalo gourd

Botanical name	*Cucurbita foetidissima*
Family	Cucurbitaceae
Other names	Stinking gourd, wild gourd, wild pumpkin

Botanical name	*Cucurbita foetidissima*
Botany	Herbaceous perennial vine, bearing 1–30 stems, usually 1–5 m long and 1.5 cm girth, from a perennial taproot
Origin and distribution	North America, central America and Mexico
Growth period	
Primary product	Fleshy tap roots
Yield and starch content	15%

Out of the total contribution of 5.4% energy from all the tuber crops (from annual consumption of 61.1 kg capita^{-1} year^{-1}), cassava contributes 1.9% energy (from 14.3 kg capita^{-1} year^{-1}); sweet potato, 1.5% (from 8.3 kg capita^{-1} year^{-1}); yam 0.3% (from 3.0 kg capita^{-1} year^{-1}); and aroids 0.2% (from 2 kg capita^{-1} year^{-1}) (Table 1.3).

The area, production and productivity of different edible TTCs are given in Table 1.3.

The Consultative Group on International Agricultural Research (CGIAR) facilitated a study conducted jointly by the International Food Policy Research Institute (IFPRI), Washington, DC, and the International Potato Centre, Lima, Peru (CIP) in 2000 to determine the present contribution of TCs to food security and to draw up a plan of action for 2020. The results indicated that in 2000, TCs contributed 5% energy towards food security and the study recommended that this might be enhanced to 6% by 2020 (IFPRI, 2000). This target has not been achieved. Table 1.4 gives the number of countries/entities in which TCs constitute a staple or major crop in 2009, compared to 1961. The data show that in the case of TCs, the number of countries where they form a staple or secondary staple has been slowly and steadily reducing over the period.

Table 1.3. Area, production and yield of tuber crops, 2016 (FAOSTAT 2016).

Crop	Area (10^6 ha)	Production (10^8 t)	Yield (t ha^{-1})
Cassava	23.4	2.77	11.8
Sweet potatoes	8.6	1.05	12.2
Yams	7.5	0.65	8.67
Taro (Cocoyam)	1.67	0.10	6.06

Table 1.4. Number of countries where tuber crops form a staple or major food, in 1961 and 2009 (FAO, 2009).

	Number of countries/entities Total: 172/236	
Commodity	Staple[a] (1961/2009)	Major[b] (1961/2009)
Cassava	18/19	14/9
Sweet potato	5/2	4/5
Yam	4/4	6/6
Roots, other (taro & cocoyam)	6/4	11/5
Tropical tuber crops	33/29	35/25
Potatoes	9/8	22/47

Note: The first figure in each case is for 1961 and the second is for 2009.
[a]Staple: when the daily consumption is 100 or more kcal head^{-1}
[b]Major: when the daily consumption is <100 kcal head^{-1}

The UN (2012) defined that food security 'exists when all people at all times have both physical and economic access to the basic food they need'. TTCs constitute staple food in three major regions of the world: tropical Africa, Oceania and the Caribbean islands. In the last five decades, another observed phenomenon is that as the income levels go up, the populations consume fewer vegetarian and more of non-vegetarian items. Further, even among the vegetarian diets, the shift in food preference is from 'inferior' vegetal items like millets and TTCs, to wheat, rice, and potato. This trend is affecting the growth of TTCs.

The world's population is expected to rise to 9.4 billion by 2050 from the present 7.2 billion (UN, 2012). Most of the population increase is taking place in the developing world. According to UN estimates, as many as one in seven – more than one billion people – is going to bed hungry each day. Almost all are in developing countries, mostly sub-Saharan Africa and south Asia. In recent years, China has made remarkable progress in eliminating the number of undernourished people. Elimination of undernourishment is one of the priority aspects of the UN's Millennium Development Goals (MDGs); but here another disturbing aspect has been observed – about a third of all food produced is lost globally – 1.3 billion tons approximately (FAO, 2011). This varies from 19 to 36% in different regions of the world (Fig. 1.1), during the various stages of production to consumption.

The losses during the production process are higher in less developed and economically backward countries and wastages during the consumption stage are highest in developed countries. The per capita food wasted in North America and Europe is estimated at 95–115 kg head^{-1} year^{-1}, while it is only 6–11 kg head^{-1} year^{-1} in sub-Saharan and south-east Asia. In tuber crops the maximum losses in potato take place in medium and high-income countries during grading and consumption, while in cassava, in sub-Saharan Africa, the losses happen during production and post-harvest handling (FAO, 2011). Avoidance of food wastage may be achieved only through serious efforts via appropriate political and social actions and political leadership. The UN recently appointed a high level Taskforce on Food Security to study this aspect. This Taskforce (UN, 2012) identified the three greatest challenges in the 21st century as: ensuring food security, mitigating climate change and adapting to climate change. Several other agencies have also examined this aspect from their various perspectives (FAO, 2009, 2012; IFPRI, 2010; NAS, 2010; CGIAR, 2012; UN, 2012). They have concluded that an average increase of 2°C in the near-term appears inevitable, and an increase of 3°C mean temperature appears possible. The just-released report of the IPCC (2014) has discussed only mitigation. It has observed that total anthropogenic greenhouse gases (GHGs) have continued to increase with larger absolute decadal increases (2.2%

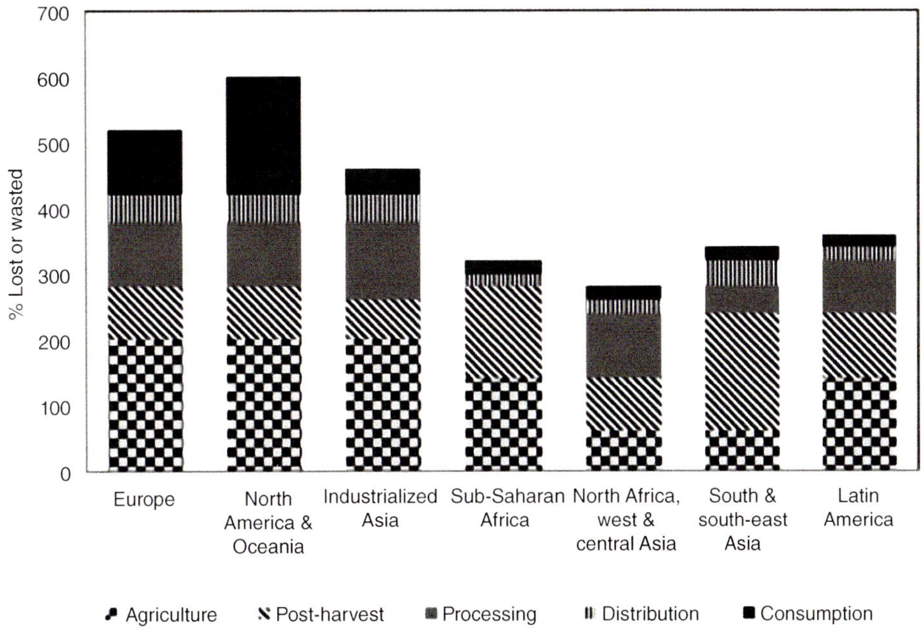

Fig. 1.1. Part of the initial production lost or wasted at different stages of the FSC for root and tuber crops in different regions (FAO, 2011).

year^{-1} during 2000–2010) compared to (1.3% year^{-1} (1970–2000). Even a 2°C increase would reduce food production in most of the lowland tropical and subtropical farmlands and will have a profoundly destabilizing effect on global food production (Foley *et al.*, 2011; Nature, 2013). TTCs supply 28.5 kg food and 75 kcals of energy head^{-1} day^{-1}. That is almost half the total contribution of TCs. Whilst the area under TTCs increased by only about a third during 1961–2011 (Table 1.5), production increased by as much as 250%.

The area/production trends in TCs, especially in TTCs, have been inconsistent during 1961–2011, except in the case of (i) in all the crops, productivity has increased by about 60%; (ii) in cassava and yams, both area and production have been steadily increasing in recent years, though they have not been making the kind of dramatic increases being shown by the two major cereal crops, wheat and rice. There are several reasons for this: (i) most of the R&D efforts are being made in only the major cereal crops; (ii) the observed increases in cassava and yam production may be attributable to the greater resilience of these two crops to climate changes in sub-Saharan Africa. The FAO identified cassava as one of the most climate-resilient crops. Very soon, it is felt by many experts that yam may also be identified as a climate-resilient crops. This will prove advantageous as it could help in attracting more funds for R&D. Until a few years ago, nearly three-quarters of all sweet potato grown in the world used to be cultivated in China. Then China took a policy decision to reduce the area under this crop substantially. At the same time, several countries in middle Africa, such as Uganda, Rwanda, and Burundi, have taken up sweet potato cultivation, which has, in effect, partially offset the major reduction in area and production of this crop globally following China's decision. The two aroid crops, taro and cocoyam, have made only moderate increases in area and production (Table 1.5). Presently, the TTCs contribute a meagre 3.4% energy to global food security and efforts should be made to double this by 2030. Narrowing the yield gap (Table 1.6) is the least difficult and fastest procedure to increase crop production. This can be achieved by providing sound technical

Table 1.5. Production and yield trends in TTCs.

Crops[a]	1961			2011		
	Area (million ha)	Yield (t ha^{-1})	Production (million t)	Area (million ha)	Yield (t ha^{-1})	Production (million t)
Cassava (105)	9.6	7.4	71.3	20.01 (+109.4%)	12.78	256.0 (+59.0%)
Sweet potatoes (120)	13.4	7.4	98.2	8.2 (–38.8%)	12.9	105.0 (+6.9%)
Yams (61)	1.2	7.2	8.3	5.0 (+316.6%)	11.5	56.8 (+584.3%)
Taro (48)	0.8	5.9	4.5	1.3 (+62.5%)	0.8	9.6 (+113.8%)
Cocoyam (IPCC, 2014)	0.023	5.1	0.117	0.043 (+6.9%)	8.4	0.363 (+17.7%)
Potato (155)	22.2	12.2	271.0	19.2 (–12.2%)	19.5	373.0 (+37.6%)
All tuber crops	47.6	9.6	445.3	54.9 (+15.3%)	14.8	810.8 (+79.8%)

Source: FAOSTAT (1961, 2011)
[a]Figures in parenthesis give the number of countries reporting information.

Table 1.6. Yield gaps in tuber crops, 2012.

Crop	Yield (t ha^{-1}) Highest	Lowest	Average
Cassava	36.4 (India)	1.4 (Burkina Faso)	12.8 (World)
Sweet potato	32.7 (Reunion)	1.5 (Bhutan)	12.9 (World)
Yam	22.1 (Japan)	3 (Bhutan)	11.5 (World)
Taro	22.9 (China)	1.2 (Côte d' Ivoire)	2.3 (World)
Yautia	10 (Cuba)	4.2 (Mexico)	8.4 (World)

Source: FAOSTAT, 2012.

advice, making available the needed inputs like seed, manure, etc., and making arrangements to buy the excess produce from the farmers. The examination of the current yields of the TTCs shows that the highest, lowest and world average yields in cassava and sweet potato are very comparable.

Cassava is a very long duration crop (10–12 months) and sweet potato a medium duration crop (3–4.5 months). This shows the unique ability of the sweet potato to produce potentially high yields. At the other extreme, the aroids (taro, yautia) show the lowest world average yields. However, the optimistic indicator here is the prevailing wide gap in yields (× 10.0) between the highest yielding country (China, 22.9t ha^{-1}) and world average yield. Further, some of the TTCs, especially the sweet potato, produce more energy per unit area than even the cereals (FAO, 1999, 2000). Though low in protein, the TTCs contain higher amounts of dietary fibre (up to 20% on dry weight basis). Taro is high in potassium. The starch granules of aroids are only about one-tenth of the size of other starch crops. This makes them very suitable for infants and invalids (FAO, 1999, 2000). There is also much variation in the GI (Glycaemic Index) values of the same crop when processed in different ways. For instance, in a gari preparation, made of cassava in Kenya, the GI value was a low 39. It also calls for developing processing techniques by which GI values can be reduced: as in gari, as stated above. Biofortification research is underway in the following crops (WHO, 2014): (i) sweet potato for enhancing iron and zinc content; (ii) carotenoid biofortification in sweet potato, and (iii) cassava for increasing pro-vitamin A content. The Biocassava Plus (BC+) programme is an ambitious project funded by the Bill and Melinda Gates Foundation engineering cassava to impart several positive attributes to it. It has the poorest tuber-keeping quality among all the TCs. Most of the cassava tubers are rendered unfit 48–72 hours after harvest. It is affected by a serious virus disease, the cassava mosaic disease (CMD). Further, it is afflicted by a serious antinutritional factor, hydrocyanic acid. The first cultivar #60 444 released under the BC+ programme contains up to 40 µg dry weight of α-carotene (1.5–2.5 µg g^{-1} in control), 40 µg g^{-1} iron (1 µg g^{-1} in control) and 10% protein in storage roots (2% in control). The shelf life of the tubers has been enhanced to 28 days (80% spoilage in seven days in control). It is also significantly resistant to the cassava mosaic disease. The studies are progressing (Fregene *et al.*, 2000; Sayre *et al.*, 2011). Among the other TTCs, biofortification has been taken up also in sweet potato (WHO, 2014). The biofortification work done on sweet potato has been to enhance its vitamin A content by increasing the anthocyanin content of the tubers and leaves and popularizing the cultivars possessing these attributes. This work is being carried out by Harvest Plus, an initiative of CGIAR. Cultivars having high carotenoid content possess deep orange or yellow flesh tubers. Aroids, mainly taro, have also been receiving some international support, mainly from Australia, the European Union (EU), and Bioversity International (BI). These programmes are TAROGEN and TANSAO (SPC, 2001). The project TAROGEN, funded by AusAid which has been concluded, was meant to collect and catalogue taro germplasm of the Pacific, including PNG. Under this project, nearly 20,000 accessions were collected and studied for their variability (TAROGEN, 2014). The International Network for Edible Aroids

(INEA) (www.ediblearoids.org) is an ongoing project with a broader mandate covering some countries in south-east and south Asia, and tropical Africa and Oceania also. It is meant to test the adaptability and performance of selected taro material for yield and field reaction to leaf blight. The future potential of TTCs lies in the ability to double their contributions to food security from the present <4% of daily energy to about 8% by 2030 and increase the utilization of starch from these crops in industrial applications. Similarly, the use of TTCs in animal feeds should be enhanced to provide added impetus to their cultivation.

References

CGIAR (2012) *Achieving Food Security in the Face of Climate Change*. Consultative Group on International Agriculture Research, Washington, DC, USA.
FAO (1999) *Taro Cultivation in Asia and the Pacific* (ed. Onwueme, I) RAP Publication 1999/16. Food and Agriculture Organization, Rome.
FAO (2000) *Edible Aroids: Post-harvest Operations* (ed. Opare, L.U.). Food and Agriculture Organization, Rome.
FAO (2009) *How to Feed the World in 2050 A.D. Executive Summary*. Food and Agriculture Organization, Rome.
FAO (2011) *Global Food Losses and Food Waste Extent, Causes, and Prevention*. Food and Agriculture Organization, Rome.
FAO (2012) *World Agriculture Towards 2030/2050. The 2012 Revision*. Food and Agriculture Organization, Rome.
FAOSTAT (2012) Food and Agriculture Organization of the United Nations, Rome.
FAOSTAT (2016) Food and Agriculture Organization of the United Nations, Rome.
Foley, J.A., Ramankutty, N., Brauman, K.A., Cassidy, E.S., Gerber, J.S. *et al.* (2011) Solutions for a cultivated planet. *Nature* 478(7369), 337–342. DOI: 10.1038/nature10452.
Fregene, M., Bernal, A., Duque, M., Dixon, A. and Tohme, J. (2000) AFLP analysis of African cassava germplasm on resistant to cassava mosaic disease. *Theoretical and Applied Genetics* 100(3), 678–685.
IFPRI (2000) *Roots and Tubers for the 21st Century*. Discussion Paper # 21. International Potato Centre, Lima/InternationalFood Policy Research Institute, Washington, DC, USA.
IFPRI (2010) *Food Security and Climate Change: Challenges to 2050 and Beyond*. Brief # 66. International Food Policy Research Institute, Washington, DC, USA.
IPCC (2014) Working group III, annual report 5. Summary for policy makers. Available at: www.ipcc.int/ (accessed 16 April 2014).
Kay, D.E. (1987) Root crops. In: *TDRI Crop and Product Digest No. 2*, 2nd edn. Natural Resources Institute, Gillingham, UK. Revised by E.B. Gooding.
NAS (2010) *Towards Sustainable Agricultural System in the 21st Century*. National Research Council/ National Academy of Sciences, Washington, DC, USA.
Nature (2013) Outlook for earth. Special issue on the IPCC 2013 report. *Nature* 2013.
Nayar, N.M. (2014) The contribution of tropical tuber crops towards food security. *Journal of Root Crops* 40(1), 1–12.
Rosling, H. (1987) *Cassava Toxicity and Food Security*. TryckKontakt, Uppsala, Sweden, pp. 3–40.
Sayre, R., Beeching, J.R., Cahoon, E.B., Egesi, C., Fauquet, C. *et al.* (2011) The BioCassava plus program: Biofortification of cassava for sub-Saharan Africa. *Annual Review of Plant Biology* 62(1), 251–272. DOI: 10.1146/annurev-arplant-042110-103751.
SPC (2001) Safeguarding Pacific Islands taro: An important root crop. Annual Report. South Pacific Commission, Suva, Fiji.
TAROGEN (2014) Available at: http://spc.int/tarogen (accessed 11 April 2014).
UN (2012) *Food and Agriculture: The Future of Sustainability*. Department of Economic and Social Development, United Nations, New York, USA.
WHO (2014) Biofortification of staple crops. Available at: http://www.who.int/ elena/ titles/ biofortification/ en (accessed 5 March 2019).

2

Starch – General Considerations

The tuber crops are rich in starch and are therefore a good source for this important food and industrial raw material. The starch content of, and the amount of extractable starch from, TTCs are given in Table 2.1.

In view of its low cost, availability and ability to impart a broad range of functional properties to food and non-food products (Bertolini, 2010; Jane, 1995; BeMiller, 2007), starch has found applications in a wide range of industries (Whistler *et al.*, 1984). Starch was originally looked upon as a staple food ingredient, but due to its attractive cost and performance, it has become one of the raw materials with the most potential for new products (Rindlav-Westling *et al.*, 1998). Starches are used mainly in the food and paper industries, with 57% of produced starch consumed in the food industries and 43% in the non-food sector (LMC, 2008). Modified and native starches form more than 85% of all hydrocolloids used in food systems (Wanous, 2004). In the food industry, the main trend in starch applications remains in syrup production and the formulation of ready meals and various sauces. From total worldwide starch production, excluding starch used to produce syrups and fermented products, more than 30 million tons of natural and modified starches are used in the production of dextrin, food ingredients, paper coatings, and adhesives (LMC, 2008). In Europe, 57% of starch is used in sweeteners and hydrolysed products, 23% as native starch and 20% as modified starch. However, in the USA, more than 73% of corn starch is used in the manufacture of products such as glycose syrups, high-fructose syrups, syrup solids, maltodextrins and fructose (Röper and Elvers, 2008). In addition to the demand for starch to produce syrups and dextrin, the demand for modified starches has shown strong growth. Though modified starches are produced mainly in North America and Europe, the latter is still one of the leading producers of modified starches and its paper and food industries tend to be at the forefront in their use of starch in innovative new products (LMC, 2008). Asian markets, however, exhibit large demand for starches in food and fermentation. India is unique in using a large portion of cassava starch for the production of sago. Some nations, including the US, have focused on starch for biofuel production. Generally, there has been a trend for demand by non-food industries to grow faster compared to food industries.

2.1 Food Applications

Texturing, thickening/gelling, retrogradation and freeze–thaw stabilizer, emulsifying agent, gel strength and clarity, and cold-water swelling are the main factors deciding the use of starch in food applications. Native and modified starches

Table 2.1. Maturity and starch content of different tuber crops.

Sl. no.	Crop	Maturity (months)	Starch content (%)	Extractable starch (%)
1	Cassava	6–10	20–35	15–28
2	Sweet potato	3–4	18–28	10–16
3	*Colocasia*	7–10	6–15	3–10
4	*Xanthosoma*	7–10	15–25	6–18
5	Arrow root	10–11	15–26	8–24
6	*D. esculenta*	7–8	12–25	8–18
7	*D. alata*	8–10	12–26	6–15
8	*D. rotundata*	9–10	15–26	10–20
9	*Amorphophallus*	10–12	12–22	5–15
10	*Canna edulis*	10–12	15–26	10–22

are used frequently in foods because of their thickening and gelling properties. Typically, native starch cannot be used at a rate higher than 6% in solids because it imparts too high a viscosity and therefore difficulty in handling. In addition, for several applications, native starches exhibit undesirable properties under certain process conditions such as temperature, pH or pressure, including low resistance to high shear rates, thermal decomposition, high retrogradation and syneresis. Improvement in these starch properties can be achieved by modifications of starch structure, using chemical, physical and enzymatic methods. Pregelatinized starches (a form of physically modified starch) which do not require cooking, are very popular in the food industry. Swollen but intact starch granules are usually desired in most food starch applications to maintain rheological properties (Kaur *et al.*, 2006; Miyazaki *et al.*, 2006). They are used to control the cold viscosity and can be cross-linked and stabilized to provide shear and freeze–thaw stability. Granular cold-water soluble starches show important functional properties desirable in several instant foods, such as a higher viscosity and a soft texture (Béllo-Perez *et al.*, 2002). Because they develop viscosity immediately, they are widely used in instant dry mixes such as instant puddings, baby foods, frozen foods and convenience products. In some applications such as salad dressings, starch can be incorporated into the oil phase to slow down hydration when water is added later, which is useful in sauce and puree formulation. Oxidized starches possess reduced swelling power, increased solubility, and lower pasting temperature than native starch (Kuakpetoon and Wang, 2001) and they are used in products demanding softness and transparency, such as gum sweets or Arabic gums (Chattopadhyaya *et al.*, 1998). Acetylated starches are used to prevent retrogradation by interrupting the linearity of the amylose or segments of the amylopectin branches, reducing syneresis and loss of texture and consistency during storage of the finished product (Filer, 1998). Acetylation diminishes bond strength thereby increasing the swelling power and solubility of the starch granule and improving freeze–thaw stability (Rutenberg and Solarek, 1984). Consequently, these starches have physicochemical characteristics such as low gelatinization temperature, high solubility and good cooking and storage stabilities, and are widely used in the food industry to overcome undesirable changes in product texture and appearance caused by retrogradation during food processing and storage. They are used extensively in a wide range of foods including baked goods, canned pie fillings, sauces, condensed soups, frozen foods, baby foods, salad dressings and snack foods. Modified starches derived by succinate derivatization are characterized by the tendency to swell in cold water and to show higher viscosity and lower gelatinization temperature than native starch. In food areas, succinate derivatives have been recommended as binders and thickening agents in soups, snacks and canned and refrigerated food products (Wurzburg, 1986), encapsulating flavour oils and emulsion stabilizers, and enhancing oxidation resistance

(Trubiano, 1995). Hydroxypropylated starches improve the shelf life, freeze–thaw stability, cold-storage stability, clarity and texture properties of starch pastes. They are very effective in retarding bread staling and improving bread-crumb texture, enhancing breadmaking functionality and bread quality (Miyazaki *et al.*, 2006). Hydroxypropylated starches are also effective thickening and stabilizing agents in low-fat or low-calorie dairy products due to their greater compatibility with milk protein. The structure of starch can be reinforced by cross-linking in order to improve the product's texture and its resistance to rupture during processing. Cross-linked starches are more resistant to low pH, rupture and high temperatures and show high capacity for holding water (Jarowenko, 1986). However, cross-linking also reduces paste clarity and stability against cold storage. Because cross-linking reduces starch granule rupture, the loss of viscosity and the formation of a stringy paste during cooking, cross-linked starches are widely used as thickeners in foods, particularly where a high and stable viscosity is desirable. Therefore, secondary modifications such as hydroxypropylation and acetylation are used to change undesirable characteristics of cross-linked starch (Gunaratne and Corke, 2007). Cross-linking of hydroxypropylated starch, for example, imparts viscosity stability and short-textured properties to the paste.

2.1.1 Dextrins, sweeteners and syrups

Starch hydrolysis has been used in the food industry to convert starches, and to obtain more versatile ingredients and they are used in a wide range of food products due to both their functional and nutritional properties. They can be obtained through acid catalysis or specific enzyme action. The starch-processing industry is the second largest segment of the industrial enzyme market, representing approximately 30% of world enzyme production. The major enzymes are amylases, glucoamylases and glucose isomerases, which are used in starch liquefaction, saccharification and isomerization. Products from starch hydrolysis, such as glucose or maltose syrup and maltodextrins, and their derivates by isomerization (iso-glucose and fructose), are used in candy, sweets, chocolate, cake, dessert, dairy and pastry products due to their anticrystallizing, sweetening, and water-holding properties. Sugar syrups and their derivatives are the most popular hydrolysed starch used in fermentation processes, beverage formulation, dietetic products, ice creams, soft gum, dairy products, desserts and candies. Maltodextrins have desirable functional properties and are employed for their thickening effects (Blanshard and Katz, 1995; Loret *et al.*, 2004).

2.1.2 Health and functional foods

In the food industry, starch in formulations of health, nutritional and functional products has been considered one of the most promising markets and it has motivated research in the development of new products. This trend supports the development of modern products such as frozen, chilled, low-fat and gluten-free foods, where starch can be used as the main or co-ingredient. For example, health nutrition trends for enhanced fibre consumption, with emphasis on low glycemic food, have pushed the growth of high amylose starch as a source of resistant starch (RS). Resistant starches result from the formation of glycosidic bonds, other than the α–(1–4) or α–(1–6) links by treatments involving heating, cross-linking, hydroxypropylation or phosphatation, which can reduce the digestibility of the starch (Englyst *et al.*, 1996; Woo and Seib, 2002). Amylose content, water content, storage time and temperature of the starch gels, number of autoclaving/cooling cycles and the presence of lipids and proteins are some processing factors that can affect the formation of RS (Gidley *et al.*, 1995; Kale *et al.*, 2002). Those starches are resistant to degradation or are not digested/absorbed in the small intestine of healthy individuals, and are fermented in the large intestine (Champ, 1992; Eerlingen and Delcour, 1995), simulating fibre action in the intestinal tract (Englyst *et al.*, 1996). They can replace fibres in food formulation without changing their sensory properties (Lee and Shin, 2006). Another trend in health is the concern to reduce fat ingestion. Fat, however, is an indispensable nutrient and plays a functional and sensory role in food: it is responsible for the stability, palatability, flavour, texture, structure, brightness, aroma and sensation of satiety (de Wijk and Prinz, 2006). Consequently, several starch derivatives have shown to be promising replacements for fat properties, probably due to their water association systems (Yackel and Cox, 1992; Filer, 1998). To replace fat in food systems, starch modifications must improve their stability in cold water, improve their characteristics as dough or gel texture or provide emulsifying characteristics through the addition of hydrophobic groups (Alexander, 1995). Starch has been used in several low-fat products to improve some properties such as texture, colour, palatability and stability of foods (Harris and Day, 1993; Esteller *et al.*, 2004). Hydrolysed starches have been shown to be excellent substitutes for fat, obtaining similar texture to those observed in high-fat products (Harris and Day, 1993; Alexander, 1995; Giese, 1996). Waxy starches and starches from roots and tubers are largely used in low-fat products. Cassava starch derivatives intercrossed with 0.05% $POCl_3$ and acetylated with 8% acetic anhydride have been shown to be an efficient fat replacer, improving the texture of mayonnaise (Teixeira, 2002). Esterification of starches with fatty acids of a linear chain gives the derivative hydrophobic characteristics and emulsifying properties (BeMiller, 1997) and acetylated starches improve stability in low-fat systems (Jarowenko, 1986). However, to replace fat, starches should maintain high moisture content in systems such as mayonnaise, salad dressings and low-fat spreads. Starches as fat replacers can also be used in bakery products such as cakes; however, they are not recommended for systems with a low moisture content such as cookies and crackers (Giese, 1996; Razavi *et al.*, 2007; Sudha *et al.*, 2007). Fermented starches are often employed in traditional culinary arts, and they have also been used as ingredients in the food industry. Sour cassava starch is an example of cassava starch modified by fermentation and sun drying, with its characteristic expansion property. For this reason, sour cassava starch is used in bakery products leading to volume expansion without the addition of yeast, biological baking powder or chemical agents (Bertolini *et al.*, 2001). It is used in breadmaking and biscuit formulation and it can be an alternative for gluten-free products consumed by people who are intolerant to wheat gluten.

2.1.3 Edible coatings

Biopolymers such as lipids, proteins, derivatives of cellulose, starch and other polysaccharides have been tested in the formulation of edible films and coatings. Recently, there has been increasing interest in the development of films and edible coatings that can be used to increase the shelf-life of fruits and vegetables (Pagella *et al.*, 2002). Starch gelatinized with sodium hydroxide, glycerol or sorbitol can be used as plasticizers in edible coatings and films (Larotonda *et al.*, 2004; Cyras *et al.*, 2006), and blends of pectin and high amylose maize starch have also been developed to produce food coatings and films (Dimantov *et al.*, 2004).

2.2 Non-food Applications

In non-food applications, the most traditional area of starch usage has been in the paper industry, which comes second to food applications and is estimated at more than 10 million tons annually. Recently, with the increasing cost of cellulose pulp, paper manufacturers have turned to lower-cost fillers. These fillers often require special modified starches to bind them to the pulp. Consequently, several innovations in this area have increased the share of starch in those applications (LMC, 2008). Oxidized (Lawal *et al.*, 2005) and enzymatic hydrolized (van der Maarel *et al.*, 2002) starches have wide applications in paper coating, improving the strength and printability of paper. Those starches are also often used in textile and laundry finishing products (Lawal *et al.*, 2005). Recently, development of biodegradable polymers has resulted in the use of starch in this field because of its low cost, its availability and its ability to be produced from renewable resources. Because it is a thermoplastic material, it has been used to produce blends with non-biodegradable polymers, biodegradable materials and injection moulding (Dufresne, 2007).

In recent years, most of the starch-based polymers produced are used for packaging applications, including soluble films for industrial packaging. Loose-fill containers and transparent films are widely used in the packaging industry. Biodegradable bags are used for waste collection and for disposable cutlery, whereas biodegradable films are used for agricultural mulching, pet products and hygiene, including cotton buds, packaging for toilet paper, overwrap and backing material for flushable sanitary products. Pencil sharpeners, rulers, cartridges, toys, and plant pots are among the other uses (García-Alonso *et al.*, 1999; Karim *et al.*, 2000). Starch-based plastics are highly hydrophilic and readily disintegrate on contact with water. There is increasing interest in biomaterials in the development of biodegradable products. Starches used as thermoplastic materials have some limitations, such as low moisture resistance, high viscosity and incompatibility with some hydrophobic polymers. Starches modified by chemical treatment have been utilized as an alternative to solve this problem. To enhance viscosity, texture and stability among many desired functional properties for industrial applications, starches and their derivatives are modified by chemical, physical and biotechnological means. Starch-based and biodegradable polymers have been used in the production of automotive parts (Leão *et al.*, 2007) and nanoparticle-sized maize-derived starch can also be used in a wide range of products (Dufresne, 2007), including as a biofiller to enhance tyre performance, reducing their rolling resistance and improving fuel consumption (Scott, 2002). Another potential area is the fuel market. Consequently, concerning this starch application, it is necessary to improve conversion technology in order to increase the ethanol yield.

Maize starch remains the most used starch in the world, the USA being its largest producer. However, regarding cassava starch, Thailand is emerging as the leading exporter of native starch and starch derivatives. Whereas maize starch is used to produce mainly modified starches and syrups, cassava starch is the most widely traded native starch in the world. Demand for native and modified starches closely follows growth in the food and paper industries. The largest applications of starch in the food industry include syrups, formulation of ready meals, instant food and sauces, and in the paper industry, for paper coating. However, starch application is also increasing in health, nutritional and functional foods, considered a potential food market. The increased value of modified starch remains very high due to several food and industrial applications. There are several new technologies that can improve the production of modified starches. Due to progress in plant breeding, a wide range of starch sources is available. New varieties of cereals and roots with high-resistant starch or high/low sugar content, potato with high phosphate content, and starches with enhanced or lowered amylose content and chain length through biotechnological innovations, are being developed in different laboratories around the world and thereby signal a bright future for starch.

References

Alexander, R.J. (1995) Fat replacers based on starch. *Cereal Foods World* 40(5), 366–368.

Béllo-Perez, L.A., Contreras-Ramos, S.M., Romero-Manilla, R., Solorza-Feria, J. and Jiménez-Aparicio, A. (2002) Chemical and functional properties of modified starch from banana *Musa paradisiaca* L. (Var. Macho). *Agrocencia* 36(2), 169–180.

BeMiller, J.N. (1997) Structure of the starch granule. *Journal of Applied Glycoscience* 44, 43–49.

BeMiller, J.N. (2007) *Carbohydrate Chemistry for Food Scientists*. AACC International, Inc, St. Paul, USA.

Bertolini, A.C., Mestres, C., Lourdin, D., Valle, G.D. and Colonna, P. (2001) Relationship between thermomechanical properties and baking expansion of sour cassava starch (Polvilho azedo). *Journal of the Science of Food and Agriculture* 81(4), 429–435. DOI: 10.1002/1097-0010(200103)81:4<429::AID-JSFA833>3.0.CO;2-2.

Bertolini, A.C. (2010) Trends in starch applications. In: Bertolini, A.C. (ed.) *Starches Characterization, Properties and Applications*. CRC Press, Boca Raton, USA, pp. 1–19.

Blanshard, P.H. and Katz, F.R. (1995) Starch hydrolysates. In: Stephen, A.M. (ed.) *Food Polysaccharides and Their Applications*. Marcel Dekker, New York, pp. 99–153.

Champ, M. (1992) Determination of resistant starch in foods and food products: Interlaboratory study. *European Journal of Clinical Nutrition* 46(2), 51–62.

Chattopadhyaya, S., Singhal, R.S. and Kulkarni, P.R. (1998) Oxidised starch as gum arabic substitute for encapsulation of flavours. *Carbohydrate Polymers* 37(2), 143–144. DOI: 10.1016/S0144-8617(98)00054-X.

Cyras, V.P., Tolosa Zenklusen, M.C. and Vazquez, A. (2006) Relationship between structure and properties of modified potato starch biodegradable films. *Journal of Applied Polymer Science* 101(6), 4313–4319. DOI: 10.1002/app.23924.

de Wijk, R.A. and Prinz, J.O.N.F. (2006) Mechanisms underlying the role of friction in oral texture. *Journal of Texture Studies* 37(4), 413–427. DOI: 10.1111/j.1745-4603.2006.00060.x.

Dimantov, A., Kesselman, E. and Shimoni, E. (2004) Surface characterization and dissolution properties of high amylose corn starch–pectin coatings. *Food Hydrocolloids* 18(1), 29–37. DOI: 10.1016/S0268-005X(03)00039-0.

Dufresne, A. (2007) Biopolymers on nanocomposities. In: Bertolini, A.C. (ed.) *Biopolymers Technology*. Cultura Academica, São Paulo, pp. 59–84.

Eerlingen, R.C. and Delcour, J.A. (1995) Formation, analysis, structure and properties of type III enzyme resistant starch. *Journal of Cereal Science* 22(2), 129–138. DOI: 10.1016/0733-5210(95)90042-X.

Englyst, H.N., Kingman, S.M., Hudson, G.J. and Cummings, J.H. (1996) Measurement of resistant starch *in vitro* and *in vivo*. *British Journal of Nutrition* 75(5), 749–755. DOI: 10.1079/BJN19960178.

Esteller, M.S., Amaral, R.L. and Lannes, S.C.D.A.S. (2004) Effect of sugar and fat replacers on the texture of baked goods. *Journal of Texture Studies* 35(4), 383–393. DOI: 10.1111/j.1745-4603.2004.tb00602.x.

Filer, J.L. (1998) Modified food starch: An update. *Journal of the American Dietetic Association* 88(3), 342–344.

García-Alonso, A., Escrig, A.J., Carrón, N.M. and Bravo, L. (1999) Assessment of some parameters involved in the gelatinization and retrogration of starch. *Food Chemistry* 66(2), 181–187. DOI: 10.1016/S0308-8146(98)00261-1.

Gidley, M.J., Cooke, D., Darke, A.H., Hoffmann, R.A., Russell, A.L. *et al.* (1995) Molecular order and structure in enzyme-resistant retrograded starch. *Carbohydrate Polymers* 28(1), 23–31. DOI: 10.1016/0144-8617(96)81387-7.

Giese, J. (1996) Fat, oils, and fat replacers. *Food Technology* 50(4), 77–83.

Gunaratne, A. and Corke, H. (2007) Functional properties of hydroxypropylated, cross-linked, and hydroxypropylated cross-linked tuber and root starches. *Cereal Chemistry Journal* 84(1), 30–37. DOI: 10.1094/CCHEM-84-1-0030.

Harris, D.W. and Day, G.A. (1993) Structure versus functional relationships of a new starch-based fat replacer. *Starch/Stärke* 45(7), 221–226. DOI: 10.1002/star.19930450703.

Jane, J. (1995) Starch properties, modifications, and applications. *Journal of Macromolecular Science, Part A* 32(4), 751–757. DOI: 10.1080/10601329508010286.

Jarowenko, W. (1986) Acetylated starch and miscellaneous organic esters. In: Wurzburg, O.B. (ed.) *Modified Starches: Properties and Uses*. CRC Press, Florida, Boca Raton, pp. 55–57.

Kale, C.K., Kotecha, P.M., Chavan, J.K. and Kadam, S.S. (2002) Effect of processing conditions of bakery products on formation of resistant starch. *Journal of Food Science and Technology* 39(5), 520–524.

Karim, A., Norziah, M.H. and Seow, C.C. (2000) Methods for the study of starch retrogradation. *Food Chemistry* 71(1), 9–36. DOI: 10.1016/S0308-8146(00)00130-8.

Kaur, L., Singh, J. and Singh, N. (2006) Effect of cross-linking on some properties of potato (*Solanum tuberosum* L.) starches. *Journal of the Science of Food and Agriculture* 86(12), 1945–1954. DOI: 10.1002/jsfa.2568.

Kuakpetoon, D. and Wang, Y.-J. (2001) Characterization of different starches oxidized by hypochlorite. *Starch/Stärke* 53(5), 211–218. DOI: 10.1002/1521-379X(200105)53:5<211::AID-STAR211>%203.0.CO;2-M.

Larotonda, F.D.S., Matsui, K.N., Soldi, V. and Laurindo, J.B. (2004) Biodegradable films made from raw and acetylated cassava starch. *Brazilian Archives of Biology and Technology* 47(3), 477–484. DOI: 10.1590/S1516-89132004000300019.

Lawal, O.S., Adebowale, K.O., Ogunsanwo, B.M., Barba, L.L. and Ilo, N.S. (2005) Oxidized and acid thinned starch derivatives of hybrid maize: Functional characteristics, wide-angle X-ray diffractometry and thermal properties. *International Journal of Biological Macromolecules* 35(1-2), 71–79. DOI: 10.1016/j.ijbiomac.2004.12.004.

Leão, A.L., Ferrão, P.C., Teixeira, R. and Sator, S. (2007) Biopolymers applications on automotive industry. In: Bertolini, A.C. (ed.) *Biopolymers Technology*. Cultura Academica, São Paulo, pp. 165–195.

Lee, H. and Shin, M. (2006) Comparison of the properties of wheat flours supplemented with various dietary fibers. *Food Science and Biotechnology* 15(5), 746–751.

LMC (2008) LMC International's Global Markets for Starch and Fermentation [NKE2] Products Database 2008 report. Available at: http://www.lmc.co.uk/Report.aspx?id=6&repID=2983&flag=1537 (accessed 16 October 2018).

Loret, C., Meunier, V., Frith, W.J. and Fryer, P.J. (2004) Rheological characterisation of the gelation behaviour of maltodextrin aqueous solutions. *Carbohydrate Polymers* 57(2), 153–163. DOI: 10.1016/j.carbpol.2004.03.026.

Miyazaki, M., Van Hung, P., Maeda, T. and Morita, N. (2006) Recent advances in application of modified starches for breadmaking. *Trends in Food Science & Technology* 17(11), 591–599. DOI: 10.1016/j.tifs.2006.05.002.

Pagella, C., Spigno, G. and De Faveri, D.M. (2002) Characterization of starch based edible coatings. *Food and Bioproducts Processing* 80(3), 193–198. DOI: 10.1205/096030802760309214.

Razavi, S.M.A., Habibi Najafi, M.B. and Alaee, Z. (2007) The time independent rheological properties of low fat sesame paste/date syrup blends as a function of fat substitutes and temperature. *Food Hydrocolloids* 21(2), 198–202. DOI: 10.1016/j.foodhyd.2006.03.008.

Rindlav-Westling, A˚sa., Stading, M., Hermansson, A.-M. and Gatenholm, P. (1998) Structure, mechanical and barrier properties of amylose and amylopectin films. *Carbohydrate Polymers* 36(2-3), 217–224. DOI: 10.1016/S0144-8617(98)00025-3.

Röper, H. and Elvers, B. (2008) Starch. 3. Economic Aspects. In: *Ullmann's Encyclopedia of Industrial Chemistry*, 7th edn. John Wiley & Sons, New York, pp. 21–22.

Rutenberg, M.W. and Solarek, D. (1984) Starch derivatives: production and uses. In: Whistler, R.L., BeMiller, J.N. and Pascha, E.F. (eds) *Starch Chemistry and Technology*. Academic Press, San Diego, pp. 311–388.

Scott, A. (2002) Nanoscale starch challenges silica in fuel-efficient 'green' tires. *Chemical Week* 164(15), 15–26.

Sudha, M.L., Srivastava, A.K., Vetrimani, R. and Leelavathi, K. (2007) Fat replacement in soft dough biscuits: Its implications on dough rheology and biscuit quality. *Journal of Food Engineering* 80(3), 922–930. DOI: 10.1016/j.jfoodeng.2006.08.006.

Teixeira, M.A.V. (2002) *Amidos quimicamente modificados empregados na substituição de gordura em alimentos*. Food Technology doctoral thesis. Universidade Estadual de Campinas, São Paulo, Brazil. Available at: http://bdtd.ibict.br/vufind/Record/CAMP_58b2c2bc2f702e07b19bed0a895baadf (accessed 22 May 2019).

Trubiano, P.C. (1995) The role of speciality food starches in flavor encapsulation. *Flavor Technology* 610, 244–253.

van der Maarel, M.J.E.C., van der Veen, B., Uitdehaag, J.C.M., Leemhuis, H. and Dijkhuizen, L. (2002) Properties and applications of starch-converting enzymes of the α-amylase family. *Journal of Biotechnology* 94(2), 137–155. DOI: 10.1016/S0168-1656(01)00407-2.

Wanous, M.P. (2004) Texturizing and stabilizing, by gum. *Prepared Foods* 173(1), 108–118.

Whistler, L., BeMiller, J.M. and Paschall, E.P. (eds) (1984) *Starch: Chemistry and Technology*, 2nd edn. Academic Press, New York.

Woo, K.S. and Seib, P.A. (2002) Cross-linked resistant starch: preparation and properties. *Cereal Chemistry Journal* 79(6), 819–825. DOI: 10.1094/CCHEM.2002.79.6.819.

Wurzburg, O.B. (ed.) (1986) *Modified Starches: Properties and Uses*. CRC Press, Boca Raton, Florida.

Yackel, W.C. and Cox, C. (1992) Application of starch-based fat replacers. *Food Technology* 46(6), 146–148.

3

Cassava

3.1 Introduction

Cassava (*Manihot esculenta* Crantz) is a sturdy perennial crop grown in many parts of Asia, Africa and South America. The yield of the crop is normally around 20 t/ha and the starch yield forms nearly one quarter of the total yield (Moorthy *et al.*, 2018). Among the root and tuber crops, cassava is mainly used for starch extraction. This is due to ease of extraction and good quality of product obtained. Yield and quality of extracted starch depend on the nature of tubers used and also the method and machinery used for starch extraction. The starch has been studied in detail and it finds uses in a large range of industries.

3.2 Extraction and Purification

Breuninger *et al.* (2009) have comprehensively reviewed the industrial-scale process of starch isolation from cassava roots. A brief description of the same has been given. Washed cassava tubers were first grated in water in the extractor followed by isolation of starch fraction from the pulp by centrifugation. The obtained starch was dewatered before drying. The undried starch extract can be preserved from microbial growth by storing it in sodium metabisulphite solution. The storage of starch in this solution has minimal effect on starch properties (Jyothi *et al.*, 2007). (See Table 3.1.)

Yield is an important factor, especially in its industrial application. There is a wide range of starch yield from various reports (Zhu, 2015). (See Table 3.2.)

The starch yield can be even up to 80% of the dried weight of cassava root (Olomo and Ajibola, 2003; Osunsami *et al.*, 1989). Wide genetic variation in starch yield has been reported (Benesi *et al.*, 2004, 2008; Joseph *et al.*, 2004). High starch yield can be achieved by selecting suitable varieties with high dry matter content and thin peel thickness (Pérez *et al.*, 2011). The extrinsic factors that can affect the starch yield include:

1. The drying conditions of raw materials (e.g. oven-dried chips and flour provided higher yield compared to sun-dried ones) (Olomo and Ajibola, 2003);
2. Time of harvest and root storage period (longer root storage resulted in lower starch yield) (Benesi *et al.*, 2008; Osunsami *et al.*, 1989);
3. The nature of the roots used for extraction (fresh root gave better extraction efficiency and purer starch than dried chips and pellets) (Meuser *et al.*, 1978); and
4. The prevailing environmental conditions during crop growth (e.g. starch yield of a

Table 3.1. Peak viscosity (Pa s) of cassava starch during wet storage in presence of sodium metabisulphite.

	Concentration of sodium metabisulphite (%)				
Weeks after storage	0	0.5	1.0	1.5	2.0
0	2.88 ± 0.11	2.90 ± 0.10	2.85 ± 0.08	2.87 ± 0.21	2.82 ± 0.16
1	2.88 ± 0.21	2.89 ± 0.10	2.87 ± 0.11	3.47 ± 0.08	3.52 ± 0.10
2	2.79 ± 0.10	2.89 ± 0.13	2.96 ± 0.09	3.12 ± 0.16	3.22 ± 0.10
3	2.69 ± 0.09	2.92 ± 0.08	3.15 ± 0.02	3.24 ± 0.13	3.24 ± 0.07
4	2.50 ± 0.05	2.88 ± 0.07	2.80 ± 0.05	2.77 ± 0.07	2.74 ± 0.11
5	2.40 ± 0.10	2.55 ± 0.11	2.83 ± 0.08	2.67 ± 0.23	2.79 ± 0.09
6	2.18 ± 0.08	2.31 ± 0.09	2.77 ± 0.04	2.47 ± 0.11	2.59 ± 0.10
7	2.08 ± 0.02	2.03 ± 0.06	2.55 ± 0.12	2.10 ± 0.12	2.32 ± 0.15
8	2.05 ± 0.02	1.90 ± 0.10	2.50 ± 0.09	2.07 ± 0.10	2.07 ± 0.22

(From: Jyothi *et al.*, 2007)

given variety differed significantly with the growth location) (Benesi *et al.*, 2004).

Quality of extracted starch can also be influenced by the quality and quantity of water used during starch extraction, hydrogen cyanide content of tubers and type of drying employed.

By controlled and careful selection of the parameters, maximum starch yield and quality can be achieved (Radley, 1976).

Cassava starch manufacture is carried out at different levels in most of the countries. It can be small-, medium- or large-scale and employs different processes and machinery, and the quality of the resultant starch can vary considerably (Whistler and Paschall, 1967). The main steps involved in manufacture are outlined in the following steps.

Step 1: determination of starch in roots

The starch content of roots is frequently the basis used by the larger starch manufacturers for determining the root purchase price. Each truck or cart load of roots arriving at the plant from the fields is evaluated on the basis of a 3–4 kg sample. The root sample, washed free of dirt and dried with a towel, is weighed in air and then in water using a specific gravity balance (see Fig. 3.1).

Step 2: water requirements

Another prerequisite for the production of high-quality tapioca starch is an uninterrupted supply of pure, soft water. A factory which grinds 100 tons of roots per day will require at least 500 litres of water. If the water is hard and contains iron, a lime-alum treatment followed by filtration is recommended.

Step 3: industrial-grade tapioca starch manufacture

From the washer, the roots fall into a grinder. This is often a crude machine consisting of a short rotating wooden log about 25–30 cm in diameter with nails projecting a few millimetres out of the surface. A stationary wooden plate, also studded with sharp nails, is located to provide a small clearance with the rotating log. Roots are torn apart to form a coarse pulp which is mixed with a little water and allowed to ferment in a cement tank or wooden tub for a day or two in order to break up the root cells and release the starch trapped in the fibrous material. The starch milk is then channelled to a large cement or wooden tank and allowed to settle. The waste coarse pulp, deposited outside the factory, is carted away or sometimes is spread on large

Table 3.2. Starch yield and chemical composition.

No	Yield (%)	Amylose (%)	Ash (%)	Protein (%)	Lipid (%)	Phosphorus (%)	Fibre (%)	Reference
1	24.4		0.24	0.49	0.13		0.15	Osunsami *et al.* (1989)
4[a]		16–20	0.12–0.33	0.05–0.28	0.04–0.38	0.009–0.021		Asaoka *et al.* (1991)
1[b]		20.6–21	0.32–0.66	0.59–0.63		0.0026		Padmanabhan and Lonsane (1992)
1				0.15	0.05			Gorinstein and Lii (1992)
3*2[c]		17.4–20						Asaoka *et al.* (1993)
5		22.6–26.2			0.11–0.22		0.112–0.15	Moorthy *et al.* (1996)
4*2*5[d]		17.9–23.6						Defloor *et al.* (1998)
4*4[e]		19.6–24.1	0.08–0.15	0.15–0.30	0.01			Sriroth *et al.* (1999)
3*3[f]	20.72–27.8	22.5–23.8	0.7–0.16	0.25–0.36	0.03–0.15		0.17–0.42	Abera and Rakshit (2003)
1*8[g]	48.8–79.9						0.3–1.9	Olomo and Ajibola (2003)
1		23	0.09	0.1	1.2		0.5–0.9	Freitas *et al.* (2004)
23	11.3–26.0	17.5–28.9						Joseph *et al.* (2004)
20*2[h]	26.25–52.19		0.05–0.09					Benesi *et al.* (2004)
4*2		15.9–23.1	0.7–0.9	0.75–1.1	0.1			Charles *et al.* (2004)
10		1.6–019.2				0.008–0.0095		Raemakers *et al.* (2005)
4		17.14–19.11[m]						Charoenkul *et al.* (2006)
1		16.3	0.2	0.51	0.5	0.00756		Mishra and Rai (2006)
40		15.2–30.2	0.4–0.49	0.23–0.7				Onitilo *et al.* (2007)
39		16.2–22.1	0.03–0.77	0.06–0.7				Eke *et al.* (2009)
3		0–20.4			0.16–0.28	0.0029–0.0048		Ceballos *et al.* (2007)

Continued

Table 3.2. Continued

No	Yield (%)	Amylose (%)	Ash (%)	Protein (%)	Lipid (%)	Phosphorus (%)	Fibre (%)	Reference
33[i]		16–28.1						Teerawanichpan *et al.* (2008)
4050		15.2–26.5				0.0073–0.0078 [n]		Sánchez *et al.* (2010)
3		14.1–24.3	0.11–0.23	0.15–0.34				Mbougueng *et al.* (2009)
1		29.3	0.29					Nwokocha *et al.* (2009)
17		17.1–21.3						Anggraini *et al.* (2009)
2		6–18.4						Gomand *et al.* (2010)
9		2–24.9						Koehorst-van Putten *et al.* (2012)
8		0–30.3						Rolland-Sabaté *et al.* (2012)
8		0–16.5						Rolland-Sabaté *et al.* (2013)

No. = number of genotypes/samples; the yield was based on wet weight except stated. DSC = differential scanning calorimetry-based method for amylose measurement; SEC = size-exclusion chromatography-based method. SEC-AD = size-exclusion chromatography after debranching. SEC-WD = size-exclusion chromatography without debranching. Iodine-S = iodine-spectrophotometry/colorimetry-based method. Iodine-A, iodine-amperometry-based method; Con A = concanavalin A-based precipitation method. N.A. = not available.

From: (Zhu, 2015).

[a]3–4 different month harvest; [b]2 extraction methods; [c]with/without pruning aerial growth; [d]5 different harvesting months and two different planting seasons; [e]4 different harvesting months; [f]3 different ways of extraction; [g]starches extracted from sun-dried and oven-dried chips and flour; [h]2 different growth locations; [i]33 samples and genotypes with variations in the layers of roots, harvesting month, and planting season; [j]dry base; [k]fresh root; [l]2 different measuring methods; [m]3 different measuring methods, and 2 variations for SEC after debranching methods with/without amylopectin isoamylolyzate involvement; [n]7 genotypes.

Fig. 3.1. Specific gravity balance for detecting starch content in cassava.

patios to sun-dry and is often used as raw material for cattle feed.

The free water remaining on the surface of the starch cake is blotted up with rags. Starch cake scraped out from the settling tubs is crumbled and spread on bamboo mats to sun-dry. Sometimes the starch is dried on a large cement surface heated by a wood fire underneath. A heavy wooden roller may be drawn over the lumps of starch to pulverize them and facilitate drying. Wooden rakes are then used to turn the starch over and expose new surfaces for drying.

Step 4: semi-mechanized tapioca starch manufacture

Another method of tapioca starch manufacture is more mechanized but again involves use of some crude wooden equipment. The roots are washed by tumbling in a water-filled U-shaped trough fitted with paddle arms which move the roots along to the peeler. Peeling is accomplished in a slotted rotating drum with a provision for an internal water spray, or in a U-shaped slotted trough with paddle arms which toss the roots and carry them through the peeler. The roots with 75–80% of the thin outer peel removed are fed into a Jahn rasper for grinding. The rasper disintegrates the roots into a relatively fine pulp, which collects in a tank. A piston pump is used to deliver the pulp, which has been mixed with a small amount of fresh water, to a series of vibrating screens fitted with 50–60-mesh phosphor bronze gauze. Usually a small spray of fresh water is applied to assist in the separation of starch and coarse fibre and at the same time keep the screen meshes clean.

The starch milk having a density of 2–3° Baume is led into a cement settling basin which is 100ft long, 25ft wide and 6–8ft in depth. The settling of the starch creates a very gradual inclined slope to the exit channel. When the basin is full, the surface of the starch cake is cleared of fibre. The starch cake is transferred to open yards and dried and stored in airtight sacks.

Step 5: Modern methods of tapioca starch manufacture

The most modern mills are mechanized with electrical equipments. Diesel-electric generators are used to furnish the power either as a complete source of energy or as stand-by power in case of failure in the municipal supply. Approximately 200 H.P are needed for the processing of 50 tons of roots per 24 hours. Water is supplied from wells or cisterns and is usually treated with lime and alum and filtered before use.

Roots with too much peel as well as those with woody ends or other defects are removed, trimmed and returned to the head end of the peeler. Grinding in the primary stage is done in Jahn Rasps or hammer mills having retaining screens with circular holes of 3.5 mm diameter. The ground pulp is pumped to brush washers which extract the freed starch. These brush washers are troughs of 20ft length with finely perforated U-shaped copper plates fitted to the bottom. A gentle spray of water from a pipe mounted above the machine aids in leaching out the starch. The resulting starch milk, having a density of 2–3° Baume, is passed over vibrating screens fitted with 120-mesh gauze which removes fine fibre. The pulp discharged from the first brush washer undergoes a second grinding in a second rasp or hammer mill in which the retaining screens have 1 mm pores. The reground pulp is pumped to a second brush washer from which the starch milk passes to the first vibrating screen, while the spent pulp is discharged from the factory as waste or to a pulp drying section for use in cattle feed. The first vibrating screen produces starch milk which is pumped to a second set of vibrating screens fitted with a 200-mesh phosphor bronze gauze. Just prior to the final screening a 0.2% sulfur dioxide solution is often added at the rate of approximately 0.6 litre per kg of dry starch. The addition of sulfur dioxide improves starch colour, aids in the subsequent settling steps and prevents deterioration of the starch from the action of bacteria and mould.

The thin inner core of the centrifuge cake containing impure starch is scraped, crumbled and sent to vacuum batch driers. These dryers are generally heated with steam, taking care to prevent formation of lumps of cooked starch or charred starch.

In the centrifugal separator process, the washed and peeled roots are comminuted in a chopping machine which facilitates the feeding of roots at a regulated rate to the rasping steps. The finely ground pulp is introduced into a small end of a vertically revolving, conical refining screen. The centrifugal action forces the starch through the screen, while the pulp migrates into the layer tank from which it is discharged. These refiners are arranged in two batteries of two or three screens in series, with a second rasping of the pulp in between. The starch milk comes out with a density of 2–3° Baume.

It has been recognized in recent years that the finest quality tapioca starch results when the enzyme bearing 'fruit waters', that is root juices, are removed early in the process. Nozzle centrifuges perform this function effectively while giving the starch a preliminary washing and removing some fine fibre. The starch milk is introduced into the top of the centrifuge and enters the base of the rapidly revolving bowl (3500–5100 rpm) near the central shaft. A series of nozzles of carefully selected diameter are arranged around the periphery of the bowl.

The starch milk is treated with sulfur dioxide solution and diluted to 8° Baume. It is then passed over vibrating screens of 120–150 mesh and subsequently over 200-mesh phosphor bronze gauze to remove fine fibre.

Final drying of tapioca starch, to 12% moisture, is carried out in most factories by a flash dryer. Air at high velocity is passed through a heat exchanger to raise the temperature to 150°–160°C. Starch is drawn into the riser duct through an air lock and in a matter of seconds, the excess moisture is 'flashed' off. The dried starch is stored in sacks free from moisture(see Fig. 3.2a, Fig. 3.2b).

CTCRI Trivandrum has developed a mobile starch extraction unit. The mobile starch extraction plant is cost-effective equipment permitting on-farm starch extraction by the producers and thus avoids the exploitation of the farmers by middlemen. The major components of the machine are a hopper to feed the tubers, a crushing disc or cylinder with nail-punched protrusions rotating inside the crushing chamber to crush the tubers, a sieving tray to remove the fibrous and other cellulosic materials, stainless steel or plastic tank to collect the sieved starch suspension, tuber storage chamber, a handle and wheels for easy transportation from place to place, and a frame to support these components. An electric motor (¾ HP) or a generator (kerosene/petrol) attached to the frame can be used as the energy source to operate the machine. The size of the equipment is 1350 × 1800 × 1320 mm and the capacity of the machine is 120–200 kg h^{-1}. (See Fig. 3.3)

The features of the unit include:

- suitable for starch extraction from different tuber crops
- provision for easy transportation
- provision to store peeled tubers
- provision for easy removal of rasper blades
- easy to fabricate, operate and maintain.

3.3 Physiochemical properties

3.3.1 Biochemical content

Though extracted starch appears 100% pure, it is actually accompanied by a number of minor constituents which can influence the physical, chemical and functional properties of the starch. The major contaminants are lipids, proteins, fibre and micronutrients. Different studies indicate considerable variation in the chemical composition (Table 3.2). It has been reported that the starch extract can

Fig. 3.2a. Cassava manufacturing units.

contain 0.01–1.2% lipids, 0.0029–0.0095% phosphorus, 0.11–1.9% fibre, 0.03–0.29% ash and 0.06–0.75% protein. Cassava starch contains lower lipid content relative to cereal starches. Palmitic, oleic, linoleic, and linolenic acids are the major types of fatty acids found in starch granules (Lehrman, 1932). Phosphorus is mainly found in starches as phosphate monoester (0.0065%) as detected by ^{31}P nuclear magnetic resonance spectroscopy (Kasemsuwan and Jane, 1996). Moreover, C3-phosphate is present in significantly higher amounts than C6-phosphate, with the latter present in only trace amounts (Anggraini *et al.*, 2009; Gomand *et al.*, 2010). The phosphorus content of cassava starch has been found to be lower than potato, wheat, mung bean, normal and high-amylose starches (Kasemsuwan and Jane, 1996). Various non-genetic factors like planting season, year and biomass form (fresh versus dry) can also affect starch composition (Abera and Rakshit, 2003; Asaoka *et al.*, 1991; Eke *et al.*, 2009). Abera and Rakshit (2003) reported lower fibre content in starch obtained from fresh roots compared to dried roots, and the type of milling process (wet versus dry milling) had only a small effect on starch composition. There was no major effect of planting seasons on starch composition in the study reported by Asaoka *et al.* (1991). Moorthy *et al.* (1993) observed that the starchy flour obtained by inoculum provided fermentation of cassava tubers had higher fibre content.

3.3.2 Amylose and amylopectin content

Starch is made up of two major molecular components, amylose and amylopectin, and cassava starch also has the normal ratio of 1:3 for amylose and amylopectin content, observed in most starches. Amylose is mostly linear with a few side chains through α-(1-6)-linkage branched from the α-(1–4)-linked d-glucose backbones (Pérez and Bertoft, 2010). The amylose can interact with iodine to form inclusion complex that has a characteristic bluish colour.The ability of the iodine inclusion formation reflects the size and structure of the amylose and has been quantified by iodine affinity (%) (18.8–19.3%) and blue value (1.43–1.5) (four genotypes) (Charoenkul *et al.*, 2006). The amylose content is one of the most important quality attributes of starch and determines the diverse properties and eventual end-use; hence it has been subjected to a great deal of investigation (Srichuwong

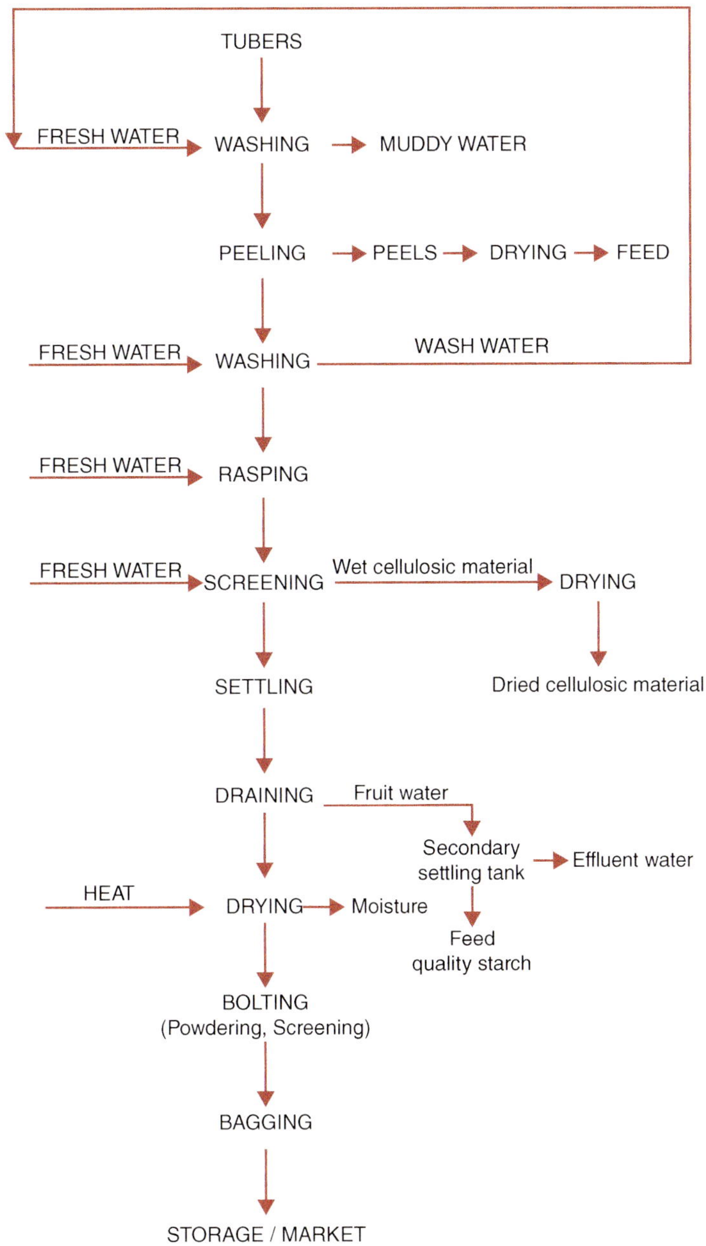

Fig. 3.2b. Cassava processing machine.

and Jane, 2007). For the quantification of amylose content in starch, there are diverse methods based on differential scanning calorimetry (DSC), iodine-binding, chromatography, and so on. Different methods can provide diverse results for the same sample (Charoenkul *et al.*, 2006; Zhao *et al.*, 2011). For example, it has been observed that diverse results are obtained for the following pairwise methods:

- gel permeation chromatography in Sepharose CL-2B versus

Fig. 3.3. Mobile starch echo unit (CTCRI Trivandrum).

iodine-binding–amperometry (Charles *et al.*, 2004);

- two iodine-binding-based methods (spectrophotometry/colorimetry versus amperometry) (Rolland-Sabaté *et al.*, 2013);
- DSC versus iodine-binding–amperometry method (Rolland-Sabaté *et al.*, 2012);
- Iodine-binding–amperometry versus high-performance size-exclusion chromatography after debranching the starch (Charoenkul *et al.*, 2006);
- the iodine-binding–spectrophotometry versus DSC method for the same sample (Ceballos *et al.*, 2007).

Thus it is obvious that the specific method for amylose measurement should be considered when comparing data from different studies.

The values reported for amylose content vary considerably from 14 to 30% (see Table 3.2; Zhu, 2015). The majority of reports of amylose content in starch state it is in the range 20–27%, similar to most other starches. Wickramasinghe *et al.* (2009) obtained values of 25.4–28.8% for starch of some cassava varieties. The amylose content shows variation among varieties, but the age of the crop and environmental factors may or may not affect the amylose content to any large extent. GPC analysis of debranched starch of different varieties did not reveal much variation among the varieties (Fig. 3.4).

Gu *et al.* (2013) have found that variation in the amylose content was statistically insignificant with rainfall and temperature, but genetic factors influence the amylose content (Fig. 3.5).

Similarly, the amylose content ranged from 15.2 to 26.5% in 4050 cassava genotypes from a world collection (Sanchez *et al.*, 2010). Influence of agronomic factors on the amylose content has also been documented (Defloor *et al.*, 1998; Moorthy and Ramanujam, 1986; Sriroth *et al.*, 1999). Amylose content remained similar at various growth stages up to 6 month (Moorthy, 1985) (see Table 3.3).

However, the amylose content dropped during the months 6 to 16 (Sriroth *et al.*, 1999). This showed a significant effect of harvesting time but only a small effect of planting season on amylose content (Asaoka *et al.*, 1991; Defloor *et al.*, 1998). It is interesting to note that amylose content in the outermost layers of the root does not differ significantly from the innermost layers (Teerawanichpan *et al.*, 2008).

A diversity of amylose content in cassava starches can be obtained using traditional

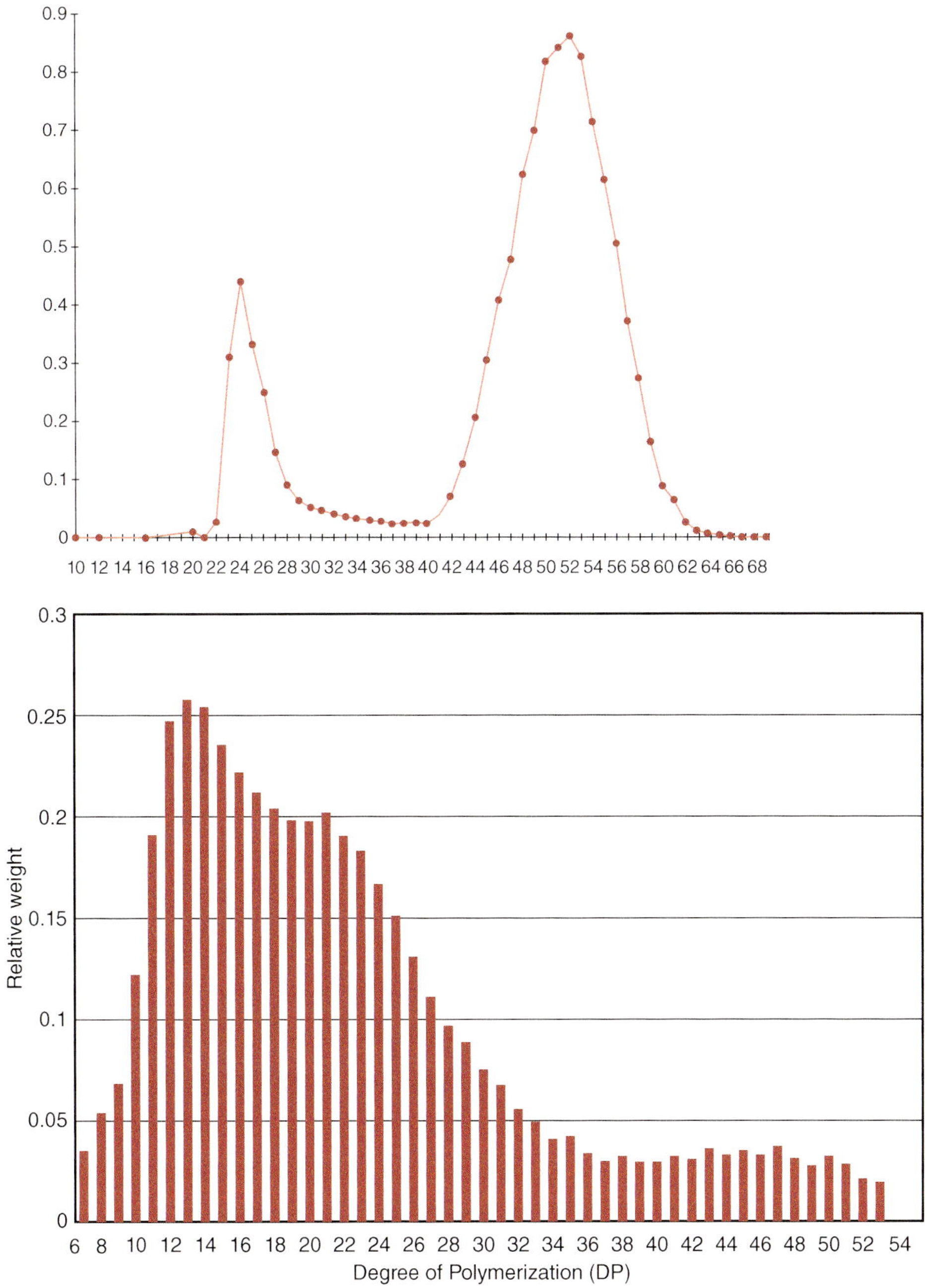

Fig. 3.4. GPC and HPSE pattern for debranched cassava starch (Moorthy *et al.* unpublished data).

plant breeding as well as modern biotechnology techniques. A waxy clone of cassava with minute amylose content was obtained by selection of natural mutations that reduce the presence of granule-bound starch synthase (GBSS), an enzyme responsible for amylose biosynthesis (Ceballos *et al.*, 2007). Joseph *et al.* (2004) employed a cyclic somatic embryogenic system and

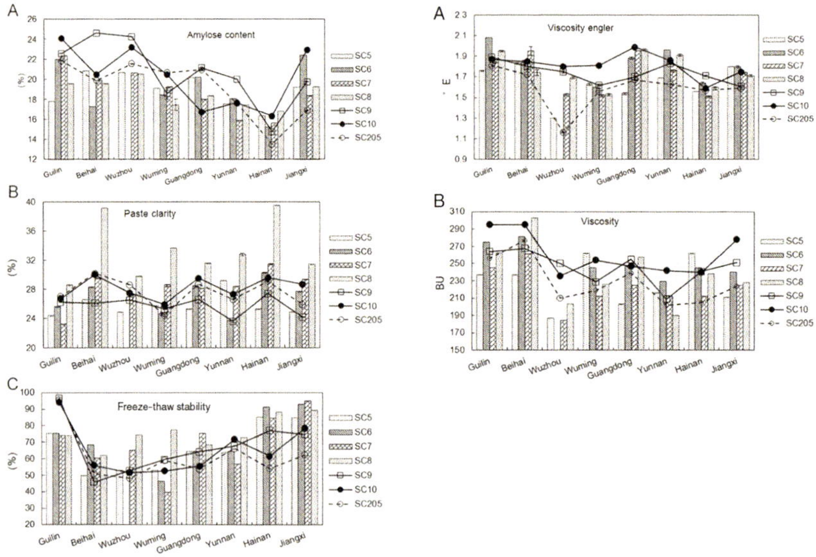

Fig. 3.5. Histogram illustrating distribution of amylose content (A), paste clarity (B), freeze–thaw stability (C) of starches extracted from seven genotypes grown at eight locations. Histogram illustrating distribution of viscosity Engler (A) and viscosity (B) of starches extracted from seven genotypes grown at eight locations. The bars indicate the calculated SD.

Source: (Gu *et al*., 2013).

λ-irradiation to generate waxy mutants with minimal starch content. A genotype with amylose-free starch was obtained by Raemakers *et al.* (2005) using antisense inhibition of granule-bound starch synthase 1 (GBSS1). Similarly Zhao *et al.* (2011) produced low-amylose and amylose-free genotypes by suppressing the expression of GBSS1. Field testing of low/nil amylose varieties of cassava in Indonesia showed great commercial potential owing to good root yield and high starch quality (Koehorst-van Putten *et al.*, 2012). On the other hand, cassava mutants with relatively high amylose content (~30%) have been obtained using λ-irradiation (Ceballos *et al.*, 2007; Rolland-Sabaté *et al.*, 2012). Other genotypes with higher amylose content (e.g. >50%) have been obtained which are highly suitable for certain applications.

Soluble amylose is that fraction of amylose that easily gets leached out into water medium during gelatinization and thus has been considered responsible for the cohesive texture of many soluble-based products. The soluble amylose fraction forms about 40% of total amylose of cassava starch (Waterschoot *et al.*, 2015). It has been found that complex formation with emulsifying surfactants can reduce soluble amylose and amylose content (Table 3.4).

3.4 Structural Properties

3.4.1 Granular morphology

The molecular components, namely amylose and amylopectin, are naturally assembled in the form of semi-crystalline granules and contribute to granular characteristics. There exists wide variation in the reported morphological features of starch from different sources. The morphological features of cassava starch, including the size and shape, have been summarized in Table 3.5.

The shape of cassava granules has been described as oval, truncated and rounded as observed by various microscopic techniques (i.e. scanning electron microscopy and light microscopy). Under a light microscope the granules appear mostly round with a flat surface on one side containing a conical pit, which extends to

Table 3.3. Amylose content (%) in relation to age of the crop.

Variety	Age in months															
	2	3	4	5	6	7	8	9	10	11	12	13	14	15	16	17
M4	21.07	19.84	20.44	19.72	18.73	19.15	21.07	20.8	22.04	-	21.85	21.17	21.17	21.2	23.5	22.05
H 2304	20.58	20.58	20.58	19.64	20.60	19.15	20.58	23.11	-	-	19.15	22.0	21.17	21.58	22.0	20.58
H 1687	20.50	21.17	20.58	19.12	21.17	20.0	20.58	22.0	19.11	21.15	22.15	20.0	20.0	22.0	24.0	20.58
H 165	17.04	18.23	19.12	18.27	19.12	21.17	17.64	20.0	20.58	20.58	19.15	21.0	22.0	22.0	20.58	20.58
H 226	20.58	20.58	18.7	20.0	19.12	20.58	20.58	22.04	21.17	21.25	20.58	17.45	20.58	23.08	20.28	18.15
H 97	19.12	20.58	18.3	19.12	20.58	17.64	17.64	20.58	20.58	-	21.17	22.0	20.58	20.58	20.0	20.58

Source: (Moorthy and Ramanujam, 1986).

Table 3.4. Total and soluble amylose content of surfactant-incorporated starch.

Starch	Blue value	
	Total amylose:0.58	Soluble amylose:0.35
Starch (100 g)+0.02 mol of potassium stearate	0.45	0.18
Starch (100 g)+0.04 mol of potassium stearate	0.41	0.11
Starch (100 g)+0.06 mol of potassium stearate	0.41	0.13
Starch (100 g)+0.02 mol of potassium palmitate	0.41	0.17
Starch (100 g)+0.04 mol of potassium palmitate	0.38	0.11
Starch (100 g)+0.06 mol of potassium palmitate	0.37	0.10
Starch (100 g)+0.02 mol GMS	0.42	0.16
Starch (100 g)+0.04 mol GMS	0.41	0.16
Starch (100 g)+0.06 mol GMS	0.41	0.14
Starch (100 g)+0.02 mol of sodium lauryl sulphate	0.44	0.22
Starch (100 g)+0.04 mol of sodium lauryl sulphate	0.41	0.20
Starch (100 g)+0.06 mol of sodium lauryl sulphate	0.39	0.18
Starch (100 g)+0.02 mol of cetyltrimethyl ammonium bromide	0.37	0.05
Starch (100 g)+0.04 mol of cetyltrimethyl ammonium bromide	0.35	0.02
Starch (100 g)+0.06 mol of cetyltrimethyl ammonium bromide	0.34	0.01
CD (5%) for interactin	0.021	0.014
CD (5%) for concn	0.009	0.006
CD (5%) for reagents	0.011	0.006

Source: (Moorthy, 1985).

a well-defined eccentric hilum. Under polarized light, a well-defined cross is observed.

The study of starch granule surface with atomic force microscopy (AFM) in non-contact mode revealed blocklets, depressions and protrusions on the surface of starch granules as illustrated in Fig. 3.6. However, the surface of cassava starch granules has been reported to be smoother than that for potato (Juszczak *et al.*, 2003).

The granule size can be observed to vary between 2 and 32 μm (Table 3.5) but lies most frequently between 7 and 20 μm (Rolland-Sabaté *et al.*, 2012). A significant variation in average granule size of normal genotypes can be observed across studies (Defloor *et al.*, 1998; Rolland-Sabaté *et al.*, 2012). The size can have unimodal as well as bimodal distribution (centred at ~15 and 12 μm) with the latter becoming more evident with root ageing up to 16 months (Sriroth *et al.*, 1999). It should be noted that the size analysis of the same sample with different techniques (e.g. laser scattering based techniques versus manual measurement from scanning electron microscopic photos) may yield different results.

Studies have reported that various environmental/controllable factors can influence granule morphology. The storage roots had larger starch granules than fibrous roots. The innermost layers of storage root contained starch with higher granule size than that in the outermost layers. Wet planting season favoured higher granule size (Teerawanichpan *et al.*, 2008). Moorthy and Ramanujam (1986) reported variation in granule in five varieties of cassava. The granule size increased mostly in the first six months after cultivation and remained almost similar thereafter (see Table 3.6).

Table 3.5. Granular morphology and X-ray diffraction (XRD) pattern.

No	Pattern	Crystallinity %	Granule size µm	Shape	Reference
4[a]	C_a	15.3–17.8	2.4–31.1		Asaoka *et al*. (1991)
6			6.96–11.21		Safokantanka and Owusunipah (1992)
1[b]	C_a			Round, cylindrical and polygonal	Padmanabhan and Lonsane (1992)
1	A				Gorinstein and Lii (1992)
3×2[c]	C_a	15.4–17.8	2.4–31.1		Asaoka *et al*. (1993)
1	A				Francisco *et al*. (1996)
1	A		15av.		Garcia *et al*. (1996)
5			3–32		Moorthy *et al*. (1996)
4×2×5[d]	A		3–32		Defloor *et al*. (1998)
4×4[e]			8–22	Irregular, oval, round, truncated	Sriroth *et al*. (1999)
11			9–20	Oval, truncated, rounded	Niba, Bokanga, Jackson, Schlimme, and Li (2002)
5			12.9–17.2		Charles *et al*. (2005)
40			12.5–23.8		Onitilo *et al*. (2007)
1			16		Tetchi *et al*. (2007)
4			5.8–18.73		Ceballos *et al*. (2008)
39			12.5–22.5		Eke *et al*. (2009)
1			2.81–14.03		Nwokocha *et al*. (2009)
17			7.3–9.7		Anggraini *et al*. (2009)
2	A	40–49	16.9–18	Oval to round	Gomand *et al*. (2010)
12	A		4.6–22.8	Round and truncated	Charoenkul *et al*. (2011)
8	C_a	25–40	8–17	Spherical, polyhedral, truncated	Rolland-Sabaté *et al*. (2012)
8	C_a	35–40	12.1–16.1		Rolland-Sabaté *et al*. (2013)
1			7.1–25	Spherical, truncated	Mishra and Rai (2006)

No., number of genotypes/samples.
Source: (Zhu, 2015).
[a]3–4 different harvesting months; [b]2 extraction methods; [c]with/without pruning aerial growth; [d]5 different harvesting month, and two different planting seasons; [e]4 different harvesting months; peak and average in the parenthesis denote the peak and average size of the granules, respectively.

Crops under water stress had lower granule size and this effect was more prominent for stress during earlier growth stage (Santisopasri *et al*., 2001). Granule sizes are larger for the crop planted in the rainy season than those planted in the dry season (Defloor *et al*., 1998). Asaoka *et al*. (1993) found that pruning also affected the granule size. The maximum granule size was

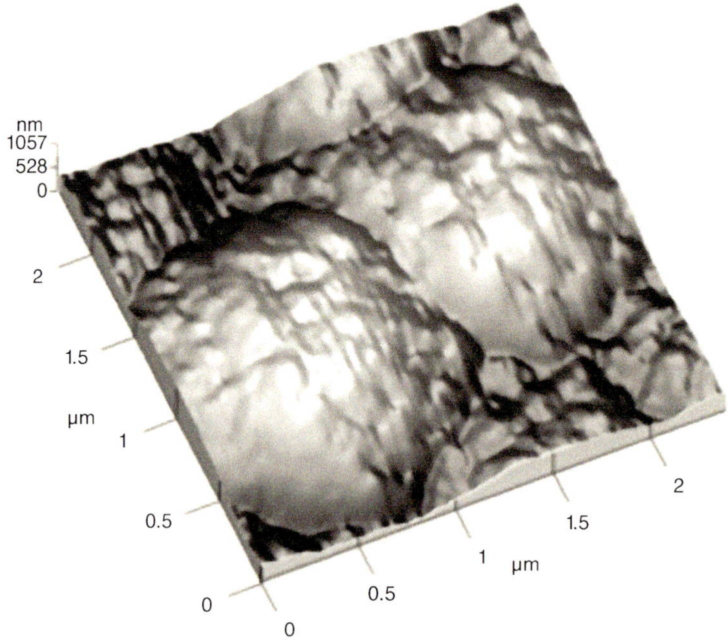

Fig. 3.6. Atomic force microscopy (AFM) image of the surface of cassava granule.

Source: Juszczak *et al*., 2003.

lower for pruned plants but the effect varied with variety of cassava. According to Gu *et al.* (2013), the granule shape was little affected by rainfall and temperature.

The crop genotypes with different amylose compositions can have wide differences in granule morphology. A high amylose mutant (~30% amylose), induced by λ-irradiation, was found to have smaller granules (average size 5.80 µm) with higher irregularity in shape as compared with a commercial normal genotype (13.97–18.73 µm). However, a natural waxy genotype and a normal genotype were found to be not significantly different (Ceballos *et al.*, 2008).

3.4.2 XRD and starch crystallinity

The structure of a starch granule involves alternating amorphous and semi-crystalline shells, each 100–400 nm thick. Using small-angle X-ray scattering analysis (SAXS), the semi-crystalline shells have been found to contain a further periodicity of 9–10 nm thickness. The specific arrangement of crystalline portions in starch granules gives rise to birefringence under polarized light (Pérez and Bertoft, 2010).

Amylopectin chains form molecular double helices with each other by intermolecular interaction and interaction with intrinsic water molecules. These helices form crystalline structures which are further arranged to give rise to two fundamentally different polymorphs: A- and B-type. This polymorphism was revealed by Pérez and Bertoft (2010) using wide-angle X-ray diffraction (XRD) analysis. A mixture of these two forms can also occur and is called C-type (Pérez and Bertoft, 2010). Cassava has been reported to contain A- or C_a-type polymorphism (Table 3.5). Detailed studies by Rolland-Sabaté *et al.* (Rolland-Sabaté *et al.*, 2013, 2012) showed that C_a-type starch contained 10–20% B-type crystallinity. Unlike granule size, the diffraction pattern (polymorphism) does not vary with environmental factors and stage of crop development. In a study by Defloor *et al.* (1998), crop

Table 3.6. Granule size (μm) in relation to age.

Variety	Age in months														
	2	3	4	5	6	7	8	9	10	11	12	13	14	15	16	17
M4	10.0	11.25	12.5	13.75	13.50	15.2	15.0	13.0	13.0	12.6	12.3	12.0	12.5	14.2	14.7	15.2
H 2304	9.0	11.7	12.5	17.0	15.75	14.0	13.75	16.25	14.75	13.75	15.0	13.75	14.4	14.0	15.0	16.25
H 1687	7.5	9.0	12.20	13.75	16.55	14.75	13.75	12.75	15.75	14.5	13.5	13.75	14.25	15.75	14.2	14.75
H 165	7.0	9.0	10.50	12.75	15.75	15.50	15.0	15.75	15.75	14.5	12.25	13.25	15.0	14.25	14.0	12.75
H 226	8.4	11	11.0	11.6	13.6	12.44	12.80	12.16	13.2	11.8	11.6	12.8	13.8	12.7	12.6	12.9
H 97	9.04	9.60	9.88	11.56	11.84	14.04	14.84	14.5	16.50	14.10	14.5	16.1	13.7	13.7	14.8	13.6

Source: (Moorthy and Ramanujam, 1986).

age and planting season did not affect crystallinity as all samples had A-type polymorphism. Another study found that harvesting season did not affect wide-angle X-ray diffraction pattern and crystallinity percentage (Asaoka *et al.*, 1991). Similarly, partial or total deactivation of GBSS1 using crop genetics did not affect granular structure (Rolland-Sabaté *et al.*, 2013). It should be noted that the final results can be affected by the method used to calculate crystallinity (Lopez-Rubio *et al.*, 2008) and the sample treatment used for wide-angle X-ray diffraction analysis. The diffraction pattern shifted from A-type to C-type due to increase in sample moisture content (Francisco *et al.*, 1996). The discrepancy in the percentage crystallinity values reported by Asaoka *et al.* (1991, 1993) (~15–18%) and other studies (Gomand *et al.*, 2010; Rolland-Sabaté *et al.*, 2012, 2013) (Table 3.5) (25–50%) has also been attributed to the use of different analytical and calculation methods.

3.4.3 Structure of amylose and amylopectin

The structural properties of amylose and amylopectin have been studied in detail by various techniques and compared in Table 3.7.

This observed variation in values can be attributed to the analytical method employed. Amylopectin, the branched fraction, is generally calculated by subtracting the amylose content in starch and makes up almost three times the amylose content. Amylopectin is mostly around 70–75% in cassava starch with about 5–6% branches. Charoenkul *et al.* (2006) reported that cassava starch amylopectin (four genotypes) had iodine affinity value (%) and blue values of 0.18–0.23% and 0.080–0.091, respectively. The apparent average radius of gyration was 87 nm (Juna and Huber, 2012) or 188 nm (Tetchi *et al.*, 2007). The ranges of the area percentages

Table 3.7. Structural properties of amylose and amylopectin.

Characters	Amylose	Amylopectin
DP (degree of polymerization)	1035–1202 (four genotypes harvested at four different months) measured by high-performance size-exclusion chromatography (HPSEC) (Srichuwong and Jane, 2007)	6000–17,100 (four genotypes) by HPSEC coupled with fluorescent labelling technique (Charoenkul *et al.*, 2006)
	2.70×10^5 (five genotypes) by HPSEC coupled with multi-angle laser light scattering and refractive index detectors (HPSEC-MALLS-RI) (Charles *et al.*, 2004)	1.88×10^8 by HPSEC coupled with MALLS and differential refractometric index detectors(HPSEC/MALLS/DRI) (Tetchi *et al.*, 2007)
	DPn2050–439 (four genotypes) by HPSEC coupled with fluorescent labelling technique (Chang *et al.*, 2000)	$5–41.8 \times 10^7$ (eight genotypes) by A4F coupled with MALLS and quasi-elastic laser light scattering (A4F/MALLS/QELS) (Rolland-Sabaté *et al.*, 2012)
	2.1×10^6 by asymmetrical flow field flow fractionation (AF4) coupled with multi-angle light scattering and refractive index detectors (AF4/MALS/RI) (one genotype) (Juna and Huber, 2012)	19×10^6 (one genotype) by AF4/MALS/RI (Juna and Huber, 2012)
Average chain lengths	450–550 glucosyl residues	a) 20–21 glucosyl residues (four genotypes) (Charoenkul *et al.*, 2006) b) 18.4–19.5 glucosyl residues (three genotypes) (Rolland-Sabaté *et al.*, 2012) c) 18.0–19.1 glucosyl residues (two genotypes)
Number of chains	4.7–7.5 (Chang *et al.*, 2000)	300–855 (four genotypes) (Charoenkul *et al.*, 2006)

Table 3.8. Distribution of amylopectin molecular weight.

Number of chains per amylopectin	Area percentages (%) of unit chains
DP 6–9	9.3–11.6
DP 6–12	28.8–30.9
DP 13–24	47.9–49.0
DP 25–36	12.3–13%
DP >37	8.6–9.4%

Source: (Rolland-Sabaté *et al.*, 2013).

of unit chains for four genotypes are shown in (Table 3.8) (Rolland-Sabaté *et al.*, 2013).

Some reports also indicated the influence of crop genetics and environmental factors on the unit chain profile of cassava amylopectin (Asaoka *et al.*, 1991; Raemakers *et al.*, 2005). Amylopectin from amylose-free starch generated by antisense inhibition of GBSS1 had similar chain length distribution compared to the amylose-containing ones (Raemakers *et al.*, 2005). In a comparative study, chain length of cassava amylopectin was shorter than that of potato, which had a B-type polymorph (Gomand *et al.*, 2010).

The heterogeneity of amylopectin chain length can be quantified by dispersity index, which is the ratio of weight average molar mass to number average molar mass (M_w / M_n). Rolland-Sabaté *et al.* (2012) tested eight genotypes of cassava and found the dispersity value to be 1.25–2.23. High-amylose mutants (induced by γ-irradiation) possessed starch with high dispersity index and lower average number of branching points than normal genotypes (Rolland-Sabaté *et al.*, 2012). High-pressure size exclusion chromatography (HPSEC) was employed to analyse debranched samples of amylopectin to determine the unit chain length. The study also found a lower average number of glucosyl units per branching point in these mutants. High-performance anion-exchange chromatography coupled with pulsed amperometry detection (HPAEC-PAD) could also be used to determine the proportion of unit chains that have a defined chain length. Asaoka *et al.* (1991) found minimal effect of planting seasons on the size distribution of isoamylase-debranched starch. The chain length distribution of amylopectin in waxy starches produced by antisense inhibition of GBSS1 was similar to that in starch with amylose (Raemakers *et al.*, 2005). The unit chain on an amylopectin molecule that carries its sole reducing end is called the C-chain. Charoenkul *et al.* (2006) used HPSEC coupled with fluorescent labelling technique to analyse the C-chain length distribution. The asymmetrical distribution that was found had a peak at DP 41–42 and a shoulder at DP 24. The weight percentage of super-long unit chain lengths in amylopectin has been found to be 0.24–1.78% (12 genotypes) by Charoenkul *et al.* (2006), while it was found to be 0.8–3.0% (two genotypes) by Laohaphatanaleart *et al.* (2009). The above results were obtained by using different methods i.e. HPSEC versus low pressure gel-permeation chromatography, and two different types of gel for the latter (Sepharose CL 6B versus Superdex 75).

The branches in the amylopectin molecule are systematically arranged as clusters in a systematic way (Pérez and Bertoft, 2010). The external chain segments of amylopectin are cleaved by exoacting amylases such as β-amylase or phosphorylase-α and β-amylase to give rise to limit dextrins. The β-limit value (%) (reflecting the hydrolysed part) of amylopectin was 59–65% (four genotypes), and ψ- β-limit value (%) was 50.6–59.7% (two genotypes) (Laohaphatanaleart *et al.*, 2009). The unit chain length distribution of β- or , ψ-β-limit dextrins, reflecting the internal part of amylopectin, could be obtained by chromatography after debranching.

The molar amounts of A-chain in cassava starch vary from 52.9 to 53.6% (two genotypes) (Laohaphatanaleart *et al.*, 2009). Two populations (weight-based) of B-chains with peaks at DP 10 (major) and 35 (minor) and a shoulder at DP 6 were observed. By comparing the unit chain profiles of amylopectin and the, ψ-β-limit dextrins, various structural parameters have been calculated. The average values of external chain length (ECL), Internal chain length (ICL) and total internal chain length were found to be 10.6–12.9, 5.2–6.4, and 13.1–15.7 respectively (Laohaphatanaleart *et al.*, 2009). Bertoft *et al.* (2008) found a high correlation between the internal molecular structure of amylopectin and the polymorphism of granular starch by analysing a variety of botanical origins. The cluster and building-block structure of amylopectin in granules could also be closely associated to polymorphism (Bertoft *et al.*, 2012).

The structural composition of clusters and building blocks has been characterized by diverse chromatographic techniques coupled with debranching. It was found that the number of chains per cluster (NC) was 8.2–11.1, the molar amount of chains of clusters was 58.2–61.0%, DP was 58–78, and the internal chain length was 4.4–4.8 glucosyl residues for five fractions (Laohaphatanaleart *et al.*, 2010; Bertoft *et al.*, 2010). Each cluster had 7.3–8.9 blocks that were each separated 6.8–7.5 glucosyl residues (five fractions) (Bertoft *et al.*, 2010). A visual model of amylopectin with two as well as three dimensions has been developed based on this data (Fig. 3.7). The model depicts the assembly of building blocks in clusters, amylopectin and granules with a 9 nm repeating distance. This model may be just one of the ways of visualizing amylopectin. Moreover, this model is simple as it is based on averaged data while these values can vary in amylopectin molecules, as is shown by the dispersity index values (Rolland-Sabaté *et al.*, 2012). The exact structure of amylopectin still remains a subject of investigation for the generations to come.

Schmitz *et al.* (2017) have studied 10 land races from Brazil and found considerable variability in the branch chain length among them (Table 3.9). They have different viscosity and thermal properties. They point out the possibility of genetic races with desirable starch properties.

3.5 Functional Properties

Starch granules when heated in the presence of water, start to absorb water and swell. During this process some of the amylose, soluble amylopectin and other low-molecular components leach out of the granules and solubilize. The increased temperature and higher absorption of water by the swollen granule disturbs its molecular order and causes it to rupture. This process is called gelatinization of starch. It can widely vary in terms of the degree of swelling, the starting temperature and the amount of granule disruption among different starches. The cooling of gelatinized starch allows the disordered chain to have increased intermolecular interaction in the form of hydrogen bonding, which caused partial re-association. This process is called retrogradation. The diverse applications of starch in food and other industries depend on its gelatinization and retrogradation properties (Zhu and Wang, 2014). In order to determine these fundamental characteristics, various techniques have been used. The relation of these properties to the structure of starch granules has also been investigated.

3.5.1 Swelling pattern and solubility

Great diversity in both swelling power (SP) and solubility (S) of starch granules at a wide range of temperatures (30–95°C) has been observed (Table 3.10).

For example, a study on starch from 4050 genotypes that occur over the world showed large differences in the values of SP (0.8–15.5) and S (0.2–16.6%). Genetically modified starches have also been found to show different swelling and solubilization properties. Waxy mutants produced by GBSS1 suppression or breeding have both been found to have higher swelling power and reduced water solubility (Ceballos *et al.*, 2007; Zhao *et al.*, 2011). On the other hand, a high-amylose (30%) mutant (produced by γ-irradiation) with smaller granules showed lower SP and comparable S values relative to commercial ones (Ceballos *et al.*, 2008).

The influence of crop physiological maturity and harvesting month on the swelling properties of starch has been reported to be variety or genotype dependent (Moorthy and Ramanujam, 1986; Sriroth *et al.*, 1999). Santisopasri *et al.* (2001) showed that water stress, irrespective of crop growth stage, decreased SP of starch. In another study, starch obtained from fresh roots has showed lower SP than starch from dried roots (Abera and Rakshit, 2003). The proteins associated with starch granules also affect the swelling process (Hongsprabhas *et al.*, 2007; Israkarn *et al.*, 2007). These proteins have been found to develop envelopes around starch content during heating which can slow down the swelling process (Hongsprabhas *et al.*, 2007). By encasing granular starch, the protein envelopes can significantly influence pasting behaviour and other thermal properties of starch (Israkarn *et al.*, 2007).

Cassava starch has medium swelling power compared to potato and cereal starches, a property in conformity with its observed viscosity. It was observed that during the growth period, starch of two varieties of cassava maintained

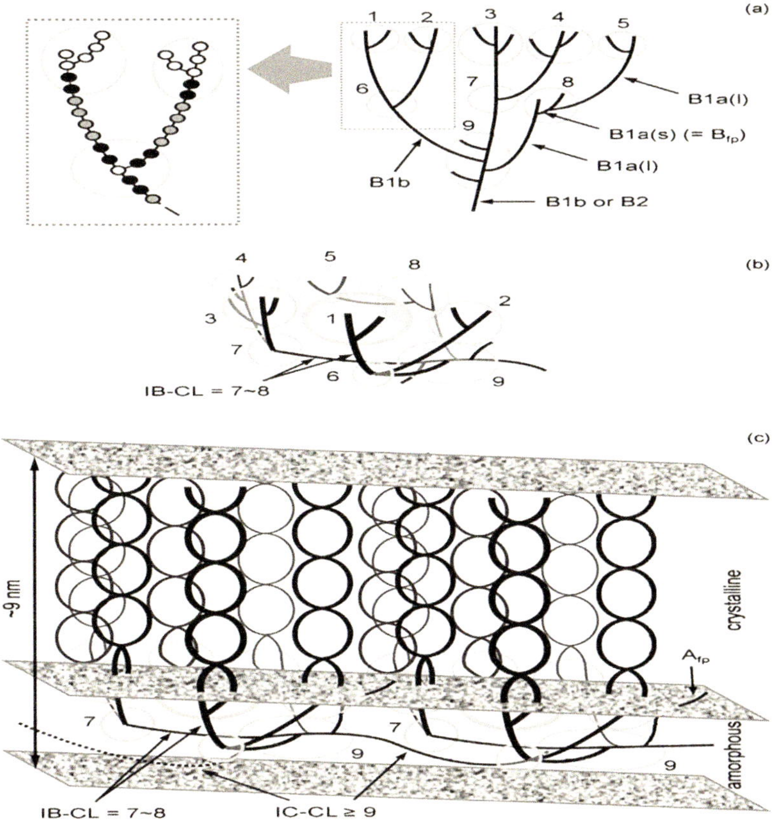

Fig. 3.7. Amylopectin 3D model.

A model of how the building blocks can be organized into a cluster of amylopectin. (a) Black lines symbolize the unit chains of diverse B-chains categories indicated;building blocks are encircled in light grey and numbered 1–9; a cluster segment is enlarged to illustrate how the glucose units are involved in building blocks (white and black circles) and in the inter-block segments (grey circles); the segmental length between branches of two adjacent blocks (inter-block chain length, IB-CL = ~7–8 glucosyl residues) involves additional residues (~4) (black circles inside the blocks). (b) A three-dimensional illustration of the cluster; numbers correspond to the building blocks from the 2D drawing in (a). IB-CL is similar between all building blocks. (c) A larger structure including two clusters shown in (b) with an amorphous lamella of the granules. The clusters are interconnected through blocks no. 9 (IC-CL, the inter-cluster chain length). Dotted line denotes the interconnection to more clusters. External chains in the form of double helices extend from the building blocks at the interfacial area between the lamellae into the crystalline part. Very short A-chains (Afp) are connected to building block no. 9 in the amorphous area and may protrude into the crystals, causing structural defects (Bertoft et al., 2010) (with permission from Elsevier). Note the nomenclature: A- and B-chains of clusters were later changed into a- and b-chains to distinguish them from the names of amylopectin chains.

Source: (Bertoft *et al.*, 2012).

their swelling volumes within small ranges while that for some varieties expressed wide variations, which indicate that these varieties are very much susceptible to environmental influences and also point to possible relationships between cooking quality and swelling volumes (Moorthy and Ramanujam, 1986). Nuwamanya *et al.* (2011) have studied the swelling power and solubility at different temperatures and the values increased from 2.43 to 10.19 (g/g) for swelling

Table 3.9. Branch chain length distribution of amylopectin of cassava starches.

	Branch chain length distribution (%)					
Variety	DP 6–12	DP 13–24	DP 25–36	DP ≥37	Average DP	Highest detectable DP
Ciganapreta	31.4 ± 0.1	46.6 ± 0.3	11.9 ± 0.1	10.1 ± 0.2	19.3 ± 0.1	81
IAC 13	31.5 ± 0.2	46.1 ± 0.2	12.1 ± 0.1	10.2 ± 0.3	19.4 ± 0.1	86
IAC 12	32.1 ± 0.2	45.9 ± 0.1	12.2 ± 0.1	9.8 ± 0.2	19.2 ± 0.1	79
IAC 14	31.3 ± 0.2	46.4 ± 0.2	12.1 ± 0.1	10.3 ± 0.3	19.4 ± 0.1	82
Caapora	30.9 ± 0.2	46.9 ± 0.1	12.2 ± 0.0	10.0 ± 0.4	19.4 ± 0.1	79
IAC 90	30.3 ± 0.2	46.9 ± 0.1	12.3 ± 0.1	10.5 ± 0.2	19.6 ± 0.1	78
Olho junto	31.9 ± 0.3	45.7 ± 0.1	12.6 ± 0.1	9.9 ± 0.2	19.3 ± 0.1	76
IAC 06–01	31.9 ± 0.3	46.0 ± 0.0	12.0 ± 0.1	10.1 ± 0.2	19.3 ± 0.1	79
IAC 576–70	31.0 ± 0.2	46.3 ± 0.1	12.4 ± 0.1	10.2 ± 0.2	19.5 ± 0.1	81
VassourinhaPaulista	31.2 ± 0.2	46.0 ± 0.1	12.4 ± 0.0	10.3 ± 0.2	19.5 ± 0.1	82

DP – degree of polymerization.
Source: (Schmitz *et al.*, 2017)

power when temperature was raised from 60 to 90°C. Soni *et al.* (1985) have reported a two-stage swelling for cassava starch and attributed it to the two types of forces that require different energy input to cause relaxation of the starch molecules. Asaoka *et al.* (1991) have determined the swelling volume of four varieties of cassava harvested during two seasons and observed that swelling power was higher in samples harvested in November compared to those harvested in August. Sanchez *et al.* (2010) observed wide variation in swelling power (0.8–15.5) among different varieties

Cassava starch has a higher solubility compared to the other tuber crop starches and the higher solubility can be attributed partly to the high swelling it undergoes during gelatinization and the reported values ranged from 25 to 48% (Table 3.10). The solubility of starch of different cassava varieties varied from 17.2 to 27.2%. However, no direct correlation between swelling and solubility could be observed. The values for solubility of starch from different varieties of cassava during the growth period also indicated that starch of two varieties had good stability in their solubility, whereas the others had medium or poor stability (Moorthy and Ramanujam, 1986). The effect of temperature studied by Nuwamanya *et al.* (2011) showed 0.19–1.57 g/g for solubility from 60–90°C. Sanchez *et al.* (2009) found values for solubility 0.2–16.6% among a number of varieties. Among a few non-aqueous solvents studied for solubility of starch, maximum solubility was obtained in DMSO and formalin, while in glycerol it was moderate. Starch was insoluble in anisole and methyl cellosolve. The solubility data indicate that starch is more soluble in polar solvents or solvents with affinity towards water (Table 3.11).

The minor components (e.g. protein) in the granules can also influence swelling and solubility as shown by the effect of fibre in the fermented starchy flour (Moorthy *et al.*, 1993). Anomalous behaviour was observed for swelling and solubility of cassava starch in presence of sodium sulphite. At a concentration of 0.01% there was drastic decrease in swelling with the corresponding increase in solubility and this has been explained due to oxidative-reductive depolymerizaton of starch (Fig. 3.8; Mat Hashim *et al.*, 1992).

3.5.2 Viscosity

Viscosity of starch suspensions has been among the most reported characteristics of starch (Table 3.12). Rapid Viscoanalyzer (RVA) and Brabender Viscoamylograph (BVA) are

Table 3.10. Swelling properties of cassava starch.

No	Parameter	Temperature °C 50	60	70	80	90	Reference
4a	Sp		17.0–22.6	27.7–32	34.8–43.6		Asaoka *et al*. (1992)
	S		7.5–11.3	14.4–17.1	18.4–19.9		Asaoka *et al*. (1992)
1	Sp	2.88					Gorinstein and Lii (1992)
	S	1.16					Gorinstein and Lii (1992)
1×4b	Sp		16.7–19.5	24.6–27.1	41.1–46.8		Asaoka *et al*. (1993)
	S		10.8–11	15.6–17	20.7–22.8		Asaoka *et al*. (1993)
1	Sp				28.7		Numfor *et al*. (1996)
	S				29.7		Numfor *et al*. (1996)
4×5c	Sp					45–77	Sriroth *et al*. (1999)
6×2×4d	Sp				33.3–73		Santisopasri *et al*. (2001)
3×3+1e	Sp				18.6–26		Abera and Rakshit (2003)
5	Sp				11.8–19.9	27.2–42.3	Charles *et al*. (2005)
	S				4.4–11.8	10.6–20.2	Charles *et al*. (2005)
40	Sp				9.0–16.9		Onitilo *et al*. (2007)
	S				1.03–2.1		Onitilo *et al*. (2007)
3	Sp			30.8–55.7			Ceballos *et al*. (2007)
	S			6–14.1			Ceballos *et al*. (2007)
4	Sp			9.25–27.92			Ceballos *et al*. (2008)
	S			8.34–36.84			Ceballos *et al*. (2008)
4050	Sp			0.8–15.5			Sanchez *et al*. (2010)
	S			0.2–16.6			Sanchez *et al*. (2010)
8	Sp		16–32	32–44	36–56		Anggraini *et al*. (2009)
7	Sp			20–48.2		35.2–54.7	Zhao *et al*. (2011)
	S			4.1–11.7		8–24.1	Zhao *et al*. (2011)

SP – swelling power; S – solubility
Source: (Zhu, 2015).

commonly used to monitor starch-pasting properties. However, RVA is quicker and requires lower sample size. The viscosity of a mixture of starch and water are measured as it undergoes programmed heating and cooling while under constant shearing force. The solid content can influence pasting behaviour and should be noted when comparing data

Table 3.11. Solubility, gelatinization and viscosity pattern of cassava starch in various solvents.

Solvent			Viscosity of 2% solution (Redwood) (in seconds)			Clarity (Absorbance at 650 nm; relative to water = 0)		
	Maximum solubility (g/100 ml)	Gelatinization temperature (°C) range	At highest gelatinization temperature	After cooling to room temperature	After standing overnight	At highest gelatinization temperature	After cooling to room temperature	After standing overnight
Glycerol	10	130–145	175	1800	3700	0.26	0.22	0.212
Ethanediol	2	110–125	17.5	38	-	0.23	0.23	2.25
DMSO	25	70–85	85	130	153	0.15	0.15	0.13
Formalin	25	70–85	78	139	170	0.11	0.10	0.13
DMF	Insoluble							
Methyl cellosolv	Insoluble							
Anisole	Insoluble							

Source: (Moorthy, 1982).

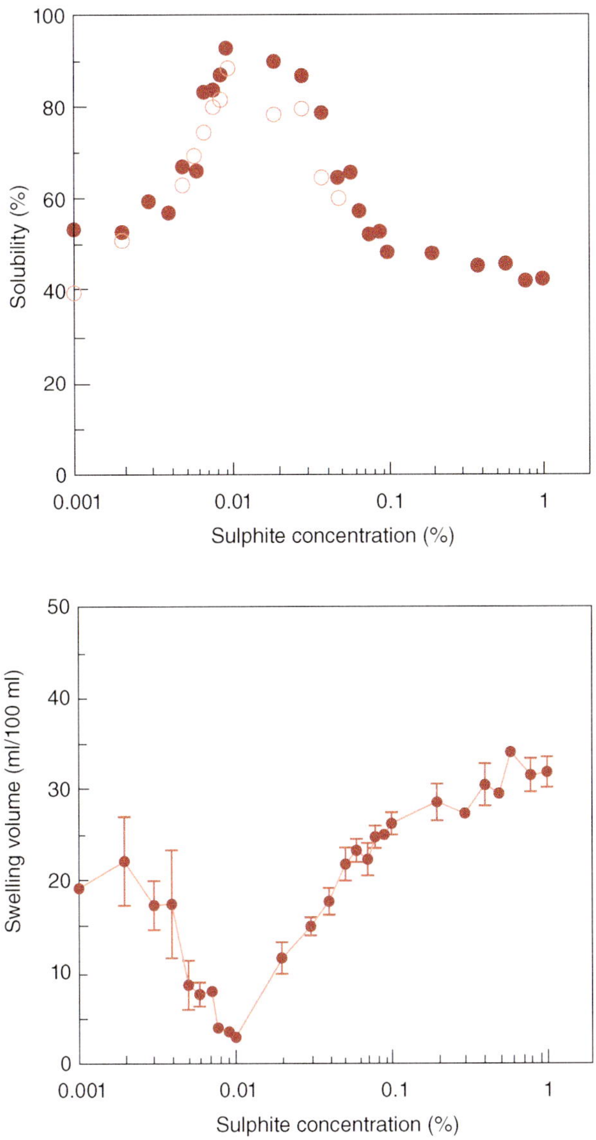

Fig. 3.8. The effect of sodium sulphite addition on the solubility and swelling volume of cassava starches.

Source: (Mat Hashim *et al.*, 1992).

from different studies (Caesar and Moore, 1935). The heating and cooling programs can also affect pasting data. Pasting properties of starches can vary widely with plant source. The extensive study by Sanchez *et al.* (2009) found broad variations in properties like peak viscosity (PV) (146–1505 cP), breakdown viscosity (BD) (28.1–859 cP), setback viscosity (SB) (−702–273 cP), and pasting temperature (PT) (58.8–71.2°C). Similar variations were observed in a study on starches from 40 Nigerian genotypes: PV = 261–593 RVU; BD = 141.2–329.0 RVU; SB = 19–79.9 RVU; and PT = 63.8–65.7°C (Onitilo *et al.*, 2007). The genes

Table 3.12. Pasting property of cassava starch.

No	Method	Starch(%)	PV	BD	SB	Pt(°C)	Ref:
4a	BA	5	635–820			58–62.3	Asaoka *et al.* (1992)
	RVA	5	487–629	344–479	124–145		Asaoka *et al.* (1992)
1b	BV	7	690–730	220–240	70–100	71	Padmanabhan and Lonsane (1992)
1	BVA	5	287				Gorinstein and Lii (1992)
1×3 c	BVA	5	660–714	420–475	160–205	60.7–62	Asaoka *et al.* (1993)
1×4b	RVA	5	493–571	354–420	137–150		Asaoka *et al.* (1993)
1	BVA	4.1	50	10	10	68	Pérez *et al.* (1998)
1	RVA	7.4				73	
4×5e	RVA		304–531	150–369	23–125	69.5–73.2	Sriroth *et al.* (1999)
6×2×4 f	RVA		317–455			64.9–72.9	Santisopasri *et al.* (2001)
11	RVA	10.7	414–515		53.6–114	73.6–75.3	Niba *et al.* (2002)
3×3+1 g	RVA	9.1	289–375	156–234	60–91	70–76	Abera and Rakshit (2003)
1×8 hour	RVA		521–615			76–77	Olomo and Ajibola (2003)
2×2×11i	RVA	9.1	222–331	112–196	51–91	68.8–74.2	Abera and Rakshit (2004)
5	RVA		88.9–142.6	55.8–118	8.8–15.6	71.3–74.6	Charles *et al.* (2004, 2005)
40	RVA	10.7	261–593	141.2–329	19–79.9	63.8–64.7	Onitilo *et al.* (2007)
1	RVA	10	3417cP			72	Tetchi *et al.* (2007)
3	RVA	5	577–890	249–491	161–400	65.4–68.3	Ceballos *et al.* (2007)
4	RVA	5	22–1080 cP			60.17–64.23	Ceballos *et al.* (2008)
4050	RVA	5	146–1505 cP	28.1–859 cP	–702 to 273 cP	58.8–71.2	Sánchez *et al.* (2010)
39×2 j	RVA	16.7	232–485	129–285.9	17–98.4	3.2–4.2	Eke *et al.* (2009)
13	RVA	9.1		56.5–107.8	9.8–14.3	63–64.7	Ikegwu, Nwobasi, Odoh, and Oledinma (2009)

Continued

Table 3.12. Continued

No	Method	Starch(%)	PV	BD	SB	Pt(°C)	Ref:
7	RVA	5	701–1094 cP	226–699 cP	3–659 cP	65.2–69.2	Zhao *et al*. (2011)
12	RVA	8	221–249		99–112	67.4–70.4	Charoenkul *et al*. (2011)
1	RVA	10.7	1769 cP	177 cP	859 cP	66.2	Mishra and Rai (2006)

Source: (Zhu, 2015).

Table 3.13. Major viscosity properties of starches from wild type and transgenic waxy cassava analysed using a rapid viscosity analyser (RVA).

Parameter	WT	A8	A19	B9	B10	B23	AM206-5
Clarity (%)	35.5	63.7	48	53.5	45.6	54	57.6
Pasting temperature (°C)	65.2	67.6	69.2	65.7	68	67.6	68.3
Peak viscosity ($_c$P)	885	783	701	1094	999	757	890
Hot paste viscosity ($_c$P)	662	356	413	395	469	337	399
Cold paste viscosity ($_c$P)	882	336	511	435	548	343	490
Breakdown ($_c$P)	223	427	288	699	530	420	491
Setback ($_c$P)	-3	–447	–190	–659	–451	–414	–400
Consistency ($_c$P)	220	–20	98	40	79	6	91
Solubility at 75°C (% db)	7.25	11.73	5.99	4.12	5.78	5.91	5.21
Solubility at 90°C (% db)	14.92	24.09	13.16	8.00	8.84	18.94	8.74
Swelling Index at 75°C (g/g)	19.98	32.58	23.32	31.96	31.08	29.78	48.23
Swelling Index at 90°C (g/g)	35.22	49.90	38.68	47.13	40.05	46.62	54.73

Source: (Zhao *et al*., 2011).
cP: centipoises, db: dry basis.

that influence the pasting properties, called quantitative trait loci (QTL), are being mapped. The identified loci can be targeted in efforts to be breed cassava with enhanced starch properties (Thanyasiriwat *et al*., 2014).

For normal non-waxy genotypes (five genotypes), amylose content was positively related to PV, BD, SB and PT of pasting (Charles *et al*., 2005). Amylopectin structure property correlation studies showed that proportion of longer chains (DP >35) was positively related to PV, BD and SB (Charles *et al*., 2005).

The variation of pasting properties in starches with different compositions has been reported. The high-amylose mutant investigated by Ceballos *et al*., 2008 had lower pasting viscosity than normal genotypes while the waxy starches showed PV, BD and SB values that were higher than normal genotypes (Ceballos *et al*., 2007). The reduced- and no-amylose genotypes studies by Zhao *et al*. (2011) were found to have higher BD and SB values and different PV values (Table 3.13).

Agronomic factors and processing conditions have also been important to pasting properties. Application of water stress to the crop at an early or late stage led to lower PV and higher PT (Santisopasri *et al*., 2001). The time of harvesting can greatly affect pasting properties (Asaoka *et al*., 1992). The highest PV and BD, as well as the lowest PT values were determined for the crop harvested at the 16th month when compared to those harvested from the 6th to 14th months (Sriroth *et al*., 1999). The pasting properties were minimally affected by aerial pruning (Asaoka *et al*., 1993). Abera and Rakshit (2003) found that the fresh root starch showed higher PV, BD and SB values and lower PT values than dry root starch. The storage of dried chips at 5°C and 35°C up to 8 months led to a reduced PV, with significantly lower PV values at 35°C. Olomo and Ajibola (2003) showed that starch extracted from flour and chips had comparable PT values. Dry chips obtained by wet milling and dry milling showed similar pasting behaviour (Abera and Rakshit, 2003). The starch from the extraction method with enzyme integration was, however, markedly different from that obtained by conventional method, with the former showing lower PV (Padmanabhan and Lonsane, 1992).

The microstructural changes during pasting have been studied by observing the effect of heating on freeze-etched starch. The water was found to infiltrate the granules from the truncated ends. Starch-water reticulum formed in central regions and grew through the granules. Swelling occurred when the peripheral regions of the granules became reticulate.

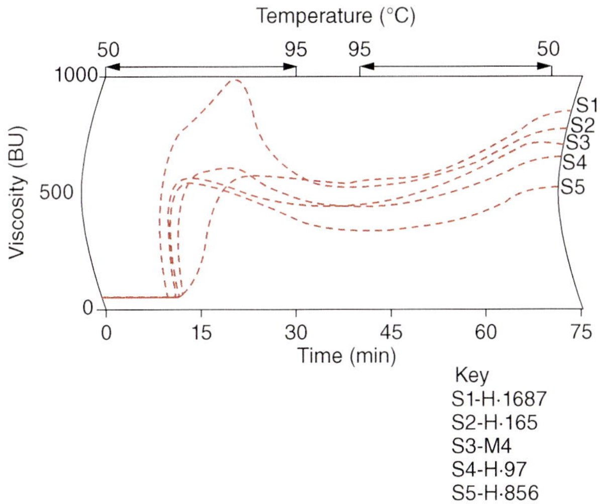

Fig. 3.9. Viscosity patterns of starch of different cassava varieties.

Source: (Moorthy *et al.*, 1996).

The reticulate bands were observable until the granule integrity was lost and the viscosity peaked (Allen *et al.*, 1977).

The four Brabender viscographs of different varieties of starch showed diverse peak patterns for, e.g. single stage gelatinization with high peak viscosity and high-viscosity breakdown, two-stage gelatinization with high-peak viscosity and breakdown, and broad two-stage gelatinization with medium viscosity and medium breakdown. These patterns seem to be genetically controlled as the patterns were maintained by these starches irrespective of the environmental factors, though there was variation in the viscosity values. In a comparative study of five cassava varieties having different cooking quality, it was observed that viscosity patterns varied considerably in terms of peak, breakdown and set back viscosities Fig. 3.9. The result indicated a possible relationship between cooking quality and starch rheology, as different tuber varieties with reasonably good cooking quality had totally different patterns compared with those having poor quality. Olorunda *et al.* (2007) found that mealier cassava varieties had slightly higher peak viscosity for their starches (Fig. 3.9).

Padmanabhan and Lonsane (1992) observed a slight reduction in peak viscosity and breakdown in cassava starch extracted by an enzymatic method. Olusola *et al.* (2015) have examined four cassava varieties for their viscosity properties (Table 3.14).

The viscosity of cassava starch extracted from inoculum provided fermentation was lowered due to presence of fibrous matter (Table 3.15).

However, the breakdown was also correspondingly reduced due to the cementing of the granules by fibrous materials and its rheology was similar to cassava flour. Another factor related to viscosity is setback which is attributed to the retrogradation of starch during cooling. Cassava starch has relatively low setback and this may be due to lower amylose content and also the structure of the amylopectin.

3.5.3 Rheological properties

Rheological tests on cassava starch such as flow behaviours and dynamic rheology with small deformation of starch gel development have been reported (Aguilera and Rojas, 1996; Charles *et al.*, 2004; Freitas *et al.*, 2004; Rao and Tattiyakul, 1999).

The flow behaviour of starch dispersions is determined by recording viscosity at different levels of shear forces and modelling it mathematically to gain structural insights. Shear

Table 3.14. Viscosity properties of four different varieties of cassava starches.

Paste properties	Red stick	Butter stick	Maracas black stick	M Col22
Peak viscosity (RVU)	393	333	344	297
Trough (lowest) Visc. (RVU)	131	118	116	101
Breakdown (RVU)	263	214	228	195
Final viscosity (RVU)	174	168	168	154
Setback (RVU)	43	49	52	52
Peak time (min)	3.6	3.8	3.6	3.9
Pasting temp (°C)	68.7	71.9	70.3	73.5

Source: (Olusola *et al.*, 2015).

Table 3.15. Viscosity data (BU) of flour from fermented and non-fermented tubers.

	Fermentation time (hr)							
Variety	0		24		48		72	
	PV	V_{97}	PV	V_{97}	PV	V_{97}	PV	V_{97}
M4	580	500	480	480	400	400	400	400
H.165	910	500	700	640	660	660	660	660
H.2304	1000	640	740	540	620	580	420	380
T.300	980	540	780	540	440	420	460	420
H.1687	860	400	640	480	450	390	500	480
CI.468	810	460	620	480	570	500	630	580

Source: (Moorthy *et al.*, 1993).

tests have been used to determine the effect of temperature, concentration, pH and cooking time on starch's rheological properties. A power law equation was found to model the flow behaviour of 2–6% starch solution well (Chen and Ramaswamy, 1999). Rao and Tattiyakul (1999) have studied the shear thinning behaviour in relation to granular size. They found that the power law index was positively correlated to granular size with an exponential relationship. The shear thickening behaviour of a 2.6% heated starch dispersion turned to shear thinning after prolonged heating and at higher temperatures. At a concentration below the overlap value, the flow behaviour of starch solutions indicated that the chains do not entangle with each other (Che *et al.*, 2008). At 0.2%, starch suspension has Newtonian shear behaviour, which turned to shear thinning over 0.4%. From 0.2 to 1%, the value of flow behaviour indices fell while the consistency indices increased (Che *et al.*, 2008).

Storage modulus (G') and loss modulus (G") are commonly used properties to determine viscoelastic properties of starch gels for dynamic or small deformation studies. They quantify the amount of stored energy that can be recovered per cycle and the energy that is lost per cycle, respectively. Dynamic rheological properties were shown to vary with genotype (Charles *et al.*, 2004). Aguilera and Rojas (1996) used oscillatory rheology to study the kinetics of cassava starch gelation in the presence of whey protein, and thereafter proposed a two-step reaction to model starch gelation. In the first step, the native starch granules (state A) irreversibly absorbed water until they swelled to a maximum packing condition (state R). The swollen granules undergo softening in a reversible second step (Abera and Rakshit, 2003).

The rheological properties of cassava studied using a Bohlin rheometer are presented in Fig. 3.10.

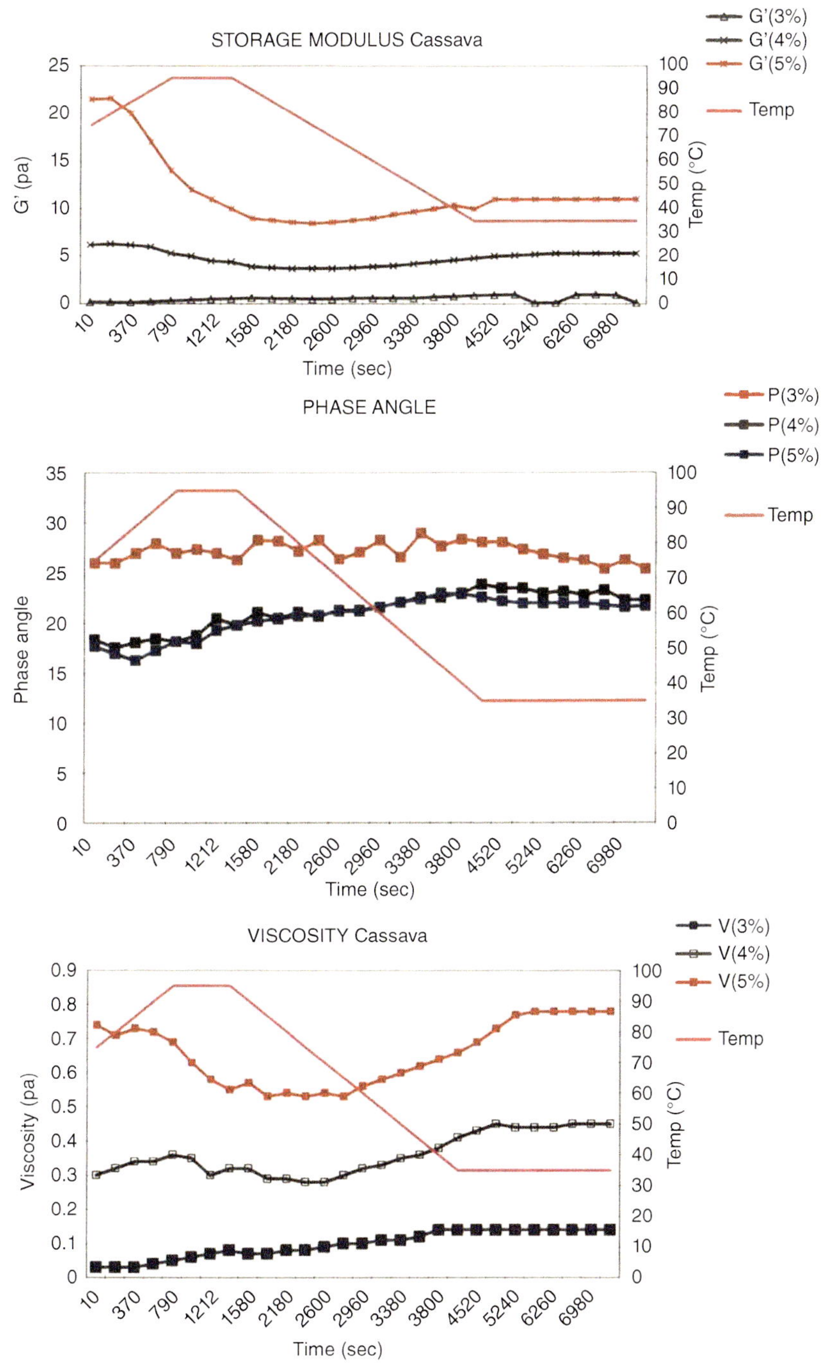

Fig. 3.10. Rheological properties of cassava.

Source: (Moorthy *et al*., 1998).

Table 3.16. Adhesive forces and adhesiveness.

	Adhesive forces (g)		Adhesiveness (gs)	
Sample	70°C	80°C	70°C	80°C
2 days after production				
Red stick	–77	–79	–1952	–1872
Butter stick	–146	–90	–3075	–2098
Maracas black stick	–132	–134	–2999	–2799
M Col 22	–84	–69	–2032	–1404
2 weeks after production				
Red stick	–71	–48	–1718	–1106
Butter stick	–104	–82	–2226	–1313
Maracas black stick	–181	–101	–3588	–2309
M Col 22 (top)	–36	–67	–936	–1344
M Col 22 (bottom)	–60	–77	–1446	–1797
3 weeks after production				
Red stick	–64	–17	–1650	–408
Butter stick	–118	–105	2,494	–2,265
Maracas black stick	–101	–135	–2,207	–2,834
M Col 22	–10	–14	–95	–286

Source: (Olusola *et al*., 2015).

The rheological properties of starch were shown to be significantly influenced by amylose content (Freitas *et al*., 2004). The proportion of longer unit chains (DP >35) in the amylopectin structure had a positive correlation to the resistance to shearing and the frequency independence of rheological properties (Charles *et al*., 2005). Asaoka *et al*. (1992) also showed that the thermal characteristics of starch measured by DSC overlapped with its dynamic rheological properties. The onset temperature of gelatinization (DSC) and the temperature of maximum G′ (oscillation rheology) were found to change together, suggesting a common structural basis for the two sources of data. Like the pasting behaviour, the rheological properties of starch can be affected by environmental factors and agronomic practices to which the crop was treated. For example, harvesting month affected rheological properties (Asaoka *et al*., 1992) while pruning aerial growth had a minor effect (Asaoka *et al*., 1993). These external factors can therefore be used as non-genetic ways to control starch rheology. Olusola *et al*. (2015) has used a texture analyser to examine the adhesive property of cassava starch for possible application in adhesives (Table 3.16).

3.5.4 Retrogradation

As mentioned before, when a gelatinized starch solution is cooled, the starch chains re-establish molecular interactions which lead to reordering of starch molecules and partial crystallization. Amylose reassociation is quicker and is largely responsible for the initial retrogradation, while the amylopectin molecules recrystallize at a much slower rate and take a long time (Hoover, 1995). DSC, syneresis, and NMR are mostly used to study various aspects of starch retrogradation.

Due to low amylose content in cassava starch, it has a lower tendency to retrograde. Abera and Rakshit (2003) studied retrogradation of starch gels obtained from three genotypes (different extraction methods) along with a commercial sample using DCS. The retrograded starch had the following properies:

T_{onset} = 48.5–52.1°C, T_{peak} = 54.7–59.7°C, T_{end} = 61.6–67.6°C and ΔH = 1.2–2.1 J/g. The degree of retrogradation (%) i.e. the ratio of ΔH of retrograded starch to that of gelatinization ranged from 7.8–13.9% (Abera and Rakshit, 2003). The starch extracted by dry milling had lower retrogradation than that obtained from wet milling. The effect was attributed to the mechanical damage and degradation caused during dry milling. This suggested that both natural and artificial factors can be controlled to obtain required retrogradation properties. The fitting of starch retrogradation kinetics in the Avrami equation and the value of the Avrami exponent (n) was found to be unity. This suggested that the mechanism of starch crystallization involves instantaneous nucleation followed by rod-like growth of crystals (Kim *et al.*, 1976). Sanchez *et al.* (2010) showed negligible syneresis in waxy cassava starch after five weeks at −20°C showing its superior potential in formulation of frozen or refrigerated foods than normal and waxy starches from maize, rice and potato. Amylose-free starch had enhanced gel stability (Raemakers *et al.*, 2005).

Another commonly used parameter reflecting starch retrogradation is the syneresis of pasted gel. During retrogradation, the formation of crystals leads to the leaching of water from the gel. Low syneresis during storage or freeze-thaw cycle is preferable for the stability and shelflife of frozen foods. Cassava starch paste possessed lower stability and had poor gelling properties compared with maize and wheat (Dickinson *et al.*, 1982). The low syneresis of waxy starch may be attributed not only to the absence of amylose but also to the amylopectin structure. External factors have been evaluated to control the starch syneresis.

Freezing and thawing history had a great influence on paste stability. Paste showed lower syneresis and better stability when frozen by cryogenic quick freezing and when thawed at higher temperatures (Varavinit *et al.*, 2000). Gel retrogradation was significantly reduced when the freezing rate was increased from 0.06 or 0.9°C/min to 2.3°C/min. Abera and Rakshit (2003) found that starch extracted from fresh roots showed lower syneresis that than from dried roots (Table 3.17).

The DSC parameters of retrograded starch of cassava from 10 land races in Brazil exhibited considerable variability (see Table 3.18).

3.6 Thermal Properties

Specific heat, thermal conductivity, thermal diffusivity and glass transition are important

Table 3.17. Differential scanning calorimetry parameters of TS gels containing different NaCl concentrations after gelatinization and 5 FTCs.

Concentration of NaCl (mol/L)	Gelatinization				
	T_o (°C)	T_p (°C)	T_c (°C)	ΔH (J/g)	
0	64.50 ± 0.06	68.68 ± 0.04	80.62 ± 0.44	14.38 ± 0.28	
0.1	68.68 ± 0.04	71.67 ± 0.18	82.20 ± 0.11	14.65 ± 0.04	
0.3	69.81 ± 0.01	73.87 ± 0.03	85.06 ± 0.04	15.14 ± 0.20	
0.5	70.87 ± 0.14	74.97 ± 0.01	86.07 ± 0.04	16.44 ± 0.01	
1.0	72.70 ± 0.12	76.85 ± 0.11	89.03 ± 0.06	16.35 ± 0.01	
Concentration of NaCl (mol/L)		5 FTCs			
	T_o (°C)	T_p (°C)	T_c (°C)	ΔH (J/g)	R%
0	51.60 ± 0.07	57.06 ± 0.08	65.42 ± 0.01	3.70 ± 0.26	25.76 ± 2.29
0.1	52.38 ± 0.12	57.68 ± 0.13	65.59 ± 0.08	3.08 ± 0.01	21.04 ± 0.12
0.3	52.76 ± 0.05	58.00 ± 0.01	65.26 ± 0.06	2.95 ± 0.04	19.50 ± 0.07
0.5	53.70 ± 0.23	58.82 ± 0.09	66.08 ± 0.21	2.60 ± 0.03	15.80 ± 0.16
1.0	55.65 ± 0.20	60.12 ± 0.16	67.76 ± 0.11	1.21 ± 0.03	7.36 ± 0.65

Source: (Wang *et al.*, 2015).

Table 3.18. Thermal properties of retrograded cassava starches.

	Retrogradation temperatures (°C)				
Variety	T_o	T_p	T_c	ΔHret (J/g)	**%R**
Ciganapreta	41.2 ± 0.5	52.8 ± 0.0	59.6 ± 0.0	3.2 ± 0.2	23
IAC 13	45.0 ± 1.4	53.5 ± 0.5	60.3 ± 0.2	3.6 ± 0.4	26
IAC 12	47.4 ± 0.1	53.1 ± 0.0	59.4 ± 0.3	2.6 ± 0.3	20
IAC 14	47.6 ± 0.9	53.8 ± 0.0	59.8 ± 0.2	2.1 ± 0.2	16
Caapora	49.9 ± 0.3	56.0 ± 0.2	60.5 ± 0.1	3.0 ± 0.2	21
IAC 90	41.8 ± 0.3	52.8 ± 0.0	59.8 ± 0.2	4.4 ± 0.3	32
Olho junto	46.1 ± 0.2	53.8 ± 0.0	59.4 ± 0.3	2.7 ± 0.1	19
IAC 06–01	48.5 ± 0.7	52.8 ± 0.0	59.9 ± 0.1	2.2 ± 0.0	15
IAC 576–70	50.1 ± 0.5	55.6 ± 0.5	60.6 ± 0.3	2.5 ± 0.2	19
Vassourinha Paulista	49.5 ± 0.1	53.2 ± 0.1	59.8 ± 0.2	2.3 ± 0.4	19

T_o – onset temperature. T_p - peak temperature, T_c – conclusion temperature, ΔHret – enthalpy change, %R – (Δhret/ΔHgel) x 100
Source: (Schmitz *et al.*, 2017).

parameters useful in thermal processing of starchy foods. Specific heat is defined as the amount of heat required to raise the temperature of a unit mass of a material by one degree and Differential Scanning Calorimety (DSC) can be conveniently used to determine this property. Thermal diffusivity is the rate at which heat is diffused into different materials and may be calculated from experimental values of specific heat, thermal conductivity, and bulk density. The glass transition temperature (T_g) is the temperature at which any material changes from the glassy to the rubbery state at a given heating rate, and plays an important role in contributing to the food stability and safety (Slade and Levine, 1991).

Thermal diffusivity of cassava starch at various temperatures and water content has been determined (Lan *et al.*, 2000). It was found that specific heat increased with the rise in temperature and moisture while thermal conductivity increased with the increase in moisture content. Models have been worked out to predict the conductivity, specific heat, and diffusivity of cassava starch (Lan *et al.*, 2000). The T_g of amorphous starch with water content of 2–35% was determined by DSC and dynamic mechanical thermal analysis (DMTA) (Perdomo *et al.*, 2009) Water anti-plasticized starch when moisture content was less than 11%, while a plasticizing effect was observed at higher levels (Perdomo *et al.*, 2009). But in another study, the anti-plasticizing effect of water on T_g was, however, not observed (Chang *et al.*, 2000) and this difference may be attributed to the variation among the cassava genotypes (Perdomo *et al.*, 2009). Typical DSC curves on the T_g as a function of moisture content have been obtained by earlier researchers (Aichayawanich *et al.*, 2011; Chang *et al.*, 2000). When the starch was in rubbery state, the particles tended to agglomerate (Aichayawanich *et al.*, 2011).

3.6.1 DSC

One most commonly used method for quantifying starch gelatinization characteristics is differential scanning calorimetry (DSC). The technique can provide parameters like onset (T_o), peak (T_p) and conclusion (T_c) temperatures, and enthalpy change (ΔH). The techniques can give different results for the same sample depending on its water content and the heating regime applied. If the amount of water is too low, it can lead to peak multiplicity, peak broadening, and elevated gelatinization temperatures (Garcia *et al.*, 1996; Francisco *et al.*, 1996). At intermediate water content (45%), a higher heating rate of DSC (from 0.2 to 10°C/min) led to disappearance of the first transitional peak. Faster heating rate also resulted in a higher T_p and lower ΔH (Freitas *et al.*, 2004).

Thermal characteristics of cassava starch have been studied in detail using differential scanning calorimetry (DSC) which has become very important in the characterization of starch gelatinization. The values reported for T_{onset} and T_{end} exhibited wide variation, showing that varietal differences, environmental conditions and experimental conditions used for DSC measurement all influence these factors.

For example, analysis of 12 genotypes grown in Thailand showed the following ranges of parameters: T_{onset} = 61.1–71.3°C ,T_{peak} = 66.8–74.9°C , T_{end} = 78.4–85°C , and ΔH = 15.1–16.4 J/g (Charoenkul, 2011). Waxy genotypes showed higher gelatinization temperatures irrespective of the method of genotype development (Ceballos *et al.*, 2007; Raemakers *et al.*, 2005). The low amylose mutant produced by down regulation of GBSSI showed lower ΔH (Zhao *et al.*, 2011). Moreover, a correlation study of five non-waxy genotypes showed that ΔH increased while the peak temperature fell with an increase in amylose content (Charles *et al.*, 2005). The study also showed a positive correlation between weight percentage of amylopectin and ΔH (Charles *et al.*, 2005). Shorter chains can cause structural defects in the crystals and cause a reduced T_{peak} and ΔH (Genkina *et al.*, 2007).

Non-genetic factors (environmental conditions, agronomic practice and processing parameters) also influence starch gelatinization. Fresh root starch showed lower gelatinization temperatures than dried root starch (Abera and Rakshit, 2003). Harvesting month had a significant effect on DSC parameter (Asaoka *et al.*, 1992) but the practice of pruning aerial growth did not (Asaoka *et al.*, 1993). Water stress on the crop increased the peak temperature (Santisopasri *et al.*, 2001). The cassava planted in dry season produced starch with peak multiplicity in DSC with the second peak temperature between 71.4 and 76.1°C (Defloor *et al.*, 1998). Ample irrigation led to looser granule packing of amylose and amylopectin chains as shown by proton NMR relaxation analysis (Chatakanonda *et al.*, 2003).

Asaoka *et al.* (1991) have examined the gelatinization characteristics of four cultivars harvested at different seasons and their results indicated that both genetic constitution and environmental protocols affected the DSC parameters. Starch of one cultivar consistently displayed the highest gelatinization temperature through all the harvesting seasons. Moorthy *et al.* (1996) examined the DSC characteristics of five cassava varieties and DSC data on native, defatted and ethanol-extracted starch and flour, shown in Table 3.20, found only slight variation.

Values for gelatinization enthalpy depend on a number of factors like crystallinity, intermolecular bonding, rate of heating of the starch suspension, presence of other chemicals, etc. For cassava starch, the values reported in the literature range all the way from 4.8 to 16 J/g. Genetic and environmental factors also affect the gelatinization enthalpy as illustrated from studies using different varieties, time of harvest and seasonal variations (Asaoka *et al.*, 1991; Defloor *et al.*, 1998).

Defloor *et al.* (1998) studied the DSC characteristics of five varieties harvested during dry and rainy seasons and found that although within each planting season for each genotype and for each harvest time, significant differences in gelatinization temperatures were noticed, no systematic changes as a function of genotype or harvest time were noticed. Starch samples harvested at the six month stage in dry season had higher onset, peak and conclusion temperatures compared to those from rainy season, but the reverse was true for the tubers harvested at 12, 15 and 18 months. Even the temperature of drying affected the DSC gelatinization temperatures (Table 3.19). SO_2 treatment slightly enhanced the gelatinization enthalpy of cassava starch from 18.1 to 19.1 J/g (Radley, 1976). The gelatinization parameters of cassava starch extracted using SO_2 incorporated water were enhanced from 59.6 to 62.0°C for T_{onset} and 84.7 to 87.2°C for T_{peak} Wickramasinghe *et al.* (2009) obtained wide range in the DSC values (Table 3.21) of four cassava varieties: T_{onset} 54.5–65.9°C, T_{peak} 59.7–69.7°C and ΔH 13.0–16.8 J/g. Torres (2011) gave the values 58.6°C, 65.0°C and 11.7 J/g, respectively, for T_{onset}, T_{peak} and ΔH.

Among different tuber starches, cassava starch possesses the lowest gelatinization temperatures. Varietal difference is also evident

Table 3.19. Gelatinization measured by differential scanning calorimetry (DSC).

No	Methods	Starch : Water ratio	Heating Rate	T_0	T_p	T_c	ΔH	Reference
4[a]	DSC	1:25	5	50.7–52.7	54.7–61.3	59.8–67.2	1.7–2.2 (cal/g)	Asaoka *et al.*, 1992
1	DSC		10	52.3	63.1	71.1	3.3 (cal/g)	Gorinstein and Lii, 1992
3 x 2[b]	DSC	1:2.5	5	53.3–55.8	58.2–59.5	64.2–66.2	1.7–2.0 (cal/g)	Asaoka *et al.*, 1993
1	DSC	~1 : 3	10	65.6	70.8	75.2	15.6	Francisco *et al.*, 1996
1	DSC	1:9	10		66.0			Aguilera and Rojas, 1996
1	DSC	1:4		67.7			12.8	Numfor *et al.*, 1996
5	DSC	1:3	10	65.4–69.4	69.2–73.2	74.9–78.5	2.7–3.4	Moorthy *et al.*, 1996
1	DSC	1:2	10	62.4	69.3	84.1	4.8 (J/g)	Perez et al., 1998
4 x 2 x 5 [c]	DSC	1:2	2	53.9–62.1	62.9–68.7[g]	76.3–83.2		Defloor *et al.*, 1998
4 x 4[d]	DSC	~1:2	10		69–71			Sriroth *et al.*, 1999
6 x 2 x 4[e]	DSC	~1: 2	10		58.3–67.8			Santisopasri *et al.*, 2001
3 x 3 + 1[f]	DSC	1:3	10	63.9–69.1	68.9–75.5	80.3–85.6	14.2–16.2	Abera and Rakshit, 2003
1	DSC	~1: 2	2.5, 3, 4, 5	50.5–52.0	60.5–63.5	69.5–71.0	6.5–8.5	Freitas *et al.*, 2004
4	DSC	1:3	1.2	59.7–64.4	64.4–69.9	71.6–75.7	12.8–13.7	Charles *et al.*, 2004
1	DSC	1:5	3	61.5	69.0		15.0	Tetchi *et al.*, 2007
3	DSC	~1: 5	10	60.4–63.1				Ceballos *et al.*, 2007
1	DSC	1:9	0.1	60.1	66.2	72.7	14.4	Nwokocha *et al.*, 2009
17	DSC	4:1	10	63.5–66.1	68–69.5[h]		11.1–14.3[h]	Anggraini *et al.*, 2009
12	DSC	1:2	5	61.1–71.3	66.8–74.9	78.4–85.5	15.1–16.4	Charoenkul *et al.*, 2001
8	DSC	1:4	3	50.3–61.2	56.2–68.4	66.6–75.7	8.9–16.9	Rolland-Sabaté *et al.*, 2012
8	DSC	1:4	10	54.0–61.2	63.3–69.4	72.7–78.3	12.9–16.9	Rolland-Sabaté *et al.*, 2013

No., number of genotypes/samples; DSC, differential scanning calorimetry; KHSM, Kofler hot-stage microscopy; To, Tp, Tc, and H denote onset, peak, and conclusion temperatures, and enthalpy change of gelatinization, respectively.
Source: (Zhu, 2015).
[a]3–4 different harvesting months; [b]with/without pruning aerial growth; [c]different harvesting months and two different planting seasons; [d]4 different harvesting months; [e]4 harvesting months with or without water stress; [f]rch sample; [g]some genotypes had double peaks and the major one is recorded here; [h]Tp and H were from analysis of 7 genotypes.

Table 3.20. DSC data on native, defatted and ethanol extracted starch and flour.

	Starch				Flour			
Variety	Tint (°C)	Tmax (°C)	Tend (°C)	ΔH (cal/g)	Tint (°C)	Tmax (°C)	Tend (°C)	ΔH (cal/g)
H-1687								
Pl	67.1	71.4	75.4	3.2	70.0	73.9	79.1	2.2
De	67.3	70.9	75.1	3.4	69.5	73.3	78.1	2.0
EE	67.5	71.3	75.4	3.1	69.4	73.0	77.0	2.0
H-165								
Pl	65.4	69.2	74.9	3.3	68.7	72.0	77.2	2.1
De	66.0	69.4	73.9	3.3	69.9	73.1	78.1	2.0
EE	65.0	68.9	73.7	3.2	68.9	73.0	77.9	2.1
H-97								
Pl	69.4	72.3	77.1	3.4	71.8	75.0	79.9	2.3
De	69.0	71.3	76.9	2.8	72.0	75.4	80.0	2.1
EE	68.0	71.0	76.1	2.8	71.7	75.2	79.8	2.0
S-856								
Pl	65.6	70.1	74.9	2.7	69.1	74.1	79.2	2.4
De	65.2	70.3	75.1	2.8	68.1	73.2	78.3	2.4
EE	65.1	70.7	74.8	2.7	69.3	73.4	79.4	2.3
M-4								
Pl	68.2	73.2	78.5	3.0	71.1	75.7	81.2	2.0
De	67.9	72.8	77.3	2.9	70.1	75.1	82.1	2.0
EE	67.8	73.0	77.8	3.1	70.1	74.8	81.5	2.0

Source: (Moorthy *et al*., 1996).

for the reported data from various sources. Correlation between granule size and gelatinization temperature was not observed. The gelatinization temperatures of cassava starch determined microscopically by various workers ranged from 49–64°C to 62–73°C, quite similar to DSC values. Gu *et al*. (2013) could find effect of both genetic and environmental factors on pasting temperature of cassava starch. In a Brabender viscographic study involving starch of different varieties, pasting temperature of starch of variety H165 was slightly lower than that for most of the other varieties, and M4 starch had the highest range of pasting temperature. The values were quite close to DSC values of 66 and 78°C for T_{onset} and T_{peak} respectively. SO_2 treatment lowered the pasting temperature of cassava starch from 92 to 89°C. Li *et al*. (2014) obtained much higher values for T_{onset}, T_{peak} of 86°C and 91°C, and ΔH as 1.47 J/g

Schmitz *et al*. (2017) examined the thermal property of 11 varieties of cassava starches and found considerable variability in thermal property (Table 3.22).

3.7 Digestibility

The nutritional quality of starch is a subject of much discussion in view of the high glycaemic index associated with starch. The *in vitro* enzyme susceptibility of cassava starch to amylases and/or amyloglucosidase, reflecting the *in vivo* digestibility, has been studied (Gorinstein, 1993; Charles *et al*., 2005; Asaoka *et al*., 1991). The digestibility of cassava starch in native and gelatinized forms is quite high in both *in vivo* and *in vitro* studies. It was over 60% in *in vivo* studies on small animals. In fact the use of sago,

Table 3.21. DSC characteristics of five varieties harvested during dry and rainy seasons.

Cultivar	Harvested month	T onset	T end	ΔH	Swelling power			Solubility			RVA data (SNU)			Brabender data (BU)		
		°C	**°C**	**J/g**	**60°C**	**70°C**	**80°C**	**60°C**	**70°C**	**80°C**	**Pv**	**Breakdown**	**Setback**	**PT°C**	**PV**	**V95**
HMC-1	**March**	53.2	63.8	8.4												
	August	53.6	63.5	6.8	19.8	30.6	39.4	9.1	15.8	19.0	567	405	132	59–69	680	340
	November	50.7	60.5	7.6	22.6	32.0	43.0	9.6	17	19.9	559	403	137	58–62	820	310
CM 489–1	**March**	55.4	64.6	8.8												
	August	54.8	63.7	7.2	17.6	27.7	38.7	7.5	14.4	19.6	590	437	137	60–70	720	360
	November	50.8	59.8	7.6	20.6	27.7	34.8	9.5	15.9	19.3	629	479	140	58–66	724	340
CM6812	**March**	57.6	67.2	8.4												
	August	57.7	67.2	7.6	17.0	31.1	40.9	10.0	17.4	19.9	487	344	142	62–72	635	320
	November	54.1	63.9	8.0	19.0	27.7	35.5	11.3	17.8	19.6	557	416	124	61–72	660	335
CM15595	**March**	55.0	64.6	8.4												
	August	54.6	64.1	6.8	20.2	30.3	40.9	9.3	15.3	18.4	567	412	144	60–70	750	370
	November	51.0	60.2	7.6	21.0	29.3	43.6	8.9	15.6	18.8	624	472	145	58–95	795	335

Source: (Asaoka *et al.*, 1991).

Table 3.22. Thermal properties of cassava starches.

	Gelatinization temperatures (°C)				
Variety	T_o	T_p	T_c	ΔT	ΔHgel(J/g)
Ciganapreta	57.2 ± 0.0	61.7 ± 0.2	66.9 ± 0.3	9.7	14.2 ± 0.2
IAC 13	57.2 ± 0.1	61.2 ± 0.1	66.5 ± 0.1	9.3	13.6 ± 0.3
IAC 12	52.0 ± 0.5	57.9 ± 0.2	64.0 ± 0.1	12.0	13.4 ± 0.1
IAC 14	56.4 ± 0.3	61.4 ± 0.0	66.4 ± 0.0	10.0	13.5 ± 0.2
Caapora	54.3 ± 0.1	61.0 ± 0.1	67.4 ± 0.2	13.1	14.0 ± 0.1
IAC 90	58.2 ± 0.2	63.3 ± 0.1	69.2 ± 0.2	11.0	13.6 ± 0.2
Olho junto	49.2 ± 0.5	58.8 ± 0.1	65.1 ± 0.1	15.9	14.9 ± 0.1
IAC 06–01	52.0 ± 0.1	61.3 ± 0.1	67.0 ± 0.3	14.9	15.1 ± 0.2
IAC 576–70	53.5 ± 0.7	62.3 ± 0.7	67.3 ± 0.7	13.9	13.0 ± 0.2
VassourinhaPaulista	55.0 ± 0.0	59.2 ± 0.1	63.5 ± 0.0	8.5	11.9 ± 0.0

Source: (Schmitz *et al.*, 2017).

made using cassava starch, as food during fasting in India can be attributed to its easy digestibility. Wickramasinghe *et al.* (2009) compared different raw root starches and found highest digestibility values for cassava starch. Recently digestibility parameters of native and modified cassava and potato starches were studied by Remya *et al.* (2017) (Table 3.23).

Gorinstein (1993) described the hydrolysis of starch by bacterial amylase using the Michaelis–Menten kinetics model. The enzymatic susceptibility to glucoamylase was found to be similar for starches obtained from four genotypes grown and harvested under different conditions (Asaoka *et al.*, 1991). This suggested that different genotypes and agronomic factors did not affect the quality of these starches for fermentation. Chen *et al.* (2011) studied the structural changes of cassava starch after treatment with a mixture of α-amylase and glucoamylase. The hydrolysis occurred mainly at the granule surface and in the amorphous regions, resulting in increase in crystallinity (Chen *et al.*, 2011). On the other hand, the enzyme treatment of various starches at 60°C showed a decrease in degree of crystallinity and lowered gelatinization temperatures (Gorinstein and Lii, 1992). Lacerda *et al.* (2008) studied the effect of fungal amylase on starch granules and found that the thermal degradation temperatures were reduced while the heat of gelatinization increased. This indicated that the degradation was limited to the amorphous regions. The hydrolysis of starch obtained from five genotypes using pancreatic amylase and amyloglucosidase showed an enzyme-resistant starch content

Table 3.23. Digestibility parameters of native and modified cassava and potato starches.

Sample	*In vitro* starch digestibility (%)	Rapidly digestible starch (RDS) (%)	Slowly digestible starch (SDS) (%)	Resistant starch (RS) (%)	Glycaemic index
Native cassava starch	42.3 ± 1.04	78.1 ± 0.26	20.5 ± 0.41	1.5 ± 0.39	91.2 ± 1.34
S1	41.0 ± 0.12	46.8 ± 0.32	38.4 ± 0.57	14.7 ± 0.64	64.3 ± 0.94
S2	36.1 ± 3.50	40.9 ± 1.09	38.4 ± 0.88	20.8 ± 0.62	62.1 ± 1.68
S3	32.1 ± 1.57	39.5 ± 0.09	32.6 ± 1.12	27.9 ± 1.52	60.3 ± 0.44

Source: (Remya *et al.*, 2017).

ranging from 6.8 to 14% (Charles *et al.*, 2005). This suggested a difference in *in vivo* digestibility as well as structure of starch granules. A comparative study of cassava and corn starch showed that the former was less susceptible to the amylase of *Bacillus subtilis* and amyloglucosidase of *Aspergillus niger* (Franco and Ciacco, 1987). Microscopic observation showed that corn starch experienced radial degradation as well as surface degradation due to enzymes while cassava granules were corroded only at the surface (Franco *et al.*, 1988). The rheological characterization of starch hydrolysates from bacterial amylase treatment showed pseudoplastic behaviour and had reduced viscosity, firmness, brittleness, chewiness and gumminess (Sangeeta and Rai, 2008).

Langworthy and Deuel (1922) conducted *in vivo* digestibility trials on humans to compare cassava starch with other types of starch. Cassava starch digestibility was similar to taro, true arrowroot, rice, but higher than canna, potato and arrowroot starches. Moorthy and Padmaja (1991) have also found in rat-feeding trials that raw cassava starch has good digestibility. In rat feeding trials, raw starch digestibility was 99%, and was unaffected by variety, cyanide content, drying process, grinding fineness, dietary inclusion level (up to 50%), and animal age (Brough *et al.*, 1995).

References

Abera, S. and Rakshit, S.K. (2003) Processing technology comparison of physicochemical and functional properties of cassava starch extracted from fresh root and dry chips. *Starch/Stärke* 55(7), 287–296. DOI: 10.1002/star.200390072.

Aguilera, J.M. and Rojas, E. (1996) Rheological, thermal and microstructural properties of whey protein-cassava starch gels. *Journal of Food Science* 61(5), 962–966. DOI: 10.1111/j.1365-2621.1996.tb10911.x.

Aichayawanich, S., Nopharatana, M., Nopharatana, A. and Songkasiri, W. (2011) Agglomeration mechanisms of cassava starch during pneumatic conveying drying. *Carbohydrate Polymers* 84(1), 292–298. DOI: 10.1016/j.carbpol.2010.11.036.

Allen, J.E., Hood, L.F. and Chabot, J.F. (1977) Effect of heating on Freeze-Etchultrastructure of hydroxypropyl distarch phosphate and unmodified tapioca starches. *Cereal Chemistry* 54, 783–793.

Anggraini, V., Sudarmonowati, E., Hartati, N.S., Suurs, L. and Visser, R.G.F. (2009) Characterization of cassava starch attributes of different genotypes. *Starch/Stärke* 61(8), 472–481. DOI: 10.1002/star.200800121.

Asaoka, M., Blanshard, J.M.V. and Rickard, J.E. (1991) Seasonal effects on the physicochemical properties of starch from 4 cultivars of cassava. *Starch/Stärke* 43(12), 455–459.

Asaoka, M., Blanshard, J.M.V. and Rickard, J.E. (1992) Effects of cultivar and growth season on the gelatinisation properties of cassava (*Manihot esculenta*) starch. *Journal of the Science of Food and Agriculture* 59(1), 53–58. DOI: 10.1002/jsfa.2740590108.

Asaoka, M., Blanshard, J.M.V. and Rickard, J.E. (1993) The effect of pre-harvest pruning on the quality of cassava starch. *Annals of Applied Biology* 122(2), 337–344. DOI: 10.1111/j.1744-7348.1993.tb04039.x.

Benesi, I.R.M., Labuschagne, M.T., Dixon, A.G.O. and Mahungu, N.M. (2004) Stability of native starch quality parameters, starch extraction and root dry matter of cassava genotypes in different environments. *Journal of the Science of Food and Agriculture* 84(11), 1381–1388. DOI: 10.1002/jsfa.1734.

Benesi, I.R.M., Labuschagne, M.T., Herselman, L., Mahungu, N.M. and Saka, J.K. (2008) The effect of genotype, location and season on cassava starch extraction. *Euphytica* 160(1), 59–74. DOI: 10.1007/s10681-007-9589-x.

Breuninger, W.F., Piyachomkwan, K. and Sriroth, K. (2009) Tapioca/cassava starch:production and use. In: BeMiller, J. and Whistler, R. (eds) *Starch: Chemistry and Technology*. Academic Press, New York.

Brough, S.H., Neale, R.J., Norton, G. and Wenham, J.E. (1995) The effects of variety, drying procedure, fineness of grinding and dietary inclusion level on the bioavailability of cassava (*Manihot esculenta* Crantz) starch. *Journal of the Science of Food and Agriculture* 67(1), 71–76.

Ceballos, H., Sánchez, T., Morante, N., Fregene, M., Dufour, D. *et al.* (2007) Discovery of an amylose-free starch mutant in cassava (*Manihot esculenta* Crantz). *Journal of Agricultural and Food Chemistry* 55(18), 7469–7476. DOI: 10.1021/jf070633y.

Caesar, G. V. and Moore, E.E. (1935) Consistency changes in starch pastes – tapioca, corn, wheat, potato, and sweet potato. *Industrial & Engineering Chemistry* 27(12), 1447–1451.

Ceballos, H., Sánchez, T., Denyer, K., Tofiño, A.P., Rosero, E.A. *et al.* (2008) Induction and identification of a small-granule, high-amylose mutant in cassava (*Manihot esculenta* Crantz). *Journal of Agricultural and Food Chemistry* 56(16), 7215–7222. DOI: 10.1021/jf800603p.

Chang, Y.P., Cheah, P.B. and Seow, C.C. (2000) Plasticizing-antiplasticizing effects of water on physical properties of tapioca starch films in the glassy state. *Journal of Food Science* 65(3), 445–451.

Charles, A.L., Chang, Y.-H., Ko, W.-C., Sriroth, K. and Huang, T.-C. (2004) Some physical and chemical properties of starch isolates of cassava genotypes. *Starch/Stärke* 56(9), 413–418. DOI: 10.1002/star.200300226.

Charles, A.L., Chang, Y.H., Ko, W.C., Sriroth, K. and Huang, T.C. (2005) Influence of amylopectin structure and amylose content on the gelling properties of five cultivars of cassava starches. *Journal of Agricultural and Food Chemistry* 53(7), 2717–2725.

Charoenkul, N., Uttapap, D., Pathipanawat, W. and Takeda, Y. (2006) Molecular structure of starches from cassava varieties having different cooked root textures. *Starch/Stärke* 58(9), 443–452. DOI: 10.1002/star.200600515.

Chatakanonda, P., Chinachoti, P., Sriroth, K., Piyachomkwan, K., Chotineeranat, S. *et al.* (2003) The influence of time and conditions of harvest on the functional behaviour of cassava starch—a proton NMR relaxation study. *Carbohydrate Polymers* 53(3), 233–240. DOI: 10.1016/S0144-8617(03)00047-X.

Che, L.-ming., Li, D., Wang, L.-jun., Özkan, N., Chen, X.D. *et al.* (2008) Rheological properties of dilute aqueous solutions of cassava starch. *Carbohydrate Polymers* 74(3), 385–389. DOI: 10.1016/j.carbpol.2008.03.007.

Chen, C.R. and Ramaswamy, H.S. (1999) Rheology of tapioca starch. *Food Research International* 32(5), 319–325. DOI: 10.1016/S0963-9969(99)00090-3.

Chen, Y.S., Huang, S.R., Tang, Z.F., Chen, X.W. and Zhang, Z.F. (2011) Structural changes of cassava starch granules hydrolyzed by a mixture of amylase and glucoamylase. *Carbohydrate Polymers* 85, 272–275.

Defloor, I., Dehing, I. and Delcour, J.A. (1998) Physico-chemical properties of cassava starch. *Starch/Stärke* 50, 58–64.

Dickinson, E., McKay, J.E. and Thomas, V.D. (1982) Viscometric study of the dispersion of whole and gelatinized cassava starch. *LWT—Food Science and Technology* 15, 242–244.

Eke, J., Achinewhu, S.C., Sanni, L., Barimalaa, I.S., Maziya-Dixon, B. *et al.* (2009) Pasting, color, and granular properties of starches from local and improved cassava varieties in high rainfall region of Nigeria. *International Journal of Food Properties* 12(2), 438–449. DOI: 10.1080/10942910701867665.

Francisco, J.D., Silverio, J., Eliasson, A.C. and Larsson, K. (1996) A comparative study of gelatinization of cassava and potato starch in an aqueous lipid phase (L2) compared to water. *Food Hydrocolloids* 10, 317–322.

Franco, C.M.L. and Ciacco, C.F. (1987) Studies on the susceptibility of granular cassava and corn starches to enzymatic attack. 1. study of the conditions of hydrolysis. *Starch/Stärke* 39, 432–435.

Franco, C.M.L., Ciacco, C.F. and Tavares, D.Q. (1988) Studies on the susceptibility of granular cassava and corn starches to enzymatic attack. Part 2: Study of the granular structure of starch. *Starch/Stärke* 40(1), 29–32. DOI: 10.1002/star.19880400108.

Freitas, R.A., Paula, R.C., Feitosa, J.P.A., Rocha, S. and Sierakowski, M.-R. (2004) Amylose contents, rheological properties and gelatinization kinetics of yam (*Dioscorea alata*) and cassava (*Manihot utilissima*) starches. *Carbohydrate Polymers* 55(1), 3–8. DOI: 10.1016/S0144-8617(03)00142-5.

Genkina, N.K., Wikman, J., Bertoft, E. and Yuryev, V.P. (2007) Effects of structural imperfection on gelatinization characteristics of amylopectin starches with A- and B-type crystallinity. *Biomacromolecules* 8(7), 2329–2335. DOI: 10.1021/bm070349f.

Gomand, S.V., Lamberts, L., Derde, L.J., Goesaert, H., Vandeputte, G.E. *et al.* (2010) Structural properties and gelatinisation characteristics of potato and cassava starches and mutants thereof. *Food Hydrocolloids* 24(4), 307–317. DOI: 10.1016/j.foodhyd.2009.10.008.

Gorinstein, S. (1993) Kinetic studies during enzyme hydrolysis of potato and cassava starches. *Starch/Stärke* 45, 91–95.

Gorinstein, S. and Lii, C.Y. (1992) The effects of enzyme hydrolysis on the properties of potato, cassava and amaranth starches. *Starch/Stärke* 44, 461–466.

Gu, B., Yao, Q., Li, K. and Chen, S. (2013) Change in physicochemical traits of cassava roots and starches associated with genotypes and environmental factors. *Starch/Stärke* 65(3-4), 253–263. DOI: 10.1002/star.201200028.

Hongsprabhas, P., Israkarn, K. and Rattanawattanaprakit, C. (2007) Architectural changes of heated mungbean, rice and cassava starch granules: Effects of hydrocolloids protein-containing envelope. *Carbohydrate Polymers* 67(4), 614–622. DOI: 10.1016/j.carbpol.2006.07.012.

Hoover, R. (1995) Starch retrogradation. *Food Reviews International* 11(2), 331–346. DOI: 10.1080/87559129509541044.

Israkarn, K., Hongsprabhas, P. and Hongsprabhas, P. (2007) Influences of granule-associated proteins on physicochemical properties of mungbean and cassava starches. *Carbohydrate Polymers* 68(2), 314–322. DOI: 10.1016/j.carbpol.2006.12.016.

Joseph, R., Yeoh, H.-H. and Loh, C.-S. (2004) Induced mutations in cassava using somatic embryos and the identification of mutant plants with altered starch yield and composition. *Plant Cell Reports* 23(1-2), 91–98. DOI: 10.1007/s00299-004-0798-7.

Juna, S. and Huber, A. (2012) Determination of molar mass distribution of tapioca starch using asymmetrical flow field flow fractionation. *Starch/Stärke* 64(2), 87–96. DOI: 10.1002/star.201100066.

Juszczak, L., Fortuna, T. and Krok, F. (2003) Non-contact atomic force microscopy of starch granules surface. Part I. potato and tapioca starches. *Starch/Stärke* 55(1), 1–7. DOI: 10.1002/star.200390012.

Jyothi, A.N., Kiran, K.S., Wilson, B., Moorthy, S.N. and Nambisan, B. (2007) Wet storage of cassava starch: Use of sodium metabisulphite and acetic acid and the effect on starch properties. *Starch/Stärke* 59(3–4), 141–148. DOI: 10.1002/star.200600583.

Kasemsuwan, T. and Jane, J.L. (1996) Quantitative method for the survey of starch phosphate derivatives and starch phospholipids by 31P nuclear magnetic resonance spectroscopy. *Cereal Chemistry* 73, 702–707.

Kim, S.K., Ciacco, C.F. and D'appolonia, B.L. (1976) Kinetic study of retrogradation of cassava starch gels. *Journal of Food Science* 41, 1249–1250.

Koehorst-van Putten, H.J.J., Sudarmonowati, E., Herman, M., Pereira-Bertram, I.J., Wolters, A.M.A. *et al.* (2012) Field testing and exploitation of genetically modified cassava with low-amylose or amylose-free starch in Indonesia. *Transgenic Research* 21, 39–50.

Lacerda, L.G., Azevedo, J.A.M., Carvalho, M.A.D., Demiate, I.M., Schnitzler, E. *et al.* (2008) Thermal characterization of partially hydrolyzed cassava (*Manihot esculenta*) starch granules. *Brazilian Archives of Biology and Technology* 51, 1209–1215.

Lan, Y.B., Fang, Q., Kocher, M.F. and Hanna, M.A. (2000) Thermal properties of tapioca starch. *International Journal of Food Properties* 3, 105–116.

Langworthy, C.F. and Deuel, H.J. (1922) Digestibility of raw rice, arrowroot, Canna, cassava, taro, tree-fern, and potato starches. *Journal of Biological Chemistry* 52, 251–261.

Laohaphatanaleart, K., Piyachomkwan, K., Sriroth, K. and Bertoft, E. (2010) The fine structure of cassava starch amylopectin. Part 1: Organization of clusters. *International Journal of Biological Macromolecules* 47(3), 317–324. DOI: 10.1016/j.ijbiomac.2010.01.004.

Lehrman, L. (1932) The fatty acids associated with cassava starch. *Journal of the American Chemical Society* 54(6), 2527–2530. DOI: 10.1021/ja01345a053.

Lopez-Rubio, A., Flanagan, B.M., Gilbert, E.P. and Gidley, M.J. (2008) A novel approach for calculating starch crystallinity and its correlation with double helix content: A combined XRD and NMR study. *Biopolymers* 89, 761–768.

Mat Hashim, D.B., Moorthy, S.N., Mitchell, J.R., Linfoot, S.E. and Blanshard, J.M.V. *et al.* (1992) The effect of low levels of antioxidants on the swelling and solubility of cassava starch. *Starch/Stärke* 44(12), 471–475. DOI: 10.1002/star.19920441206.

Meuser, F., Smolnik, H.-D., Rajani, C. and Giesemann, H.G. (1978) Comparison of starch extraction from Tapioca chips, pellets and roots. *Starch/Stärke* 30(9), 299–306. DOI: 10.1002/star.19780300904.

Moorthy, S.N. (1982) Behaviour of cassava starch in various solvents. *Starch/Stärke* 34, 372–374.

Moorthy, S.N. (1985) Effect of different types of surfactants on cassava starch properties. *Journal of Agricultural and Food Chemistry* 33(6), 1227–1232. DOI: 10.1021/jf00066a053.

Moorthy, S.N. and Ramanujam, T. (1986) Variation in properties of starch in cassava varieties in relation to age of the crop. *Starch/Stärke* 38(2), 58–61. DOI: 10.1002/star.19860380206.

Moorthy, S.N. and Padmaja, G. (1991) Comparative study on digestibility of raw and cooked starch of different tuber crops. *Journal of Root Crops* 17, 255–258.

Moorthy, S.N., Thankamma Pillai, P.K. and Unnikrishnan, M. (1993) Variability in starch extracted from taro. *Carbohydrate Polymers* 20(3), 169–173. DOI: 10.1016/0144-8617(93)90147-V.

Moorthy, S.N., George, M. and Padmaja, G. (1993) Functional properties of the starchy flour extracted from cassava on fermentation with a mixed culture inoculums. *Journal of the Science of Food and Agriculture* 61(4), 443–447.

Moorthy, S.N., Wenham, J.E. and Blanshard, J.M.V. (1996) Effect of solvent extraction on the gelatinisation properties of flour and starch of five cassava varieties. *Journal of the Science of Food and Agriculture* 72(3), 329–336. DOI: 10.1002/(SICI)1097-0010(199611)72:3<329::AID-JSFA663>3.0.CO;2-1.

Moorthy, S.N., Sajeev, M.S. and Anish, R.J. (2018) Functionality of tuber starches. In: *Starch in Food Structure, Function and Applications*, 2nd edn. Woodhead Publishing Series in Food Science, Technology and Nutrition, pp. 421–508.

Numfor, F.A., Walter, W.M. and Schwartz, S.J. (1996) Effect of emulsifiers on the physical properties of native and fermented cassava starches. *Journal of Agriculture and Food Chemistry* 44, 2595–2599.

Nuwamanya, E., Baguma, Y., Wembabazi, E. and Rubaihayo, P. (2011) A comparative study of the physicochemical properties of starches from root, tuber and cereal crops. *African Journal of Biotechnology* 10, 12018–12030.

Olomo, V. and Ajibola, O. (2003) Processing factors affecting the yield and physicochemical properties of starch from cassava chips and flour. *Starch/Stärke* 55(10), 476–481. DOI: 10.1002/star.200300201.

Olorunda, A.O., Aworh, O.C. and Numfor, F.A. (2007) Technical note: Predicting quality of cassava products with the aid of chemical and rheological properties. *International Journal of Food Science & Technology* 16(4), 447–450.

Olusola, A.J., Adebiyi, O.B. and Riyaad, K. (2015) Evaluation of new cassava varieties for adhesive properties. *Starch/Stärke* 67(7-8), 561–566.

Onitilo, M.O., Sanni, L.O., Daniel, I., Maziya-Dixon, B. and Dixon, A. (2007) Physico-chemical and functional properties of native starches from cassava varieties in Southwest Nigeria. *Journal of Food, Agriculture and Environment* 5, 108–114.

Osunsami, A.T., Akingbala, J.O. and Oguntimein, G.B. (1989) Effect of storage on starch content and modification of cassava starch. *Starch/Stärke* 41(2), 54–57. DOI: 10.1002/star.19890410205.

Padmanabhan, S. and Lonsane, B.K. (1992) Comparative physico-chemical and functional properties of cassava starches obtained by conventional and Enzyme-Integrated conventional techniques. *Starch/Stärke* 44(9), 328–331. DOI: 10.1002/star.19920440903.

Perdomo, J., Cova, A., Sandoval, A.J., García, L., Laredo, E. *et al.* (2009) Glass transition temperatures and water sorption isotherms of cassava starch. *Carbohydrate Polymers* 76(2), 305–313. DOI: 10.1016/j.carbpol.2008.10.023.

Pérez, E.E., Breene, W.M. and Bahnassey, Y.A. (1998) Variations in the gelatinization profiles of cassava, sagu and arrowroot native starches as measured with different thermal and mechanical methods. *Starch/Stärke* 50(2-3), 70–72.

Pérez, J.C., Lenis, J.I., Calle, F., Morante, N., Sánchez, T. *et al.* (2011) Genetic variability of root peel thickness and its influence in extractable starch from cassava (*Manihot esculenta* Crantz) roots. *Plant Breeding* 130, 688–693.

Pérez, S. and Bertoft, E. (2010) The molecular structures of starch components and their contribution to the architecture of starch granules: A comprehensive review. *Starch/Stärke* 62(8), 389–420. DOI: 10.1002/star.201000013.

Radley, J.A. (1976) *Starch Production Technology*. Applied Science Publishers, London.

Raemakers, K., Schreuder, M., Suurs, L., Vincken, J.-P., Vincken, J.-P. *et al.* (2005) Improved cassava starch by antisense inhibition of granule-bound starch synthase I. *Molecular Breeding* 16(2), 163–172. DOI: 10.1007/s11032-005-7874-8.

Rao, M.A. and Tattiyakul, J. (1999) Granule size and rheological behavior of heated tapioca starch dispersions. *Carbohydrate Polymers* 38(2), 123–132. DOI: 10.1016/S0144-8617(98)00112-X.

Remya, R., Jyothi, A.N. and Sreekumar, J. (2017) Comparative study of RS4 type resistant starches derived from cassava and potato starches via octenyl succinylation. *Starch/Stärke* 69, 1–11.

Rolland-Sabaté, A., Sánchez, T., Buléon, A., Colonna, P., Jaillais, B. *et al.* (2012) Structural characterization of novel cassava starches with low and high-amylose contents in comparison with other commercial sources. *Food Hydrocolloids* 27(1), 161–174. DOI: 10.1016/j.foodhyd.2011.07.008.

Rolland-Sabaté, A., Sanchez, T., Buléon, A., Colonna, P., Ceballos, H. *et al.* (2013) Molecular and supra-molecular structure of waxy starches developed from cassava (*Manihot esculenta* Crantz). *Carbohydrate Polymers* 92(2), 1451–1462. DOI: 10.1016/j.carbpol.2012.10.048.

Sanchez, T., Dufour, D., Moreno, I.X. and Ceballos, H. (2010) Comparison of pasting and gel stabilities of waxy and normal starches from potato, maize and rice with those of a novel waxy cassava starch under thermal, chemical and mechanical stress. *Journel of Agricultural and Food Chemistry* 58, 5093–5099.

Sangeeta, M. and Rai, T. (2008) Rheological properties of corn, potato and tapioca starch hydrolysates. *Journal of Food Science and Technology—Mysore* 45, 406–411.

Santisopasri, V., Kurotjanawong, K., Chotineeranat, S., Piyachomkwan, K., Sriroth, K. *et al.* (2001) Impact of water stress on yield and quality of cassava starch. *Industrial Crops and Products* 13(2), 115–129. DOI: 10.1016/S0926-6690(00)00058-3.

Schmitz, G.J.H., Peroni-Okita, F.H.G., Nascimento, J.R.O., Campanha, R.B., Valle, T.L. *et al.* (2017) Selected physicochemical properties of starches isolated from ten cassava varieties reveal novel industrial uses. *Starch/Stärke* 69(1-9), 1600272A3.

Slade, L. and Levine, H. (1991) Beyond water activity: Recent advances based on an alternative approach to the assessment of food quality and safety. *Critical Reviews in Food Science and Nutrition* 30, 115–360.

Soni, P.L., Sharma, H.W., Dobhal, N.P., Bisen, S.S., Srivatsava, H.C. *et al.* (1985) The starches of *Dioscorea ballophylla* and *Amorphophalus campanulatus.* Comparison with tapioca starch. *Starch/Stärke* 37(1), 6–9.

Srichuwong, S. and Jane, J.L. (2007) Physicochemical properties of starch affected by molecular composition and structures: A review. *Food Science and Biotechnology* 16, 663–674.

Sriroth, K., Santisopasri, V., Petchalanuwat, C., Kurotjanawong, K., Piyachomkwan, K. *et al.* (1999) Cassava starch granule structure-function properties: influence of time and conditions at harvest on four cultivars of cassava starch. *Carbohydrate Polymers* 38, 161–170.

Teerawanichpan, P., Lertpanyasampatha, M., Netrphan, S., Varavinit, S., Boonseng, O. *et al.* (2008) Influence of cassava storage root development and environmental conditions on starch granule size distribution. *Starch/Stärke* 60, 696–705.

Tetchi, F.A., Rolland-Sabate, A., Amani, G.N. and Colonna, P. (2007) Molecular and physicochemical characterisation of starches from yam, cocoyam, cassava, sweet potato and ginger produced in the Ivory Coast. *Journal of the Science of Food and Agriculture* 87, 1906–1916.

Thanyasiriwat, T., Sraphet, S., Whankaew, S., Boonseng, O., Bao, J. *et al.* (2014) Quantitative trait loci and candidate genes associated with starch pasting viscosity characteristics in cassava (*Manihot esculenta* Crantz). *Plant Biology* 16(1), 197–207. DOI: 10.1111/plb.12022.

Varavinit, S., Anuntavuttikul, S. and Shobsngob, S. (2000) Influence of freezing and thawing techniques on stability of sago and tapioca starch pastes. *Starch/Stärke* 52(6-7), 214–217. DOI: 10.1002/1521-379X(200007)52:6/7<214::AID-STAR214>3.0.CO;2-3.

Wang, G., Hong, Y. and Gu, Z. (2015) Effect of NaCl addition on the freeze-thaw stability of tapioca starch gels. *Starch/Stärke* 67(7–8), 604–611. DOI: 10.1002/star.201500048.

Waterschoot, J., Gomand, S.V., Fierens, E. and Delcour, J.A. (2015) Production, structure, physicochemical and functional properties of maize, cassava, wheat, potato and rice starches. *Starch/Stärke* 67(1–2), 14–29. DOI: 10.1002/star.201300238.

Whistler, R.L. and Paschall, E.F. (1967) Starch chemistry and technology: Fundamental aspects. *Starch/Stärke* 19(1), 22–25.

Wickramasinghe, H.A.M., Takigawa, S., Matsuura-Endo, C., Yamauchi, H. and Noda, T. (2009) Comparative analysis of starch properties of different root and tuber crops of Sri Lanka. *Food Chemistry* 112(1), 98–103. DOI: 10.1016/j.foodchem.2008.05.046.

Zhao, S.-S., Dufour, D., Sánchez, T., Ceballos, H. and Zhang, P. (2011) Development of waxy cassava with different biological and physico-chemical characteristics of starches for industrial applications. *Biotechnology and Bioengineering* 108(8), 1925–1935. DOI: 10.1002/bit.23120.

Zhu, F. and Wang, S. (2014) Physicochemical properties, molecular structure, and uses of sweet potato starch. *Trends in Food Science and Technology* 36(2), 68–78. DOI: 10.1016/j.tifs.2014.01.008.

Zhu, F. (2015) Composition, structure, physicochemical properties, and modifications of cassava starch. *Carbohydrate Polymers* 122, 456–480. DOI: 10.1016/j.carbpol.2014.10.063.

4

Sweet Potato

4.1 Introduction

Sweet potato (*Ipomoea batatas* Lam.), grown widely in the tropics, is a dicotyledonous herbaceous perennial plant that belongs to the family *Convolvulaceae* and can grow at altitudes from sea level to 2500 m and comes in many varieties, having skin and flesh colours that may be white, yellow, orange or deep purple. Globally, sweet potato is the sixth most important food crop after rice, wheat, potato, maize and cassava, with annual production exceeding 105 million metric tonnes (International Potato Center, 2013). In developing countries, it has pride of place as the fifth most important food crop playing a critical role in food security. The yield varies anywhere from 8 to 30 t/ha and it has been possible to raise three crops in a year, thus giving a very high annual starch yield. The starch content in the fresh tubers varies from 12% to 30% depending on variety, location and environmental conditions.

The properties of sweet potato starch are very similar to those of cassava starch. However, starch extraction from sweet potato is not a usual procedure compared to cassava. The major limiting factor is the high residence time and lower settling rate that lead to microbial growth which reduces the quality and economic value of the sweet potato starch.

Many reviews that have appeared during the last decade have summarized data on the composition, physiochemical properties, and structure of sweet potato starch (Hoover, 2001; Moorthy, 2002; Moorthy *et al.*, 2012; Tian *et al.*, 1991). The last decade witnessed a great increase in our understanding of sweet potato starch, which can be attributed to development of more advanced and precise experimental tools and newer concepts.

4.2 Extraction and Purification

Starch production from sweet potato is one of the major uses of the crop in many producer countries. In the largest producer, China, about 15–20% of annual production is converted into starch in some provinces (Gitomer, 1987), with an annual production of over 300,000 tonnes (Wang, 1984). In Japan, during 1986, starch production accounted for 35% of all uses for sweet potato with annual production of nearly 100,000 tonnes (Duell, 1990). The rapid increase in demand for industrial starch in China, not only for direct use but also for further processing into a variety of products, makes sweet potato play an increasingly important role. Strangely, the manufacture of sweet potato starch has shown a declining trend in some countries, like Taiwan and Japan, which freely import cheaper maize or cassava starches from other countries.

Sweet potato has the advantage of being a high starch-producing crop with 30%, 30% and 49% greater starch yield compared to rice, maize or wheat, respectively, under the same conditions (Wang, 1984). In addition, sweet potato can be grown with low inputs on marginal soils and can tolerate adverse natural conditions. In the devastating super-cyclone in Odissa state of eastern India in 1990, sweet potato survived, dramatically indicating the resilience of this crop. Sweet potato may be one of the crops chosen for crews that travel to Mars. It provides an important energy source, carbohydrate, as well as beta-carotene, and is projected to be 'astronauts' superfood'. The sweet potato is able to adapt to a controlled environment with artificial sunlight. Cultivars with relatively high starch content, especially suitable for processing, have been developed in many laboratories.

In countries producing sweet potato starch, processing takes place both on small- (cottage industry) and large-scale levels using completely different conditions and machinery. For instance, in the Sichuan province of China, processing capacity ranges from 300–500 to 10,000–15,000 tonnes of fresh roots per annum in 17 plants, compared to only 100–300 tonnes per annum in more than 200 village enterprises producing starch and noodles (Wiersema *et al.*, 1989). In some countries, starch extraction is carried out on a seasonal basis (usually six months or less).

While the normal extraction rate from traditional and large-scale processing is usually not more than 12% and 15%, respectively (Wiersema *et al.*, 1989), a high extraction rate of 28% has been reported elsewhere. Sweet potato starch is often less pure and of a darker colour than cassava or maize starch due to contamination by jalapin, the resin formed by the sweet potato latex on reaction with polyphenolic compounds formed during processing.

In China, starch is extracted using two types of raw materials. In one system, fresh roots are processed at the earliest point after harvest. In the other, roots are first converted into sun-dried chips. The dried chips can be stored for long periods and can also be used by large-scale processing plants. In areas where natural climatic conditions do not favour field drying, processing units are much smaller, to cut down on transport costs of fresh roots (Wiersema *et al.*, 1989). A traditional method used in rural areas of China to enhance starch extraction is one in which an acid paste, naturally formed during starch production by the action of lactic acid bacteria, is added to the starch slurry (Zhao and Jia, 1985). The treatment enhances the rate of coagulation and separation of starch, fibre and protein, helping rapid starch sedimenting and forming a bottom layer with the other constituents suspended above it. The top layer, after removal, is allowed to ferment again to form the 'acid paste' for the subsequent processing. The acidity not only helps to enhance starch separation but also inhibits the growth of undesirable microorganisms. In Vietnam, village-scale processing involves the storing of wet starch, after extraction under anaerobic conditions, in underground holes covered with a layer of salt to inhibit microbial growth (Wiersema *et al.*, 1989). After closing the hole with plastic, a cover of bricks and soil is provided. The method permits storage of wet starch for six to nine months. It was found that sweet potato during the winter season gives a starch recovery of 12–14% compared to 18–20% during the spring season.

In Japan, the starch extraction process is maintained alkaline throughout by the addition of lime (Anon, 1968). The benefits of the addition of lime, to increase the pH to 8, in terms of enhanced starch yield and improved whiteness, have been proved experimentally in Korea (Suh, 1966). Centrifugal separators are often used in Japan for starch separation in place of settling tanks. It is also frequent practice in Japan for roots to be processed up to the slurry stage in autumn, for the concentrated starch slurry to be stored in outside concrete ponds, and for the process of purification, concentration and drying to take place during the winter. The raw starch pulp, which is retained by the sieves, is used for livestock feeding or citric acid manufacture. Where processing starts with dried chips, as in some large plants in China, the chips are first sieved to remove soil, crushed and soaked in water prior to being processed like fresh roots. In the larger processing plants in China and Japan, starch slurry is concentrated by centrifuge and the wet starch dried by hot air before being packed or stored for secondary processing.

For attaining optimum yield and quality of starch, the correct time of maturity, the

Table 4.1. Properties of enzymatically separated sweet potato starch. (From Moorthy and Balagopalan, 1999)

Enzyme conc (%)	Starch content (%)	Reducing value	Swelling volume (%)	Solubility (%)
0.000	90.33	1.37	19.50	22.5
0.010	91.00	1.35	17.50	19.5
0.025	92.05	1.50	20.50	20.2
0.050	90.92	1.85	17.85	22.3
0.100	91.25	2.25	17.75	37.5
0.200	90.85	1.95	18.25	39.5

methodology used for extraction, and the processing machinery are very crucial aspects. If roots are harvested late, the starch may get converted to sugar and fibre and thus may limit the yield and quality of starch to be harvested, while delays in processing freshly harvested sweet potato can result in sucrose and sugar reduction (Heinze and Appleman, 1943) and to toxic compounds such as *Ipomea marone* (Jain *et al.*, 1951)

In the isolated starch which,, in turn, affects the properties and yield of the starch indirectly (Lii and Chang, 1978). Radley (1976) suggested that lime, dilute acetic or lactic acids have been used to improve starch yield and quality. The recovery of starch from sweet potato roots has been enhanced by over 20% by using pectinase and cellulase enzymes, which act by breaking the pectin-cellulosic matrix of cell membranes of the starch granules. The treatment up to 0.05% concentration of enzyme provided higher yields without affecting its starch properties (Kallabinski and Balagopalan, 1991; Moorthy and Balagopalan, 1999) (see Table 4.1).

4.3 Physiochemical Properties

4.3.1 Biochemical characteristics

Similar to all starches, extracted starch is invariably contaminated by various other components (see Table 4.2) – fibre, lipids, proteins and minerals, depending on a number of factors such as method of extraction, age of the crop, environmental conditions etc. The extracted starch usually contains 95% starch, but the value depends on variety, maturity and method of extraction. The recommended moisture content for safe storage of starch is 13% (ISI, 1970; Radley, 1976), but various reports indicate a large variation in values (Kay, 1987; Takeda *et al.*, 1986; Mélo *et al.*, 1994; Soni *et al.*, 1990; Zhu, 2015).

Sweet potato starch is similar to cassava starch in its lipid and phosphorus content and the values for protein, lipid, ash and phosphorus vary considerably according to various reports (Table 4.2). The approximate composition of sweet potato starch isolated by sour liquid processing, compared to commercial starch, has

Table 4.2. Approximate composition of isolated starch.

Parameters	Range (g/kg)	Reference
Moisture	139–150	Lii and Chang, 1978
Ash	0.8–1	Delpeuch *et al.*, 1978, 1979
	2.6–5.1	Lii and Chang, 1978
Fibre	0.7–1.8	Delpeuch *et al.*, 1978, 1979
	0.5–1	Lii and Chang, 1978
Crude protein	4.8–5.4	Delpeuch *et al.*, 1978, 1979
	1.3–2	Lii and Chang, 1978
Crude lipid	0.6–6	Delpeuch *et al.*, 1978, 1979
Phosphorus	0.19	Lii and Chang, 1978
Starch	980–988	Delpeuch *et al.*, 1978, 1979

Table 4.3. Approximate composition of sweet potato starch during different processings. (From Deng *et al.*, 2013)

Sample	LSLPS	LCFS	CSLPS	CCFS	Control
Protein (%)	0.11	0.078	0.19	0.17	0.088
Lipid (%)	0.16	0.1	0.11	0.07	0.12
Ash (%)	0.26	0.13	0.18	0.15	0.17
P (mg/kg)	0.023	0.010	0.029	0.011	0.014
Al (mg/kg)	2.42	30	2.8	19.82	22.65
K (mg/kg)	10.7	6	35.6	15.5	4.5
Ca (mg/kg)	238	173	827	305	140
Mg (mg/kg)	28.6	32	7.3	55	26
Fe (mg/kg)	3.9	5	4.7	6.2	5

LSLPS, Laboratory sweet potato starch by sour liquid processing; LCFS, Laboratory sweet potato starch by centrifugation; CSLPS, Commercial sweet potato by sour liquid processing; CCFS, Commercial sweet potato by centrifugation.

been compared and wide variations were observed (see Table 4.3) (Deng *et al.*, 2013).

4.3.2 Amylose and amylopectin content

Various properties like pasting, thermal parameters, swelling, solubilization and enzyme susceptibility could be correlated to amylose content (Waramboi *et al.*, 2011; Zhu *et al.*, 2011a; Srichuwong *et al.*, 20072007). Among the various methods of measuring the amylose content, the most commonly used ones are based on iodine binding, size-exclusion chromatography before or after debranching, concanavalin A precipitation and DSC. Since these diverse methods can provide different values for the same sample, the specific measuring method should be taken into consideration when comparing results from different studies (Zhu *et al.*, 2011a).

The amylose content in sweet potato starch has been found to be slightly higher than that of cassava but lower than that of wheat, maize or potato (Rickard *et al.*, 1991). The amylose content of sweet potato starch can considerably affect the cooking and textural qualities of starch-based products (Collado *et al.*, 1999). Sweet potato starch amylose content has been reported to be between 8.5% and 35% (see Table 4.4). The amylose content of starch from six varieties of sweet potato from the Philippines had a value of 29.4–32.2%, the values being lower than other tuber crop starches (Madamaba *et al.*, 1975). Uehara (1983) found an amylose content of 21.6% in sweet potato starch while Watanabe *et al.* (1982) reported 20.9%.

For some Peruvian cultivars, Garcia and Walter (1998) obtained values ranging from 20–25% (by potentiometric titration). The value for amylose from various reports has been recently compiled by Zhu and Wang, 2014 (Table 4.5). Curing of the tubers exerted either a minor effect (Bertoniere *et al.*, 1966) or a slight increase in amylose content (Hammet and Barrentin, 1961). Delpeuch and Favier (1980) reported that cultivation practices and year of harvest did not relate to the amylose content. Collado and Corke (1997) reported that peak viscosity negatively correlated with amylose content significantly.

Lately, breeding and genetic modification have been used to develop sweet potato genotypes with widely varying diversified amylose content and genetic resources, in various countries and biogeographic regions, and has been evaluated (Aina *et al.*, 2012; Katayama *et al.*, 2006; Waramboi *et al.*, 2011; Zhu *et al.*, 2011a). High-amylose and amylose-free genotypes have been generated using genetic modifications (e.g. RNA interference), and environmental alternation further contributed to the diversity in amylose content (Kitahara *et al.*, 2007; Noda *et al.*, 2001; Takahata *et al.*, 2010).

Table 4.4. Physicochemical properties of sweet potato starches.

Properties	Range	References
Amylose content (%)	23.2–26.3	Hammet and Barrentin, 1961
	29.6–32.4	Madamaba *et al*., 1975
	16.1–24.4	Madamba and San Pedro, 1976
	17.5–18.3	Delpeuch *et al*., 1979
	13.4–22.5	Shen and Sterling, 1981
	22	Watanabe *et al*., 1982
	23.6–27.6	Shin and Ahn, 1983
	17.2–19	Takeda *et al*., 1986
	25–28	Seog *et al*., 1987
	19.4–22.8	Chiang and Chen, 1988
	21.5–22	Kitada *et al*., 1988
Water-binding capacity (%)	178.9–185.5	Shin and Ahn, 1983
	66.3–211.6	Seog *et al*., 1987
Swelling volume (ml/g)	46	Woolfe, 1992
	27.5–33.3 (95°C)	Chiang and Chen, 1988
	24.5–27.4 (85°C)	Seog *et al*., 1987
	63–95 (95°C)	Seog *et al*., 1987
	32–46 (80°C)	Seog *et al*., 1987
Solubility (%)	18	Woolfe, 1992
	13.2–14.4 (95°C)	Chiang and Chen, 1988
	11.4–12.9 (85°C)	Seog *et al*., 1987
	60–79 (95°C)	Seog *et al*., 1987
	30–50 (80°C)	Seog *et al*., 1987
Digestibility (%)	14.9–43.3	Fuwa *et al*., 1977
	20.8	Ueda and Saha, 1983
Acid resistance to 16% sulphuric acid at 50°C	43.7% acid-resistant portion, 49.6% low acid- resistant portion	Nara *et al*., 1983

4.4 Structural Properties

4.4.1 Granule morphology

Many reports are available about the morphology, size, crystallinity and polymorphism in sweet potato starch granules (Hoover, 2001; Moorthy, 2002; Tian *et al*., 1991). The granule shapes were diverse: round, polygonal oval or semi-oval, and non-aggregated, with diameters ranging from around 2 μm to 45 μm (Tian *et al*., 1991; Woolfe, 1992; Shin and Ahn, 1983; Bouwkamp, 1985) and characterized by a distinct centric hilum. However, polarization crosses are comparatively less distinct and the data on size and shape of starch granules from sweet potato are given in Table 4.6.

A pomegranate structure has been proposed for sweet potato granular structure based on optical micrographic studies of gelatinization. The outer layer of the granule is compared to skin, blocklets to garnet, and the amylopectin cluster with one reducing end of the hilum is equivalent to the primary body of pomegranate fruit, respectively (Li *et al*., 2012). The exact source of the amylose chain in contributing to the structure of the granule is still to be enumerated and the exact roles of amylopectin and amylose chains in structuring the granule are to be developed molecularly. In 1983, Dreher and Berry reported that the size of sweet potato starch granules shows similarity to that of cassava but is smaller than that of potato starch.

Depending on the specific varieties, the granule size ranges from 4 μm to 43 μm with a mean size of 12.3–21.5 μm. The granule size has been reported to influence some functional properties like swelling, solubility and digestibility. The particle size of sweet potato starch negatively correlates with amylase and acid degradation (Bouwkamp, 1985). In 1971,

Table 4.5. Amylose content in sweet potato starch. (From Zhu and Wang, 2014)

No.	Amylose content (%)	Method	References
21	12.8–21.3	Iodine binding	Aina *et al.*, 2012
11	23.3–26.5	Iodine binding	Zhu *et al.*, 2011a
	17.5–23.9	SEC	
	15.9–22.5	Concanavalin A pptn	
25	20–33	Concanavalin A pptn	Waramboi *et al.*, 2011
3	16.6–23.5	Iodine binding	Wickramasinghe *et al.*, 2009
2	4.6–14.4	Concanavalin A pptn	Lockwood and King, 2008
1	24.1	SEC	Tetchi *et al.*, 2007
8	0–25.1	Iodine binding	Kitahara *et al.*, 2007
1	22.6	Iodine binding	Leelavathi *et al.*, 1987
812	11.2–15.3	Iodine binding	Katayama *et al.*, 2006
1	19.8	Iodine binding	Srichuwong *et al.*, 2008
5	13.6–16.2	Iodine binding	Kitahara *et al.*, 2007
20	14–17.7	Iodine binding	Katayama *et al.*, 2004
2	32.15–34.16	DSC	Osundahunsi *et al.*, 2003
3	19.3–20	Iodine binding	Chen *et al.*, 2003
8	16.8–21	Iodine binding	Toyama *et al.*, 2003
7	0–19.5	Iodine binding	Noda *et al.*, 2002

No. = number of genotypes tested in the specific study; SEC = size-exclusion chromatography; DSC = differential scanning calorimetry

Rasper, reported that particle size and size distribution are peculiar features that can affect the functional properties of starch. With a higher specific surface area, smaller granules show higher solubility and water absorption capacity (Georing and Dehaas, 1972).

Madamaba *et al.* (1975) found significant differences in the granule size distribution among starch of six sweet potato varieties studied (Fig. 4.1).

Observations by a polarized microscope with a hot stage showed that the crystallinity disruption initially took place at different locations around the central hilum of the granules (Cai and Wei, 2013). The disruption pattern for sweet potato was more diverse than many starches from other botanical sources. The molecular basis responsible for this difference remains to be elaborated.

Table 4.6. Size, shape and X-ray diffraction pattern of sweet potato starches.

Sl. no	Size (μm)	Shape	X-ray pattern	References
1	14–34	Round polygonal	Ca	Shin and Ahn, 1983
2	3–42	Round polygonal	–	Seog *et al.*, 1987
3	4–40	–	Ca	Delpeuch *et al.*, 1978
4	10–14	–	–	Lii and Chang, 1978
5	4–43	Polygonal oval, round non-aggregated	–	Bouwkamp, 1985
6	9–38	Non-aggregated, oval polygonal	–	Madamaba *et al.*, 1975

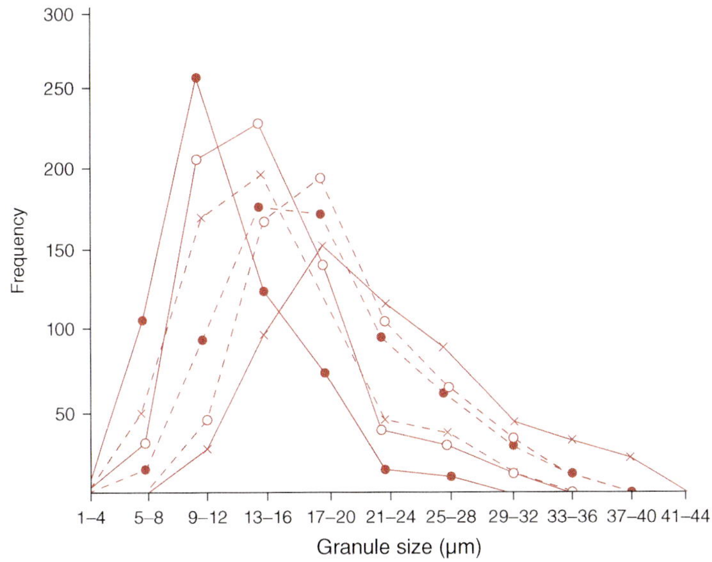

Fig. 4.1. Frequency curves of the granule sizes of starch of six sweet potato varieties grown in the Philippines. (From Madamaba *et al.*, 1975)

4.4.2 XRD and starch crystallinity

The X-ray diffraction peaks give information about the crystalline nature of a starch granule (Zobel, 1988). Figure 4.2 provides a typical X-ray pattern of sweet potato starch. Hizukuri (1969) demonstrated that mixtures of A- and B-type starches gave rise to an intermediate pattern, i.e. C-type. Sweet potato starch has a X-ray pattern between C and A, which is entirely separate from the pattern shown by cassava and cereal starches such as wheat, rice and maize, which have A- and B-type patterns of distribution by potato starch (Zobel, 1988). Sweet potato starch has also been assigned 'A' pattern (Takeda *et al.*, 1986; Szylit *et al.*, 1978; Gallant *et al.*, 1982), 'C' pattern (Shin and Ahn, 1983; Zobel, 1988; Chiang and Chen, 1988) or intermediate pattern (Tian *et al.*, 1991). Takeda *et al.* (1986) observed 'A' pattern for two varieties while it was 'C_A' for another variety that showed an absolute crystallinity of 38%.

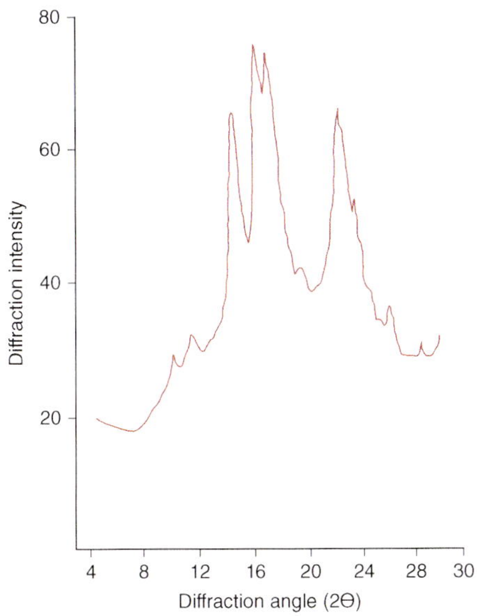

Fig. 4.2. An X-ray diffraction of sweet potato starch. (From Chiang and Chen, 1988)

4.4.3 Structure of amylose and amylopectin

The alkali number, which is a measure of the number of reducing end groups, can be related to the molecular weight (Schoch, 1964). The alkali number of six Korean sweet potato starches ranged between 7.66 and 12.13, which indicated specific difference in varieties of starch and in its molecular weights (Seog *et al.*, 1987). The amylose from sweet poatato starch has more branches per amylose molecule than that of cassava, potato, wheat or maize starches. In addition, the molecular weight is higher than maize, wheat and cassava but less than potato. Takeda *et al.* (1986) mentioned that, the low retrogradation tendency of sweet potato amylose can be attributed to this factor. The degree of polymerization and branching has been reported to have a substantial effect on the physicochemical properties of amylose and amylopectin (Zobel, 1988). The β-amylase enzyme cleaves amylose-amylopectin in a stepwise manner from the non-reducing ends until the residue; on average, up to two residues from the branching points.

Structural studies on sweet potato starch using GPC have revealed amylopectin to have peaks at DP = 12 and DP = 8. The concentrations of the peaks at DP = 6 and DP = 7 were 7.1–7.5% and 6.7–7.0%, respectively. Takeda *et al.* (1986) reported a trimodal pattern for the sweet potato amylopectin while Hizukuri (1969) reported a bimodal distribution, and results were interpreted as indicating an increased ratio of 'A' chains and short 'B' chains for the sweet potato starch compared to potato starch. Chain length has been found to vary in some varieties, as indicated by the lower viscosity and higher reducing values (Woolfe, 1992).

The molecular weight of sweet potato amylopectin can be determined as 1.77×10^8 (Tetchi *et al.*, 2007), and the degree of polymerization (DP) as 9900 (Takeda *et al.*, 2003). DP of small, medium and large proportions of amylopectin were determined to be 900, 5400 and 19,200, respectively (Takeda *et al.*, 2003). In studies on the structure of amylopectin, the external portion of amylopectin responsible for crytallinity has been cleaved enzymatically by phosphorylase, α- and β-amylase to obtain the 4, β- or β-limit dextrins (LDs) (Pérez and Bertoft, 2010). All the A-chains appear as maltose stubs, whereas the rest, other than maltose stubs, are B-chains (Peat *et al.*, 1952). The unit chain length distribution has been obtained by debranching amylopectin molecules and their 4, β- or β-LDs by isoamylase or pullulanase, and chromatography of the fractions.

The proportions of unit chains of amylopectin have been classified into four fractions by weight (f) or mole (F) (%), namely: fa (Fa) (DP 6–12); fb1 (Fb1) (DP 13–24); fb2 (Fb2) (DP 25–36); and fb3 (Fb3) (DP >36); and were found to have diverse physical properties of granular starch (Hanashiro *et al.*, 1996; Zhu *et al.*, 2011). Great diversity was found in the molecular structure of amylopectins from 11 different sweet potato genotypes (Zhu *et al.*, 2011). The ranges obtained for fa, fb1, fb2 and fb3 were 15.8–17.5%, 41.1–45.1%, 14.8–16.0% and 23.4–26.5%, respectively. The glucosyl residues range from 12.5–13.2 and 5.6–6.0 for external chain length (ECL) and internal chain length (ICL), respectively. The chain length (CL) of amylopectin and molar amounts of A-chain were 19.4–19.9 glucosyl residues and 48.8–54.3%, respectively, in contrast to the findings of Hanashiro *et al.* (2002) and may be due to the different measuring methods such as fluorescent labelling, Smith degradation method and high-performance anion exchange chromatography, whereas the difference observed in the amount of A-chain, as defined by Peat *et al.* (1952), is due to the fact that it can only be quantified from the 4, β-LDs. The C-chain length distribution showed a broad asymmetrical profile, from DP 10–130 with a peak at DP 40 (Takeda *et al.*, 2003).

Zhu *et al.* (2011) obtained interesting observations on the influence of amylopectin internal molecular structure on the physical properties of sweet potato starch. Total internal chain length (TICL) was highly related to both the pasting and thermal properties. TICL was positively related to the peak, hot paste and breakdown viscosities of pasting and melting enthalpy of DSC, while being negatively correlated to the gelatinization temperature range in DSC. Based on detailed systematic studies using starches from different sources, Bertoft (2013) proposed a backbone model, which explains the influence of internal molecular and cluster structure of amylopectin on various thermal properties of starch.

It is belived that the branches of amylopectin are clustered in the internal part, contributing to the amorphous region of the granules. Takeda *et al.* (2003) found the numbers of clusters per amylopectin were 4.8, 29 and 103 for small, medium and large molecules, respectively. The clusters of amylopectin were isolated by partial hydrolysis with α-amylase from *Bacillus amyloliquefaciens*. The α-dextrin clusters of amylopectin obtained by partial hydrolysis were further subjected to phosphorylase α- or β-amylase hydrolysis to obtain the 4, β- or β-LDs of clusters, which were characterized by chromatographic techniques (Pérez and Bertoft, 2010). In addition, the clusters have been extensively hydrolysed to isolate tightly branched 'building blocks' of α-limit dextrins using α-amylase of *B. amyloliquefaciens*. Studies have been conducted on the cluster and building-block structure of sweet potato amylopectin (Zhu *et al.*, 2011a). Clusters in the form of of 4, β-LDs had the size ranging from DP 55–86, with A-chains of 47.8–53.2% (molar basis). Each cluster had 3.6–5.8 building blocks with an inter-block chain length of 7–7.8 glucosyl residues. From the analytical data of C_A-type sweet potato starch, two models have been proposed to be representative of starch granular parts with A- and B-type X-ray diffraction patterns, respectively. Systematic studies showed that A- and B-type starches possess different cluster and building-block structures (Bertoft, 2013).

The structural defects on the granular level, resulting in difference in functional properties, have been traced to the composition of building blocks in the clusters of amylopectin and their spatial organization (Zhu *et al.*, 2011a). Longer inter-block chain length (IB-CL) of the clusters resulted in more twisted amorphous region. However, the role of cluster structure on the physicochemical properties of starch has not been completely enumerated.

The HPSE analysis of sweet potato amylopectin has been carried out to give the pattern shown in Fig. 4.3.

4.5 Functional Properties

4.5.1 Swelling pattern and solubility

Great diversity in swelling and solubilization properties was observed among different genotypes from various studies, in addition to varietal differences of composition affecting the swelling power of sweet potato starch (Tables 4.7 and 4.8). Correlation between swelling power (SP) and water soluble index (WSI) was reported (r^2¼ 0.88).

Both two-stage and one-stage swelling have been reported by Delpeuch and Favier (1980) and Madamaba *et al.* (1975), respectively (Fig. 4.4). Unlike cassava and potato starches, the higher degree of intermolecular association in sweet potato starch results in a lower swelling volume. The presence of non-carbohydrate substances in starch, such as lipid or phosphate,

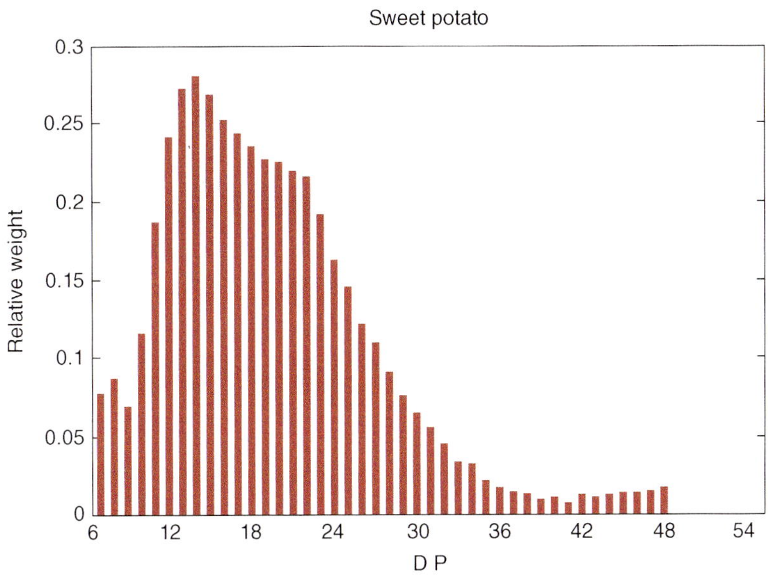

Fig. 4.3. HPSE pattern of sweet potato amylopectin. (From Moorthy, unpublished)

Table 4.7. Swelling and solubility of sweet potato starch. (From Zhu and Wang, 2014)

No.	Parameter	Temperature 55°C	65°C	75°C	85°C	References
21	Sp		7.8–31.1			Aina *et al*., 2012
21	S		1.5–9.5			Aina *et al*., 2012
2	Sp	3.5–4.1	5.0	5.0–28.1	31.4–33	Jangchud *et al*., 2003
2	S	1–1.6	2.2–2.4	3.2–11.6	20.4–20.7	Jangchud *et al*., 2003
25	Wsi	0.1–0.3 (RT)				Waramboi *et al*., 2011
11	Sp	2.0–3.2	2.8–6.5	14.5–22.1	13.3–20.4	Zhu *et al*., 2011
	Wsi	0.2–0.8	9.2–18.3	22.5–61.2	52.4–85.8	Zhu *et al*., 2011

can influence the extent of swelling (Leach *et al*., 1959; Moorthy and Ramanujam, 1986).

A number of factors such as source, inter-associative forces, swelling power and the presence of other components (lipids, surfactants, salts, sugars etc.) influence the solubility of starch. It was assumed that the bonding forces might be tenuous but comparatively extensive, immobilizing the starch within the granules even at high levels of swelling. Reported solubilities of sweet potato starch were in the range of 10–18% (Madamaba *et al*., 1975). The relatively high swelling of sweet potato is not accompanied by the expected high solubility. This characteristic was also observed by Leach *et al*. (1959) in potato. As reviewed by Tian *et al*. (1991), sweet potato amylose appears to be more branched than that from cassava.

The swelling and solubility of sweet potato starch have been shown to be less than those of potato and cassava but generally more than those of maize (Rašper, 1969b; Delpeuch and Favier, 1980). The swelling and solubility of starch extracted from 14 different varieties of sweet potato exhibited considerable differences (Table 4.8) (Moorthy *et al*., 2010).

Table 4.8. Physicochemical properties of sweet potato starches. (From Moorthy and Shanavas, 2010)

Sample	Swelling volume (ml/g)	Solubility (%)	Clarity	Amylose (blue value)
ST1	41.25	8.81	Good	0.323
ST2	46.25	10.75	Good	0.313
ST3	37.5	11.25	Good	0.317
ST4	43.75	11.81	Good	0.314
ST5	36.25	8.93	Good	0.304
ST6	33.75	7.29	Good	0.325
ST7	35	13.65	Good	0.314
ST8	43.75	12.94	Good	0.337
ST9	50	9.5	Good	0.315
ST10	40	9.6	Good	0.312
ST11	48.75	7.18	Good	0.325
ST12	37.5	10	Good	0.311
ST13	35	7.15	Good	0.327
ST 14	32.5	7.28	Good	0.339
l.s.d at 5% level	0.497	0.325		0.004

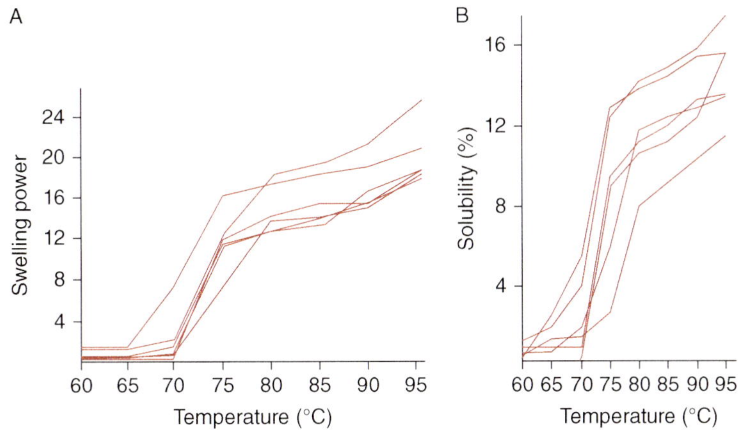

Fig. 4.4. Swelling and solubility patterns of sweet potato starches grown in the Philippines. (From Madamaba *et al*., 1975)

The water binding capacity of sweet potato gel starch determined by the method of Medcalf and Giles (1965) ranged from 66.3% to 211.6% (Table 4.4). In general, root and tuberous starches have higher water-binding capacities than those of cereal origin (Banks and Greenwood, 1975) and the majority of results indicated that sweet potato starch has a higher water-binding capacity than potato (93%) (Dreher and Berry, 1983) and cassava starches (72–92%) (Rickard *et al*., 1991).

4.5.2 Viscosity

The intrinsic viscosity depends on ability of polymer molecules to influence the solvent viscosity, in the absence of any intermolecular interactions (Young, 1981) and is directly linked to molecular size and, consequently, to the degree of polymerization (Daniels, 1966). The measured intrinsic viscosity of starches from six cultivars of sweet potato starches, which ranged from 120 ml/g to 155 ml/g, indicated that these starches are not highly polymerized in comparison with potato starch. Rašper (1969a) reported a maximum viscosity of 590 BU for sweet potato starch that was measured using a Brabender viscograph. The value indicated that sweet potato is slightly less viscous than the cassava starch but more viscous than maize starch. Reducing the concentration of starch resulted in lowering the paste viscosities and breakdown because of reduced friction, due to a lesser number of swollen granules in the medium (Fig. 4.5) (Collado *et al*., 1999). Several studies, including the findings by Tian *et al*., 1991, reported that the no-peak viscosity of sweet potato was at starch concentration of 4–6% (w/v). However, Lii and Chang (1978) reported a moderate peak and a high setback on cooling, at 7% starch concentration.

Viscosity was highly influenced by the sweet potato variety (Madamaba *et al*., 1975; Liu *et al*., 1985). The limiting viscosity of amylose from sweet potato is higher than that of amylose from wheat but lower than amylose from cassava and potato (Takeda *et al*., 1984, 1986). Similarly, amylopectin from sweet potato starch has a lower limiting viscosity number than amylopectin from potato; this may be due to the smaller spherical molecules of sweet potato starch (Suzuki *et al*., 1985; Takeda *et al*., 1986). Because of the peak viscosity value indicator of starch disintegration, cohesive forces within the granules having higher values are stronger than the granules with lower cohesive force. The breakdown viscosity, i.e. the paste consistency after maintaining at 93°C for 15 minutes, assists in estimating the resistance of the starch paste to disintegration during heating and stirring. The consistency, gel properties and gel viscosity during the pasting cycle are highly applicable for many industrial applications and influence important quality characteristics (Leelavathi *et al*., 1987).

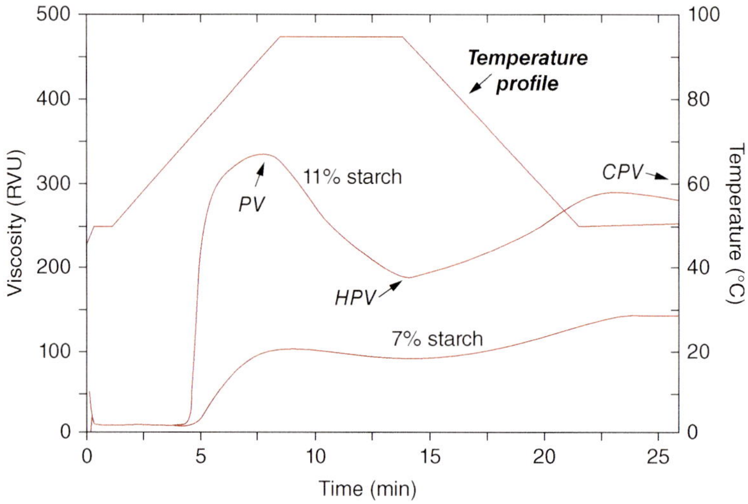

Fig. 4.5. RVA pasting curve of sweet potato starch at 7 and 11% starch concentration. (From Collado *et al*., 1999)

Sweet potato starch behaves similarly to cassava starch in its viscosity characteristics, i.e. peak viscosity, viscosity breakdown and setback viscosity. The viscosity properties of sweet potato starch measured by various methods have been reviewed by Tian *et al.* (1991) and Zhu and Wang (2014) (Table 4.9). Collado *et al.* (1999) developed Rapid Visco Analyser (RVA) parameter correlations for 44 types of sweet potato genotypes at 7% and 11% concentrations.

The authors reported a wide variation in the peak viscosity and the broadness of the peaks. In addition, they noticed a significant negative correlation between the peak viscosity and starch amylose content. The data on viscosity characteristics of a number of sweet potato varieties shows slight variation (Moorthy *et al.*, 2010) (Table 4.10). Environmental factors also could be influential to the pasting behaviours. Higher soil temperatures resulted in higher breakdown and pasting temperature, and lower setback values (Noda *et al.*, 2001).

The pasting temperature of sweet potato starch ranged from 62°C to 86°C (Tian *et al.*, 1991; Kitahara *et al.*, 1999; Collado and Corke, 1997). The pasting temperature measured using the Brabender viscograph ranged between 66.0°C and 86.3°C while microscopic evaluation indicated that the values ranged between 57–70°C to 70–90°C. Noda *et al.* (1996) found that cultivation practice like different fertilizer level affected the pasting temperatures of sweet potato starch cultivars, which were in the range of 70.8–73.9°C. The eating quality and product quality (especially the noodle quality) depend on the pasting chacteristics (Collado *et al.*, 1999). The study conducted by Katayama (Katayama *et al.*, 2002) also indicated pasting temperatures similar to previous reports.

Novel genotypes with lower pasting temperatures have been generated by the means of conventional breeding (Katayama *et al.*, 2002) or genetic manipulations (i.e. by inhibition of starch synthase II expression) (Takahata *et al.*, 2010). Increased amounts of very short amylopectin unit chains (degree of polymerization (DP) 6–12) and decreased crystalline lamellae thickness were the structural basis for lower pasting and gelatinization temperatures (Noda *et al.*, 2009).

The viscosity data of starches from enzymatic and conventional extraction recorded using the Brabender viscograph are presented in Table 4.11. The peak viscosity depended on the concentration of starch. With 5% and 6% pastes, the peak viscosity increased with increasing amounts of the enzymes, up to 0.025% or 0.05%, but with 0.25% enzyme it dropped appreciably. With the 7% paste, there was a fairly steady drop in peak viscosity as

Table 4.9. Pasting properties of sweet potato starch. (From Zhu and Wang, 2014)

No.	Method	Starch content (%)	Peak viscosity	Breakdown	Setback	Pasting temp.	References
21	RVA	10	143–288	29–163	50–79	73.5–87.7	Aina *et al.*, 2012
3	RVA	6	133–152	18–37	42–59	76.6–70.8	Wickramasinghe *et al.*, 2009
1	RVA	10	465			78	Tetchi *et al.*, 2007
1	RVA	9	281		73	72.4	Leelavathi *et al.*, 1987
20	RVA	7		85–206	106–176	53.8–56.6	Katayama *et al.*, 2004
2	RVA	10	381–433.4	197.5–237.3	125.4–176.4	74.8–80.5	Jangchud *et al.*, 2003
2	BA	10	240–275		380–405	75–79.5	Osundahunsi *et al.*, 2003
7	RVA	8	275–330	124–154	32–68	73.5–77.6	Noda *et al.*, 2002
25	RVA	10	145–1260			73.5–81.1	Waramboi *et al.*, 2011
11	RVA	8.9	268–469	149–247	51–76		Zhu *et al.*, 2011
5	BV	6	666–887	37–417	275–379	54.9–73.6	Kitahara *et al.*, 2007
8	RVA	8	340–385	120–189	69–98	74.1–77.2	Toyama *et al.*, 2003
4	RVA	7				52.6–73.6	Katayama *et al.*, 2002
		10				51.4–72.6	Katayama *et al.*, 2002
9	RVA	7	126–190	33–84	119–199	60.4–76	Takahata *et al.*, 2010

the enzyme concentration increased. This is probably due to a weakening of the associated forces rather than to a breakdown of starch granules. The breakdown viscosity was also increased with the higher levels of the enzymes. The pasting temperature did not show any definite pattern, but generally, there was a shift to lower temperatures with increasing concentrations of the enzyme (Moorthy and Balagopalan, 1999).

4.5.3 Rheological property

The sweet potato starch rheology has been investigated in detail. During heating, the storage modulus (G') and loss modulus (G'') increased while phase angle decreased, showing that the sweet potato starch transformed from sol to gel. The initial increase in G' and G'' is due to the dynamic swelling of starch granules leading to closer packing. When the starch granules became very soft, deformable and compressible, decrease in all the values has been recorded. The rheological properties of sweet potato starch have been compared using the Bohlin rheometer and wide variability in the values of G' and G'' was observed. During heating, the storage modulus (G') and loss modulus (G'') showed an increase in values and the phase angle indicated transformation from sol to gel (Fig. 4.6). Both moduli reached a maximum during heating after which their values decreased (Garcia and Walter, 1998). The starch exhibited similarity in its elastic behaviour, predominating over a viscous nature with other tuber starches.

4.5.4 Retrogradation

Sol or paste stability depends on the retrogradation characteristics of starch paste. Having low retrogradation tendency, sweet potato starch has high paste stability. When cooled, the gelatinized starch granules disperse in water and attain the consistency of gels.

Table 4.10. Viscosity properties of extracted starch. (From Moorthy and Shanavas, 2010)

	cP				
Lines	Peak viscosity	Breakdown	Final viscosity	Setback	Pasting temp (°C)
ST1	2924	1450	2098	800	68.70
ST2	3467	2246	3677	1317.5	67.48
ST3	3475	1306	3212	1182	67.85
ST4	3376	1997	4153	1683.5	65.9
ST5	3557	1069	3250	945	67.80
ST6	3714	2190	4076	1076	75.85
ST7	3498	1392	4334	1207	77.50
ST8	3902	1080	3702	872	76.50
ST9	3123	2541	2766.5	762.5	70.2
ST10	3602	1940	3031	871	70.7
ST11	3583	1484	4530	1426	75.90
ST12	3536	1571	3087	1034	71.85
ST13	4729	2092	3838	1201	68.6
ST14	3560	2180	3534	1154	70.2
l.s.d at 5% level	26.22	72.51	33.52	32.57	0.313

If the starch concentration exceeds a critical value, the swollen granules become entangled in amylose chains and diffuse out of the starch granules. The resultant composite can be viewed as an amylose gel with the swollen starch granules as filler (Gidley, 1989). Rupturing of starch granules by shearing or other thermal or mechanical damage further complicates the structural properties (Mestres *et al.*, 1988). In addition, further changes such as recrystallization (or retrogradation) of the polymer chains could occur during storage.

Retrogradation is depentent on various factors like amylose and amylopectin content, presence of sugars, salts and emulsifiers, molecular size, temperature and pH (Rosario and Pontiveros, 1983). Takeda *et al.* (1986) showed that rate of retrogradation of sweet potato amylose resembles that of cassava but is lower than potato amylose. Amylose from sweet potato retrograded at a slower rate than the amylose from cassava (Rašper (1969a). In contrast, sweet potato amylopectin retrograded at an increased rate compared to cassava amylopectin. Rosario and Pontiveros (1983) found that sweet potato starch retrograded slower than wheat, corn and cassava starches. The authors also indicated that the bread containing sweet potato flour as a substituent staled at a slower rate than other breads due to this factor.

Rheometry, DSC, nuclear magnetic resonance spectrometry, wide-angle X-ray diffractometry and turbidimetry have been employed to study retrogradation of sweet potato starch. Great genetic diversity in retrogradation properties has been obtained through DSC (Peroni *et al.*, 2006; Srichuwong *et al.*, 2005; Zhu *et al.*, 2011). Increase in firmness during gelling was observed (Ishiguro *et al.*, 2003). It has been suggested that one major factor deciding retrogradation of sweet potato starch is the proportion of short unit chains of amylopectin (Ishiguro *et al.*, 2003). Correlation studies have proved that some of the pasting parameters were distinctly related to the retrogradation behaviour (Ishiguro *et al.*, 2003; Zhu *et al.*, 2011). The retrogradation ratio calculated from the melting enthalpy data (ΔH) (DSC) was negatively related to the breakdown viscosity (Zhu *et al.*, 2011).

An optical micrographic study has brought to light the mechanism of retrogradation of sweet potato starch (Lian *et al.*, 2011). Intially, amylopectin started nucleation in the

Table 4.11. Viscosity properties of enzymatically separated sweet potato starch. (From Moorthy and Balagopalan, 1999)

Starch conc (%)	Enzyme conc (%)	Peak viscosity (BU)	Breakdown viscosity (BU)	Pasting temperature (°C)
5	0.000	260	0	87–95
	0.010	280	0	86–95
	0.025	300	40	85–95
	0.050	280	40	84–95
	0.100	280	100	82–89
	0.0200	220	40	82–90
6	0.000	440	20	88–95
	0.010	460	40	87–94
	0.025	500	40	86–95
	0.050	500	100	84–92
	0.100	460	120	82–92
	0.0200	400	60	82–88
7	0.000	780	0	88–95
	0.010	760	60	88–95
	0.025	740	100	84–95
	0.050	760	160	84–95
	0.100	680	200	82–90
	0.0200	600	120	82–89

gelatinized region containing the crystalline shells. Later on, the nucleus developed into large, rod-like crystals due to the congregation of amylose on plates that were composed of, and extend the spatial duration of, amylopectin molecules.

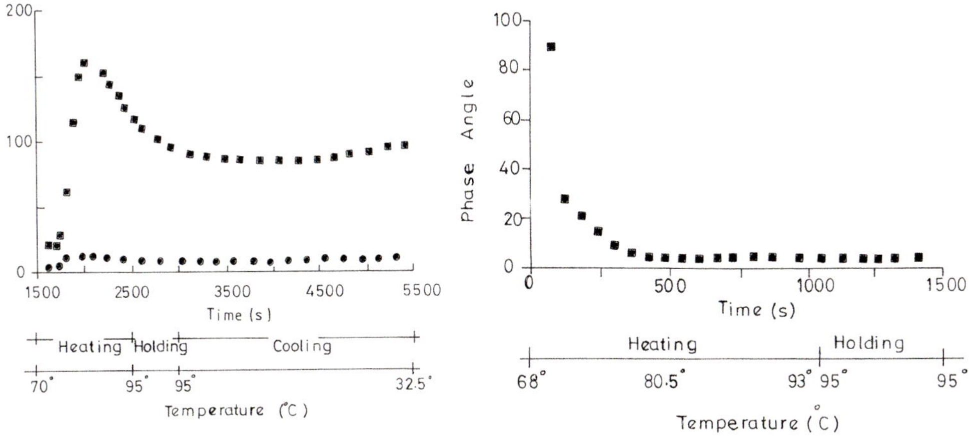

Fig. 4.6. Storage modulus, loss modulus and phase angle for Peruvian sweet potato starch. (From Garcia and Walter, 1998)

4.6 Thermal Properties

4.6.1 DSC

Gelatinization temperature can be considered as an indicator of the temperature at which the starch granules start gelatinizing. Water content and presence of salts, sugars and other small molecules in starch can modify gelatinization temperature. The typical DSC curve for sweet potato starch is given in Fig. 4.7.

Zhu and Wang (2014) have collated the DSC data for sweet potato starch from various sources (Table 4.12). Madamaba *et al.* (1975) found that the average gelatinization temperature of starch from six different cultivars of sweet potato was found to range from 63.6°C to 70.7°C.

Average gelatinization temperature and amylose content of the starch were found to be postively correlated. The range of gelatinization was normally 12–17°C. Barham *et al.* (1946) reported that average gelatinization temperatures of starch from five cultivars ranged from 69°C to 75.5°C, but after reduced curing of the roots. Rašper (1969b) reported that the onset of gelatinization in sweet potato starch took place at 77°C and the viscosity increased to 85°C. Collado *et al.* (1999) obtained considerable variation in all the DSC-derived parameters for starch from 44 different sweet potato varieties. The mean T_{onset} was 64.6°C (range 61.3–70°C), mean T_{peak} 73.9°C (range 70.2–77°C) and mean T_{end} 84.6°C (range 80.7–88.5°C), respectively. The mean gelatinization temperature was 20.1°C with values ranging from 16.1°C to 23°C. Garcia and Walter (1998) reported a T_{onset} range of 58–64°C, a T_{peak} range of 63–74°C and a T_{end} range of 78–83°C for two varieties cultivated at different locations. While the selection index did not affect the values, location played a definite role (Tian *et al.*, 1991). Kitada *et al.* (1988) found that the gelatinization temperature was affected by the region in which the sweet potato has been cultivated. Noda *et al.* (1998) reported that T_{onset}, T_{peak} and ΔH of sweet potato starch from 51 different varieties, or grown in different conditions, ranged between 55.7°C and 73.1°C, 61.3–77.6°C and 12.7–16.8 J/g, respectively. Noda *et al.* (1996) did not find any correlation with fertilization and the DSC characteristics. Zhu *et al.* (2011) obtained values for ΔH ranging from 7.6 J/g to 13.2 J/g for 11 different genotypes of sweet potato.

The increase in outer chains of amylopectin reduced the packing efficiency of double helices within the crystalline region, resulting in lower

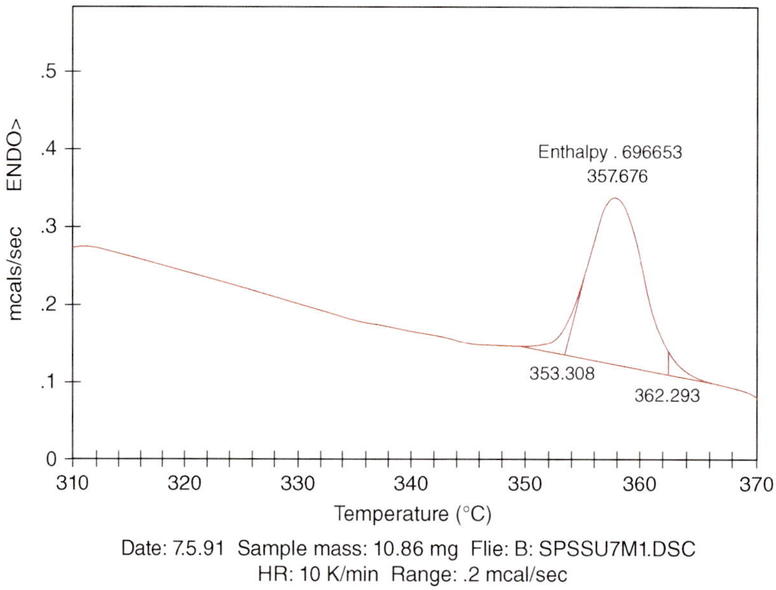

Fig. 4.7. DSC pattern of sweet potato starch (From Moorthy unpublished data.)

Table 4.12. DSC characteristics of gelatinization measured by differential scanning calorimetry and kofler hot-stage microscopy. (From Zhu and Wang, 2014)

No.	To (°C)	Tp (°C)	Tc (°C)	ΔH (J/g)	References
3	61.1–71.1	76.6–77.8		17.1–18.1	96
2	56.8–70.1	67.7–76.7	82.9–85.6	11.6–12.5	141
1	69	73.8		15.3	78
1	62.85	70.55	77.9	12.9	143
1	66.7	74.0	86.6	18.4	145
2	66.7–67.2	70.5–71.5	74.8–75.7	10.5–11	147
7	66.2–71.3	71–74.6		14.5–17.2	150
2a	51.2–76.2	57.6–79.5		12.1–17.6	177
25		68.0–78.1		11.7–27	140
11	67.2–71.1	75.4–79.4	80.4–83.4	7.6–13.6	139
4	39–64.6	46.9–71.2	64.8–84.8	8.8–13.7	178

gelatinization temperature and enthalpy. The gelatinization temperatures of sweet potato starch obtained from fresh roots and freeze-dried roots of sweet potato, as reported by Valetudie *et al.* (1995), are presented in Table 4.13. The starch from fresh tubers and freeze-dried sweet potato tubers gave nearly equal values (67–73°C), but the smaller granules gelatinized between 75°C and 88°C. Torres found DSC values of 60.2, 65.6 and 15.1 J/g'. The pasting temperature of sweet potato starch was 66.0–86.3°C by viscography and 57–70°C to 70–90°C by microscopic determination.

For sweet potato starch, the values for gelatinization enthalpy ranged from 10 J/g to 18.6 J/g (Tian *et al.*, 1991; Garcia and Walter, 1998; Collado *et al.*, 1999) (Table 4.14). These values were highly influenced by variety and environmental conditions at which the sweet potato was grown (Garcia and Walter, 1998; Noda *et al.*, 1996). During growth period, the ΔH was lowest at the earliest stage of development and the enthalpy ranged between 11.8–13.4 J/g (Noda *et al.*, 1992b).

New genotypes with rather lower gelatinization temperatures (e.g. T_{onset} 39°C) have also been reported (Katayama *et al.*, 2002).

Higher soil temperature led to higher gelatinization temperatures and melting enthalpy, indicating that environmental factors such as soil temperature can influence the gelatinization temperatures greatly (Genkina *et al.*, 2003; Noda *et al.*, 2001). The increased gelatinization temperatures and enthalpy could be traced to the distinctly lower number of short unit chains of amylopectin having DP 6 and 7 on the molecular level (Noda *et al.*, 2001), and also the lower surface energy of the crystals in the granules and the higher thickness of the crystalline lamellae at the granular level (Genkina *et al.*, 2003).

Table 4.13. DSC characteristics of the hydrothermic transition of purified starches and fresh and freeze dried and parenchyma tuber cells. (From Valetudie *et al.*, 1995)

Sweet potato	T onset (°C)	T peak (°C)	T endset (°C)	ΔH (J/g)
Starch	67.3	72.7	79.6	13.6
Fresh roots	67.4	73.5	80.1	6.8
Freeze-dried roots	67.8	73.2	81.5	9.3
Small starch granules	75.6	82.6	88.3	15.3

Table 4.14. Thermal properties of sweet potato starch (Moorthy *et al*., 2012).

No	T onset (°C)	T peak (°C)	T endset (°C)	ΔH (J/g)	Reference
1	67–75	73–79	81.4–84.8	10–12.3	Chiang and Chen, 1988
2	65.6–68.2	72.8–74.3	84.6–86.8	15.1–16.3	Kitada *et al*., 1988
3	61.3	70.2–77	80.7–88.5	15.1–16.3	Collado *et al*., 1999
4	58–64	63–74	78–83	14.8–18.6	Wankhede and Sajjan, 1981
5	67.3	72.7	79.6	13.6	Valetudie *et al*., 1995

4.7 Digestibility

Studies using *in vitro* α-amylolysis can give valuble insight, not only on the *in vivo* nutritional properties of starch but also they can reveal the molecular and physical structure of the starch. It may be difficult to compare the difference in the extent of hydrolysis from different studies due to the differences in the amylase sources and reaction conditions (Hoover, 2001; Shariffa *et al*., 2009; Yussof *et al*., 2013; Zhu *et al*., 2011). Great genetic diversity in *in vitro* digestibility (by pancreatin) of sweet potato starch range from 29.5% to 41.2%) (Zhu *et al*., 2011). Shariffa *et al*. (2009) reported that heating treatment below gelatinization temperature enhanced the degree of α-amylolysis of granular sweet potato starch.

During digestion, presence of α-amylase and glycoamylase can degrade sweet potato starch (Cerning-Beroard and Le Dividich, 1976). Digestibility of raw starch from sweet potato varieties by glycoamylase has been studied by Noda *et al*. (1992a). Gallant *et al*. (1982) reported that A-type starches have high susceptibility to α-amylase. Pelletization increased the raw starch digestibility with bacterial α-amylase from 17% to 45%. Some studies indicated that enzymatic corrosion occurs mainly at the surface of the granules. The susceptibility of sweet potato starch to α-amylase after one day's incubation was in the range of 35.7–65.5% weight loss for the six cultivars in the study.

β-amylase has been used to determine the external chain lengths and to estimate the number of branch points (Hokama *et al*., 1980; Lii *et al*., 1987; Manners, 1989). Lii *et al*. (1987) reported a β-amylase limit value of 87.9% for the amylase present in sweet potato starch, which is substantially higher than the values reported in the literature. In contrast, α-amylase is able to attack the polymers randomly at any α−1:4-linkage that is sterically accessible. The susceptibility to attack by α-amylase depends on the sweet potato variety (Madamaba *et al*., 1975).

Moorthy and Padmaja (1999) found that sweet potato starch has very good digestibility in rat-feeding experiments, 72.2% for raw starch and 91.2% for cooked starch.

References

Aina, A.J., Falade, K.O., Akingbala, J.O. and Titus, P. (2012) Physicochemical properties of Caribbean sweet potato (*Ipomoea batatas* (L) LAM*)* starches. *Food and Bioprocess Technology* 5(2), 576–583. DOI: 10.1007/s11947-009-0316-6.

Anon (1968) Production of sweet potato starch and storage of sweet potato flakes. *Journal of Food Science* 44, 1186–1190.

Banks, W. and Greenwood, C.T. (1975) *Starch and Its Components*. Edinburgh University Press, Edinburgh, UK.

Barham, H.N. and Wagoner, J.A. (1946) Effect of time and conditions of care on the carbohydrate composition of sweet potato starches. *Journal of Agricultural Research* 73, 255–280.

Bertoft, E. (2013) On the building block and backbone concepts of amylopectin structure. *Cereal Chemistry Journal* 90(4), 294–311. DOI: 10.1094/CCHEM-01-13-0004-FI.

Bertoniere, N.R., McLemore, T.A., Hasling, V.C., Catalano, E.A. and Deobald, H.J. (1966) The effect of environmental variables on the processing of sweet potatoes into flakes and on some properties of their isolated starches. *Journal of Food Science* 31(4), 574–582. DOI: 10.1111/j.1365-2621.1966.tb01907.x.

Bouwkamp, J.C. (1985) *Sweet Potato Products: A Natural Resource for the Tropics*. CRC Press, Boca Raton, Florida.

Cai, C. and Wei, C. (2013) In situ observation of crystallinity disruption patterns during starch gelatinization. *Carbohydrate Polymers* 92(1), 469–478. DOI: 10.1016/j.carbpol.2012.09.073.

Cerning-Beroard, J. and Le Dividich, J. (1976) Feeding value of some starchy tropical products: *in vitro* and *in vivo* study of sweet potato, yam, malanga, breadfruit and banana. *Annals of Zootechnology* 25, 155–168.

Chiang, W.C. and Chen, K.L. (1988) Comparison of physicochemical properties of starch and amylolytic enzyme activity of various sweet potato varieties. *Shih Pin Ko Hsueh* 15, 1–11.

Collado, L.S. and Corke, H. (1997) Properties of starch noodles as affected by sweetpotato genotype. *Cereal Chemistry Journal* 74(2), 182–187. DOI: 10.1094/CCHEM.1997.74.2.182.

Collado, L.S., Mabesa, R.C. and Corke, H. (1999) Genetic variation in the physical properties of sweet potato starch. *Journal of Agricultural and Food Chemistry* 47(10), 4195–4201. DOI: 10.1021/jf990110t.

Daniels, F. (1966) *Physical Chemistry*, 3rd edn. Wiley, New York.

Delpeuch, F. and Favier, J.C. (1980) Characteristics of starches from tropical food plants: ⟨⟨-amylase hydrolysis, swelling and solubility patterns. *Annals of Agricultural Sciences* 29(1), 53–67.

Delpeuch, F., Favier, J.C. and Chabonniere, R. (1978) Caracteristiques des amidons de plants alimentanastropicales. *Annals of Agricultural Sciences* 27(4), 809–826.

Delpeuch, F., Favier, J.C. and Chabonniere, R. (1979) Study of the physicochemical properties of starch from various Cameroon food plants. *CahOnarest* 2(3), 53–81.

Deng, F.-M., Mu, T.-M., Zhang, M. and Abegeunde, O.K. (2013) Composition, structure, and physicochemical properties of sweet potato starches isolated by sour liquid processing and centrifugation. *Starch/Stärke* 65(1–2), 162–171. DOI: 10.1002/star.201200106.

Dreher, M.L. and Berry, J.W. (1983) Buffalo gourd root starch. 1. Properties and structure. *Starch/Stärke* 35(3), 86–92.

Duell, B. (1990) Ways of eating sweet potatoes in Japan. *Proceeding of the Eighth Symposium of the International Society for Tropical Root Crops. Tropical and Tuber Crops Changing Role in Modern World, 30 October–5 November*, Bangkok, pp. 45–52.

Fuwa, H., Nakajima, M. and Hamada, A. (1977) Comparative susceptibility to amylases of starches from different plant species. *Cereal Chemistry* 54, 230–237.

Gallant, D.J., Bewa, H., Buy, Q.H., Bouchet, B., Szylit, O. *et al.* (1982) On ultrastructural and nutritional aspects of some tropical tuber starches. *Starch/Stärke* 34(8), 255–262. DOI: 10.1002/star.19820340803.

Garcia, A.M. and Walter, W.M. (1998) Physico-chemical characterization of starch from Peruvian sweet potato selections. *Starch/Stärke* 50, 331–337.

Genkina, N.K., Noda, T., Koltisheva, G.I., Wasserman, L.A., Tester, R.F. *et al.* (2003) Effects of growth temperature on some structural properties of crystalline lamellae in starches extracted from sweet potatoes (Sunnyred and *Ayamurasaki*). *Starch/Stärke* 55(8), 350–357. DOI: 10.1002/star.200300145.

Gidley, M.J. (1989) Molecular mechanisms underlying amylose aggregation and gelation. *Macromolecules* 22(1), 351–358. DOI: 10.1021/ma00191a064.

Gitomer, C.S. (1987) Sweet potato and white potato development in China. In: *A Compendium of Basic Data*. International Food Policy Research Institute, Washington, DC, pp. 105–109.

Greenwood, C.T. (1979) Observations on the structure of the starch granule. In: Blanshard, J.M.V. and Mitchell, J.R. (eds) *Polysaccharides in Food*. Butterworths and Co. (Publishers) Ltd, London, pp. 129–138.

Hammet, H.L. and Barrentin, B. (1961) Some effects of variety, curing and baking upon the carbohydrate content of sweet potatoes. *Proceedings of the American Society for Horticultural Science* 78, 421–426.

Hanashiro, I., Abe, J.-ichi. and Hizukuri, S. (1996) A periodic distribution of the chain length of amylopectin as revealed by high-performance anion-exchange chromatography. *Carbohydrate Research* 283, 151–159. DOI: 10.1016/0008-6215(95)00408-4.

Hanashiro, I., Tagawa, M., Shibahara, S., Iwata, K. and Takeda, Y. (2002) Examination of molar-based distribution of A, B and C chains of amylopectin by fluorescent labeling with 2-aminopyridine. *Carbohydrate Research* 337(13), 1211–1215. DOI: 10.1016/S0008-6215(02)00110-6.

Heinze, P.H. and Appleman, C.O. (1943) A biochemical study of curing processes in sweet potatoes. *Plant Physiology* 18(4), 548–555. DOI: 10.1104/pp.18.4.548.

Hizukuri, S. (1969) The effect of environment temperature of plants on the physicochemical properties of their starches. *Journal of the Technological Society of Starch* 17(1), 73–88. DOI: 10.5458/jag1953.17.73.

Hokama, K., Nakasone, Y., Nakama, T. and Teruya, M. (1980) Determination of external and internal chain-lengths of starch amylopectin in wet taro cultivated in Okinawa. *Bulletin of the College of Agriculture, University of the Ryukyus* 27, 163–173.

Hoover, R. (2001) Composition, molecular structure, and physicochemical properties of tuber and root starches: a review. *Carbohydrate Polymers* 45(3), 253–267. DOI: 10.1016/S0144-8617(00)00260-5.

International Potato Center. Potato Facts and Figures (2013) Available at: http://cipotato.org/potato/facts (accessed December 2019)

Ishiguro, K., Noda, T. and Yamakawa, O. (2003) Effect of cultivation conditions on retrogradation of sweet potato starch. *Starch/Stärke* 55(12), 564–568.

ISI (1970) *Specifications for Tapioca Starch for Use in Cotton Textile Industries, IS 1605–1960*. Indian Standards Institution, New Delhi.

Jain, N.L., Bhatia, B.S. and Gurdhari, L. (1951) Preparation and uses of sweet potato flour. *Journal of Scientific and Industrial Research* 10A, 332–333.

Jangchud, K., Phimolsiripol, Y. and Haruthaithanasan, V. (2003) Physicochemical properties of sweet potato flour and starch as affected by blanching and processing. *Starch - Stärke* 55(6), 258–264.

Kallabinski, J. and Balagopalan, C. (1991) Enzymatic starch extraction from tropical root and tuber crops. In: Ofori, F. and Hahn, S.K. (eds) *Symposium on Tropical Root Crops in a Developing Economy, Acta Horticulturae, 1*, pp. 83–88.

Katayama, K., Komae, K., Kohyama, K., Kato, T., Tamiya, S. *et al.* (2002) New sweet potato line having low gelatinization temperature and altered starch structure. *Starch/Stärke* 54(2), 51–57. DOI: 10.1002/1521-379X(200202)54:2<51::AID-STAR51>3.0.CO;2-6.

Katayama, K., Tamiya, S. and Ishiguro, K. (2004) Starch properties of new sweet potato lines having low pasting temperature. *Starch - Stärke* 56(12), 563–569.

Katayama, K., Komae, K., Tamiya, S., Khoyama, K., Nakatani, M. *et al.* (2006) Studies on the breeding for improving starch properties in sweet potato. *Japan Agricultural Research Quarterly: JARQ* 40(2), 115–122. DOI: 10.6090/jarq.40.115.

Kay, D.E. (1987) *TDRI Crop and Product Digest No. 2*. TDRI, London, pp. 166–173.

Kitada, Y., Sasaki, M., Yamazoe, Y. and Nakazawa, H. (1988) Measurement of thermal behavior and amylose content of Kuzu and sweet potato starches. *Nippon Shokuhin Kogyo Gakkaishi* 35(3), 135–140. DOI: 10.3136/nskkk1962.35.3_135.

Kitahara, K., Hamasuna, K., Nozuma, K., Otani, M., Hamada, T. *et al.* (2007) Physicochemical properties of amylose-free and high-amylose starches from transgenic sweetpotatoes modified by RNA interference. *Carbohydrate Polymers* 69(2), 233–240. DOI: 10.1016/j.carbpol.2006.09.025.

Kitahara, K., Ueno, J., Suganuma, T., Ishigurol, K. and Yamakawal, O. (1999) Physicochemical properties of root starches from new types of sweet potato. *Journal of Applied Glycoscience* 46(4), 391–397. DOI: 10.5458/jag.46.391.

Leach, H.W., McCoven, L.D. and Schoch, T.J. (1959) Structure of the starch granule. *Cereal Chemistry* 36, 534–544.

Leelavathi, K., Indrani, D. and Sidhu, J.S. (1987) Amylograph pasting behaviour of cereal and tuber starches. *Starch/Stärke* 39(11), 378–381. DOI: 10.1002/star.19870391103.

Lian, X.J., Zhao, S., Liu, Q. and Zhang, X. (2011) A photographic approach to the possible mechanism of retrogradation of sweet potato starch. *International Journal of Biological Macromolecules* 48(1), 125–128.

Lian, X.J., Li, L., Zhang, K.S., Xu, Y. and Lu, J. (2012) A new proposed sweet potato starch granule structure--pomegranate concept. *International Journal of Biological Macromolecules* 50(3), 471–475. DOI: 10.1016/j.ijbiomac.2012.01.014.

Lii, C.Y. and Chang, S.M. (1978) Studies of the starches in Taiwan: sweet potato, cassava, yam and arrow root starches. *Proceedings of the National Scientific Council. ROC-2 Part A: Physics Science Engineering* 2, 146–423.

Lii, C.Y., Chiou, R.W. and Chu, Y.L. (1987) The degree of branching in amylose from tuber and legume starches. Proceedings of national science Council, Republic of China. *Part A: Physics Science Engineering* 11(4), 341–345.

Liu, S.Y., Liang, C.L. and Li, L. (1985) Studies on the physicochemical properties of the tubers of new sweet potato lines. *Chung-hua Nung Yeh Yen Chiu* 34, 341–345.

Lockwood, S. and King, J.M. (2008) Amino acid additives effects on sweet potato starch thermal properties. *Starch - Stärke* 60(12), 690–695.

Madamaba, L.S.P., Bustrillos, A.R. and San Pedro, E.L. (1975) Sweet potato starch: physicochemical properties of whole starch. *Philippine Agriculturist* 58, 338–350.

Madamba, L.S.P. and San Pedro, E.L. (1976) Chemical composition of sweet potato flour. *Philippine Agriculturist* 59, 350–355.

Manners, D.J. (1989) Recent developments in our understanding of amylopectin structure. *Carbohydrate Polymers* 11(2), 87–112. DOI: 10.1016/0144-8617(89)90018-0.

Medcalf, D.G. and Giles, K.A. (1965) Wheat starches: comparison of physicochemical properties. *Cereal Chemistry* 42, 558–568.

Mélo, E. de., Krieger, N. and Stamford, T.L.M. (1994) Physicochemical properties of Jacatupé (Pachyrhizus *erosus L.* urban) starch. *Starch/Stärke* 46(7), 245–247. DOI: 10.1002/star.19940460702.

Mestres, C., Colonna, P. and Buleon, A. (1988) Gelation and crystallisation of maize starch after pasting, drum-drying or extrusion cooking. *Journal of Cereal Science* 7(2), 123–134. DOI: 10.1016/S0733-5210(88)80014-6.

Moorthy, S.N. (2002) Physicochemical and functional properties of tropical tuber starches: a review. *Starch/Stärke* 54(12), 559–592. DOI: 10.1002/1521-379X(200212)54:12<559::AID-STAR2222559>3.0.CO;2-F.

Moorthy, S.N. and Balagopalan, C. (1999) Physicochemical properties of enzymatically separated starch from sweet potato. *Tropical Science* 39, 23–27.

Moorthy, S.N. and Padmaja, G. (1999) A rapid titrimetric method for the determination of starch content of cassava tubers. *Journal of Root Crops* 28(1), 30–37.

Moorthy, S.N. and Ramanujam, T. (1986) Variation in properties of starch in cassava varieties in relation to age of the crop. *Starch/Stärke* 38(2), 58–61. DOI: 10.1002/star.19860380206.

Moorthy, S.N. and Shanavas, S. (2010) Sweet potato starch. In: Ray, R.C. and Tomlins, K.I. (eds) *Sweet Potato: Post Harvest Aspects in Food, Feed and Industry*. Nova Publishers, New York.

Moorthy, S.N., Sajeev, M.S. and Shanavas, S. (2012) Sweet potato starch: physico-chemical, functional, thermal and rheological characteristics. *Fruit, Vegetable and Cereal Science and Biotechnology* 6(1), 124–133.

Moorthy, S.N., Naskar, S.K., Shanavas, S., Radhika, G.S. and Mukherjee, A. (2010) Physicochemical characterization of sweet potato starch. *International Journal of Food Properties* 13, 1280–1289.

Nara, S., Sakakura, M. and Komiya, T. (1983) On the acid resistance of starch granules. *Starch/Stärke* 35(8), 266–270. DOI: 10.1002/star.19830350803.

Noda, T., Takahata, Y., Nagata, T. and Monma, M. (1992a) Digestibility of sweet potato raw starches by glucoamylase. *Starch/Stärke* 44(1), 32–35. DOI: 10.1002/star.19920440110.

Noda, T., Takahata, Y. and Nagata, T. (1992b) Developmental changes in properties of sweet potato starches. *Starch/Stärke* 44(11), 405–409. DOI: 10.1002/star.19920441102.

Noda, T., Kobayashi, T. and Suda, I. (2001) Effect of soil temperature on starch properties of sweet potatoes. *Carbohydrate Polymers* 44(3), 239–246. DOI: 10.1016/S0144-8617(00)00227-7.

Noda, T., Takahata, Y., Sato, T., Hisamatsu, M. and Yamada, T. (1995) Physicochemical properties of starches extracted from sweet potato roots differing in physiological age. *Journal of Agricultural and Food Chemistry* 43(12), 3016–3020. DOI: 10.1021/jf00060a005.

Noda, T., Takahata, Y., Sato, T., Ikoma, H. and Mochida, H. (1996) Physicochemical properties of starches from purple and orange- fleshed sweet potato roots at two levels of fertilizer. *Starch/Stärke* 48(11–12), 395–399. DOI: 10.1002/star.19960481103.

Noda, T., Takahata, Y., Sato, T., Suda, I., Morishita, T. (1998) Relationships between chain length distribution of amylopectin and gelatinization properties within the same botanical origin for sweet potato and buckwheat. *Carbohydrate Polymers* 37(2), 153–158. DOI: 10.1016/S0144-8617(98)00047-2.

Noda, T., Kimura, T., Otani, M., Ideta, O., Shimada, T. *et al.* (2002) Physicochemical properties of amylose-free starch from transgenic sweet potato. *Carbohydrate polymers* 49(3), 253–260.

Noda, T., Isono, N., Krivandin, A.V., Shatalova, O.V., Błaszczak, W. *et al.* (2009) Origin of defects in assembled supramolecular structures of sweet potato starches with different amylopectin chain-length distribution. *Carbohydrate Polymers* 76(3), 400–409.

Srichuwong, S. and Jane, J.L. (2007) Physicochemical properties of starch affected by molecular composition and structures: a review. *Food Science and Biotechnology* 16, 663–674.

Osundahunsi, O.F., Fagbemi, T.N., Kesselman, E. and Shimoni, E. (2003) Comparison of the physicochemical properties and pasting characteristics of flour and starch from red and white sweet potato cultivars. *Journal of Agricultural and Food Chemistry* 51(8), 2232–2236. DOI: 10.1021/jf0260139.

Peat, S., Whelan, W.J. and Thomas, G.J. (1952) The evidence of multiple branching in waxy maize starch. *Journal of the Chemical Society Resumed*, 4536–4538.

Pérez, S. and Bertoft, E. (2010) The molecular structures of starch components and their contribution to the architecture of starch granules: a comprehensive review. *Starch/Stärke* 62(8), 389–420. DOI: 10.1002/star.201000013.

Peroni, F.H.G., Rocha, T.S. and Franco, C.M.L. (2006) Some structural and physicochemical characteristics of tuber and root starches. *Food Science and Technology International* 12(6), 505–513. DOI: 10.1177/1082013206073045.

Radley, J.A. (1976) *Starch Production Technology*. Applied Science Publishers Ltd, London, pp. 189–229.

Rašper, V. (1969a) Investigations on starches from major starch crops grown in Ghana. I. Hot paste viscosity and gel forming power. *Journal of the Science of Food and Agriculture* 20(3), 165–171. DOI: 10.1002/jsfa.2740200309.

Rašper, V. (1969b) Investigations on starches from major starch crops grown in Ghana. II. Swelling and solubility patterns: amyloclastic susceptibility. *Journal of the Science of Food and Agriculture* 20(11), 642–646. DOI: 10.1002/jsfa.2740201103.

Rasper, V. (1971) Investigations on starches from major starch crops grown in Ghana.III.Particle size and size distribution. *Journal of the Science of Food and Agriculture* 20, 572–580.

Rickard, J.E., Asaoka, M. and Blanshard, J.M.V. (1991) The physicochemical properties of cassava starch. *Tropical Science* 31, 189–207.

Rosario, R.R.D. and Pontiveros, C.R. (1983) Retrogradation of some starch mixtures. *Starch/Stärke* 35(3), 86–92. DOI: 10.1002/star.19830350305.

Schoch, T.J. (1964) Swelling power and solubility of granular starches. In: Whistler, R.L. (ed.) *Methods in Carbohydrate Chemistry*. Vol. IV. Academic Press, New York, pp. 106–109.

Seog, H.M., Park, Y.K., Nam, Y.J., Shin, D.H. and Kim, J.P. (1987) Physicochemical properties of several sweet potato starches. *Applied Biological Chemistry* 30(2), 179–185.

Shariffa, Y.N., Karim, A.A., Fazilah, A. and Zaidul, I.S.M. (2009) Enzymatic hydrolysis of granular native and mildly heat-treated tapioca and sweet potato starches at sub-gelatinization temperature. *Food Hydrocolloids* 23(2), 434–440. DOI: 10.1016/j.foodhyd.2008.03.009.

Shen, M.-C. and Sterling, C. (1981) Changes in starch and other carbohydrates in baking *Ipomoea batatas*. *Starch/Stärke* 33(8), 261–268. DOI: 10.1002/star.19810330803.

Shin, M. and Ahn, S. (1983) Studies on physicochemical properties of starches from sweet potatoes of Korea cultivars. *Applied Biological Chemistry* 26(2), 137–143.

Soni, P.L., Sharma, H., Srivastava, H.C. and Gharia, M.M. (1990) Physicochemical properties of *Canna edulis* starch – comparison with maize starch. *Starch/Stärke* 42(12), 460–464. DOI: 10.1002/star.19900421203.

Srichuwong, S., Sunarti, T., Mishima, T., Isono, N. and Hisamatsu, M. (2005) Starches from different botanical sources I: contribution of amylopectin fine structure to thermal properties and enzyme digestibility. *Carbohydrate Polymers* 60(4), 529–538.

Srichuwong, S., Sunarti, T.C., Mishima, T., Isono, N. and Hisamatsu, M. (2005) Starches from different botanical sources II: contribution of starch structure to swelling and pasting properties. *Carbohydrate Polymers* 62(1), 25–34. DOI: 10.1016/j.carbpol.2005.07.003.
Suh, K.B. (1966) Effect of line treatment on pulping cooperation of current sweet potato manufacturing process. *Korean Restaurant Republic* 9, 221–230.
Suzuki, A., Takeda, Y. and Hizukuri, S. (1985) Relationship between the moleclar structures and retrogradation properties of tapioca, potato and kuzu starches. *Journal of the Japanese Society of Starch Science* 32(3), 205–212. DOI: 10.5458/jag1972.32.205.
Szylit, O., Durand, M., Borgid, L.P., Atinkpahoun, H., Prieto, F. *et al.* (1978) Raw and steam-pelleted cassava, sweet potato and yam cayenensis as starch sources for ruminant and chicken diets. *Animal Feed Science and Technology* 3(1), 73–87. DOI: 10.1016/0377-8401(78)90025-1.
Takahata, Y., Tanaka, M., Otani, M., Katayama, K., Kitahara, K. *et al.* (2010) Inhibition of the expression of the starch synthase II gene leads to lower pasting temperature in sweet potato starch. *Plant Cell Reports* 29(6), 535–543. DOI: 10.1007/s00299-010-0842-8.
Takeda, Y., Shirasaka, K. and Hizukuri, S. (1984) Examination of the purity and structure of amylose by gel-permeation chromatography. *Carbohydrate Research* 132(1), 83–92. DOI: 10.1016/0008-6215(84)85066-1.
Takeda, Y., Tokunaga, N., Takeda, C. and Hizukuri, S. (1986) Physicochemical properties of sweet potato starches. *Starch/Stärke* 38(10), 345–350. DOI: 10.1002/star.19860381006.
Takeda, Y., Shitaozono, T. and Hizukuri, S. (1988) Molecular structure of corn starch. *Starch/Stärke* 40(2), 51–54. DOI: 10.1002/star.19880400204.
Takeda, Y., Shibahara, S. and Hanashiro, I. (2003) Examination of the structure of amylopectin molecules by fluorescent labeling. *Carbohydrate Research* 338(5), 471–475. DOI: 10.1016/S0008-6215(02)00488-3.
Tetchi, F.A., Rolland-Sabaté, A., Amani, G. N.. and Colonna, P. (2007) Molecular and physicochemical characterisation of starches from yam, cocoyam, cassava, sweet potato and ginger produced in the Ivory Coast. *Journal of the Science of Food and Agriculture* 87(10), 1906–1916. DOI: 10.1002/jsfa.2928.
Tian, S.J., Rickard, J.E. and Blanshard, J.M.V. (1991) Physicochemical properties of sweet potato starch. *Journal of the Science of Food and Agriculture* 57(4), 459–491. DOI: 10.1002/jsfa.2740570402.
Toyama, J., Ishiguro, K., Noda, T., Kumagai, T. and Yamakawa, O. (2003) Influence of delayed harvest time on physico-chemical properties of sweetpotato starch. *Starch - Stärke* 55(12), 558–563.
Ueda, S. and Saha, B.C. (1983) Behaviour of *Endomycopsis fibuligera* glucoamylase towards raw starch. *Enzyme and Microbial Technology* 5(3), 196–198. DOI: 10.1016/0141-0229(83)90094-7.
Uehara, S. (1983) Amylose-amylopectin ratio of soluble and insoluble fraction of sweet potato granules treated with urea. *Journal of the Agricultural Chemical Society of Japan* 57(6), 529–533. DOI: 10.1271/nogeikagaku1924.57.529.
Valetudie, J.-C., Guadeloupe, L., Colonna, P., Bouchet, B. and Gallant, D.J. (1995) Gelatinization of sweet potato, tania and yam tuber starches. *Starch/Stärke* 47(8), 298–306. DOI: 10.1002/star.19950470804.
Wang, J. (1984) The development and utilization of starch resources from sweet potato. *Chinese Hunan Agricultural Science* 5, 44–46.
Wankhede, D.B. and Sajjan, S.U. (1981) Isolation and physico-chemical properties of starch extracted from yam, elephant (*Amorphophallus campanulatus*). *Starch/Stärke* 33(5), 153–157. DOI: 10.1002/star.19810330503.
Waramboi, J.G., Dennien, S., Gidley, M.J. and Sopade, P.A. (2011) Characterisation of sweet potato from Papua New Guinea and Australia: physicochemical, pasting and gelatinisation properties. *Food Chemistry* 126(4), 1759–1770. DOI: 10.1016/j.foodchem.2010.12.077.
Watanabe, T., Akiyama, Y., Takahashi, H., Adachi, T., Matsumoto, A. *et al.* (1982) Structural features and properties of Nageli amylodextrins. *Carbohydrate Research* 109, 221–232.
Wickramasinghe, H.A.M., Takigawa, S., Matsuura-Endo, C., Yamauchi, H. and Noda, T. (2009) Comparative analysis of starch properties of different root and tuber crops of Sri Lanka. *Food Chemistry* 112(1), 98–103.
Wiersema, S.G., Hesen, J.C. and Song, B.F. (1989) *Report on a Sweet Potato Postharvest Advisory Visit to the People's Republic of China* January 12–27. International Potato Center, Lima.
Woolfe, J.A. (1992) *Sweet Potato: An Untapped Food Resource*. Cambridge University Press, Cambridge, UK, pp. 1–5.
Young, R.J. (1981) *Introduction to Polymers*. Chapman and Hall, New York.

Yussof, N.S., Utra, U. and Alias, A.K. (2013) Hydrolysis of native and cross-linked corn, tapioca, and sweet potato starches at sub-gelatinization temperature using a mixture of amylolytic enzymes. *Starch/Stärke* 65(3–4), 285–295. DOI: 10.1002/star.201200002.

Zhao, Z. and Jia, F. (1985) *Storage and Simple Processing of Sweet Potato*. Publishing House of Agriculture, People's Republic of China.

Zhu, F. (2015) Composition, structure, physicochemical properties, and modifications of cassava starch. *Carbohydrate Polymers* 122, 456–480. DOI: 10.1016/j.carbpol.2014.10.063.

Zhu, F. and Wang, S. (2014) Physicochemical properties, molecular structure, and uses of sweet potato starch. *Trends in Food Science and Technology* 36(2), 68–78. DOI: 10.1016/j.tifs.2014.01.008.

Zhu, F., Corke, H., Åman, P. and Bertoft, E. (2011a) Structures of clusters in sweet potato amylopectin. *Carbohydrate Research* 346(9), 1112–1121. DOI: 10.1016/j.carres.2011.03.001.

Zhu, F., Corke, H., Åman, P. and Bertoft, E. (2011b) Structures of building blocks in clusters of sweet potato amylopectin. *Carbohydrate Research* 346(18), 2913–2925. DOI: 10.1016/j.carres.2011.10.009.

Zobel, H.F. (1988) Molecules to granules: a comprehensive starch review. *Starch/Stärke* 40(2), 44–50. DOI: 10.1002/star.19880400203.

5

Aroids

5.1 Introduction

Among the aroids, the world production of *Colacasia esculenta* and *Xanthosoma* spp. has been registering steady increase since the 1980s. In 2013, the top five producers of *C. esculenta* were Nigeria (3.4×10^6 tonnes), China (1.8×10^6 tonnes), Cameroon (1.55×10^6 tonnes), Ghana (1.27×10^6 tonnes) and Papua New Guinea (2.7×10^5 tonnes); and the top five producers of *Xanthosoma* spp. were Cuba (1.85×10^5 tonnes), Venezuela (1.3×10^5 tonnes), El Salvador (43,000 tonnes), Peru (30,000 tonnes) and Dominican Republic (29,104 tonnes) (FAOSTAT, 2015). However, the aroids are reported to be a neglected and underutilized family of crops (Williams and Haq, 2002). The above fact suggests that there is great potential to further develop these crops as subsistence food and as products for industrial applications.

Corm and cormels of aroids (e.g. *C. esculenta*) are rich in starch and non-starch polysaccharides. The latter can serve as dietary fibre, may contribute to favourable food texture, and be a source of nutrients beneficial for cancer prevention (Harris *et al.*, 1992; Huang *et al.*, 2010; Gaosong *et al.*, 1997; Nguimbou *et al.*, 2014). Starch constitutes over 85% of the total dry matter in the corms (Agama-Acevedo *et al.*, 2011; Lebot *et al.*, 2009, 2011; Lu *et al.*, 2005).

5.2 Extraction and Purification

Extraction of starch from the aroids has not been industrially practised because of the difficulty in the settling of starch and also the comparatively low yield of starch under normal extraction procedures. The extraction of starch from corms and cormels is similar in procedure to that of tubers and roots. Fresh corms and cormels are washed, peeled and finely cut. The pieces are then steeped and milled in a dilute solution of sodium bisulphite or sodium hydroxide. The fibres are filtered out from the slurry and the slurry is centrifuged to obtain starch. The brownish layer of the top of the starch cake is discarded. The starch obtained is washed several times before drying (Jane *et al.*, 1992; John *et al.*, 1999; Torruco-Uco *et al.*, 2006). Depending on the species, variety and extraction method, the starch yield can vary from 7% to 18.6% (wet basis) (Falade and Okafor, 2013; Moorthy *et al.*, 1994; Pérez *et al.*, 2005; Soni *et al.*, 1985). Lebot *et al.* (2011) studied 315 accessions of *C. esculenta* and found them to be a rich source of starch ranging from 56% to 89% (dry basis).

The starch yield from *Amorphophallus paeoniifolius* (10 genotypes) were 7–14.3% (Moorthy *et al.*, 1994). Ranges of starch yields from *X. sagittifolium* (two genotypes) and *C. esculenta* (three genotypes) were 10.9–18.6% and 10–14%, respectively (Falade and Okafor, 2013). Yields of starch from *X. sagittifolium* (one genotype) and *C. esculenta* (one genotype) were 10–12% (Pérez *et al.*, 2005). Extraction method can affect the yield and properties of starch (John *et al.*, 1999; Sit *et al.*, 2014a, b). Different chemical solutions have been investigated for starch extraction from *X. sagittifolium* and *A. paeoniifolius*: 1–5% sodium chloride, 1–5% sodium hexametaphosphate, 1–5% potassium metabisulphite, 1–5% ammonia and 0.025–0.125% glycerol monostearate. These studies have looked at the effect of reducing microbial growth and enzymatic browning during starch settling on the yield and purity of starch (John *et al.*, 1999; Rani *et al.*, 1998). These chemicals, except for glycerol monostearate, interact with starch to alter its solubility, lipid content, α-amylase susceptibility and intrinsic viscosity to various extents. Glycerol monostearate formed an amylose inclusion complex with starch (John *et al.*, 1999; Rani *et al.*, 1998). Sit *et al.* (2014a) successfully increased the starch yield of *C. esculenta* (from 15% to 19%) by applying ultrasound pre-treatment to the slurry (30 kHz, 100 W, up to 10 min). When combined with cellulase and xylanase, the ultrasound pre-treatment led to a yield of 20.8% (Sit *et al.*, 2014b). The ultrasound treatment affects starch properties by the phenomenon of cavitation (Zhu, 2015). The pre-treatment improved many properties of the gel-like swelling, solubility, freeze-thaw stability, firmness and consistency (Sit *et al.*, 2014a). In addition to high yield and stability, the steeping and ultrasonication parameters can affect important properties of starch that govern its application, and therefore should be chosen carefully.

5.3 Physiochemical Composition

5.3.1 Biochemical content

A great variation in chemical composition of the starches from diverse aroid species was observed. The obtained starches tend to have high protein and ash content, possibly due to inadequate purification (e.g. >0.5%, Londoño-Restrepo *et al.*, 2014). Protein and ash contents of aroid starches from diverse species ranged from 0.04–1.2% and 0.1–0.9%, respectively (Jane *et al.*, 1992; Londoño-Restrepo *et al.*, 2014; Soni *et al.*, 1985). The lipid content estimates have been found to vary with the measuring method: 0.15% lipid was detected in *C. esculenta* by an acid hydrolysis-based method while a solvent-based extraction (chloroform-methanol or hot n-propanol–water) method yield value is 0.04–0.08% (Gunaratne and Hoover, 2002; Saikia and Konwar, 2012). A phosphorus content of 0.02–0.08% was found in starches from *A. paeoniifolius*, *C. esculenta* and *X. sagittifolium* (Jane *et al.*, 1992; Moorthy *et al.*, 1993), which is lower than that in potato (Gunaratne and Hoover, 2002). Phosphorus in *C. esculenta* starch occurred mainly as phosphate monoester derivatives (Jane *et al.*, 1992). Salwa *et al.* (2010) reported values of 0.86%, 0.223%, 5.6% and 0.407% for ash, lipid, protein and phosphorus for taro starch. For amorphophallus starch, Meng *et al.* (2015) found 0.35%, 0.43%, and 0.40%, respectively, for the above constituents.

Tables 5.1 and 5.2 give the different values for the chemical composition of aroid starches (Opara, 2003; Zhu, 2016).

5.3.2 Amylose and amylopectin content

The amylose content greatly varied not only within the same species but also between different species. Lebot *et al.* (2011) analysis of the amylose content involved iodine binding colorimetry in combination with near infrared reflectance spectroscopy. For 315 genotypes of *C. esculenta* the amylose content ranged from 10.7% to 49.3%. Agama-Acevedo *et al.* (2011) reported a strain of *C. esculenta* from Mexico with 2.5% amylose content, similar to waxy starch varieties. Amylose content in *C. esculenta* (30.6%) and *X. sagittifolium* (35.3%) starches was found to be higher than cassava starch (16.9%) (Pérez *et al.*, 2005). Amylose content measurement by different techniques tends to give different values. Jane *et al.* (1992) reported that amylose contents of five strains of *C. esculenta* measured by iodine binding potentiometric titration (18.1–22.2%) were slightly lower than those measured by gel permeation chromatography (Sepharose CL-2B) of whole starch (19.2–24.3%). Similarly, amylose content values of *C. esculenta* by high-performance size exclusion chromatography (HPSEC) of

Table 5.1. Chemical composition of edible portion of corms and cormels from diverse aroids. (From Opara, 2003)

	A. campanulata	*Alocasia* spp.	*C. esculenta*	*C. chamissonis*	*Xanthosoma* spp.
Moisture (%)	77.8	70.3	69.1	75.4	67.1
Starch (%)	16.6	21.5	24.5	16.8	27.6
Sugars (%)	0.14	0.96	1.01	1.03	0.42
Dietary fibre (%)	1.45	1.85	1.46	2.78	0.99
Protein (%)	2.24	2.15	1.12	0.51	1.55
Lipids (%)	0.06	0.1	0.1	0.16	0.11
Ash (%)	1.36	0.92	0.87	0.67	1.04
Number of samples	7 samples from 1 cultivar	37 samples from 2 countries	71 samples from 3 countries	27 samples from 2 countries	37 samples from 2 countries

debranched starch exceeded those from iodine-binding potentiometric titration (e.g. by up to 5%). Endogenous lipids in starch form inclusion complexes with amylose and can limit its availability. The amylose content of starch improved by up to 4% by the removal of these lipids (Gunaratne and Hoover, 2002; Saikia and Konwar, 2012). The GPC of starch of different aroids have been studied and wide variability has been observed (Figs 5.1–5.3) and the values are given in Table 5.3. Planting season had a significant effect on the starch and amylose content of *X. sagittifolium* (Lu *et al.*, 2005). Higher contents of both starch and amylose were found in corms planted in the summer compared to those planted in winter and spring in Taiwan (Lu *et al.*, 2005). The amylose content in Xanthosoma starch extracted with different chemicals is given in Table 5.6 (John *et al.*, 1999).

5.4 Structural Properties

5.4.1 Granular morphology

Granule size and shape of aroid starches have been studied using diverse instruments like light microscopy (LM), scanning electron microscopy (SEM), laser light scattering (LLS) and Coulter counter (CC), and can vary with a variety of factors. Starch granules from *A. paeoniifolius* have been found to be polygonal in shape and ~5–15 µm in size (Moorthy *et al.*, 1994; Saikia and Konwar, 2012). The polygonal granules in *C. esculenta* are smaller, varying in the range of ~1–5 µm (Fig. 5.2) (Jane *et al.*, 1992; Moorthy *et al.*, 1993; Pérez *et al.*, 2005). The granule size of *X. sagittifolium* varied widely across studies but mostly fell in the range of 1–10 µm (Lauzon *et al.*, 1995; Perez-Rea *et al.*, 2013; Pérez *et al.*, 2005; Tetchi *et al.*, 2007). Besides the effect of genetic and varietal differences, variations in starch granule morphology can be due to misidentified species and the method of analysis. It has also been observed that granule size in the corms of *C. esculenta* was smaller (1.3 µm) in large corms compared to smaller corms (2.1 µm) (Tattiyakul *et al.*, 2007). It is, however, yet to be reported if different parts of corms and cormels differ in granule sizes. Lu *et al.* (2005) found that planting season had a significant influence. Corms from crops planted in the summer contained larger granules that those planted in spring and winter in Taiwan.

The variability in granular size of aroids is evident in Table 5.5.

5.4.2 XRD and starch crystallinity

Studies have reported a prevalence of A-type polymorphism in the starch of *A. paeoniifolius* (Table 5.5) (Moorthy *et al.*, 1994; Rani *et al.*, 1998; Saikia and Konwar, 2012), *C. esculenta* (Agama-Acevedo *et al.*, 2011; Jane *et al.*, 1992; Saikia and Konwar, 2012; Zeng *et al.*, 2014) and

Table 5.2. Chemical composition of aroid starches. (From Zhu, 2016)

Species	Genotype no.	Amylose (%)	Amylose quantification	Protein (%)	Lipid (%)	Ash (%)	Phosphorus (%)	References
A. paeoniifolius	1	24.5	IC	0.21	0.10	0.58		Wankhede and Sajjan, 1981
A. paeoniifolius	1				0.088	0.95	0.045	Soni *et al.*, 1985
A. paeoniifolius	10	21.9–23.9	IC					Moorthy *et al.*, 1994
A. paeoniifolius	1	12.5, 14.2[c]	IC		0.01–0.13		0.08	Saikia and Konwar, 2012
A. paeoniifolius	1	25.2	IP					Srichuwong *et al.*, 2012
C. esculenta	5	18.1–22.2 (IP); 19.2–24.3 (GPC)	IP, GPC[h]		0.24–0.52		0.017–0.025	Jane *et al.*, 1992
C. esculenta	10	14–19.4	IC					Moorthy *et al.*, 1993
C. esculenta	1	30.6	IC	0.53	0.27	0.31		Pérez *et al.*, 2005
C. esculenta	3[a]	8.7–14.9 (IP); 10.2–19.3 (HPSEC)	IP, HPSEC[g]	0.04–0.06	0.08–0.12			Lu *et al.*, 2008
C. esculenta	6	16.7–30.9	DSC					Njintang *et al.*, 2008
C. esculenta	1	33.3	n.a.[b]			0.09		Nwokocha *et al.*, 2009
C. esculenta	7	10.6–21	Con A	0.4–0.8	0.11–0.16	0.13–0.16		Mweta *et al.*, 2010
C. esculenta	1	2.5	IC					Agama-Acevedo *et al.*, 2011
C. esculenta	315	10.7–49.3	IC, NIR					Lebot *et al.*, 2011
C. esculenta	1	16.3	IP					Srichuwong *et al.*, 2012
C. esculenta	1	17.5–22.4[c]	IC		0.04–0.09[e]		0.07	Saikia and Konwar, 2012
C. esculenta	3	11.6–18.0	IC					Falade and Okafor, 2013
C. esculenta	1	8.4	IC	0.35		0.28		Zeng *et al.*, 2014
X. caracu	1	12.8, 16.6[c]	IC		0.02–0.2[e]		0.04	Saikia and Konwar, 2012

Continued

Table 5.2. Continued

Species	Genotype no.	Amylose (%)	Amylose quantification	Protein (%)	Lipid (%)	Ash (%)	Phosphorus (%)	References
X. robustum	1	35.4	Con A	1.2	1.2	0.18		Londoño-Restrepo *et al.*, 2014
X. sagittifolium	1	19.9	IA	0.16–0.22[f]	0.1	0.11		Valetudie *et al.*, 1993
X. sagittifolium	2	21.3–25.4	IA					Lauzon *et al.*, 1995
X. sagittifolium	1	22.3–26.4[d]	IC		0.09–0.4[e]	0.15	0.02	Gunaratne and Hoover, 2002
X. sagittifolium	2[a]	18.1–22.9	IP					Lu *et al.*, 2005
X. sagittifolium	1	35.3	IC	0.56	0.1	0.2		Pérez *et al.*, 2005
X. sagittifolium	1	26.7	DSC					Tetchi *et al.*, 2007
X. sagittifolium	1	22.5	IP					Srichuwong *et al.*, 2012
X. sagittifolium	1	18.3, 20.4[c]	IC		0.03–0.17[e]		0.05	Saikia and Konwar, 2012
X. sagittifolium	1	15	IC					Perez-Rea *et al.*, 2013
X. sagittifolium	2	16, 33.8	IC					Falade and Okafor, 2013
X. yucatanensis	1	22.4	IC	0.1	1	0.1		Torruco-Uco *et al.*, 2006; Torruco-Uco and Betancur-Ancona, 2007

[a]3 different planting seasons; [b]method according to IITA Laboratory Manual (1995) which could not be traced; [c]before and after defatting; [d]before and after removal of bound lipids; [e]3 types of extraction solvents; [f]before and after proteolysis; [g]high-performance size-exclusion chromatography of debranched starch; [h]gel-permeation chromatography on Sepharose CL-2B.

IC, Iodine binding-colorimetry; IP, Iodine binding-potentiometry; IA, Iodine binding-amperometry; NIR, near-infrared reflectance spectroscopy; Con A, Concanavalin A precipitation-based method; DSC, differential scanning calorimetry; HPSEC, high-performance size-exclusion chromatography of debranched starch;GPC, gel-permeation chromatography on SepharoseCL-2B.

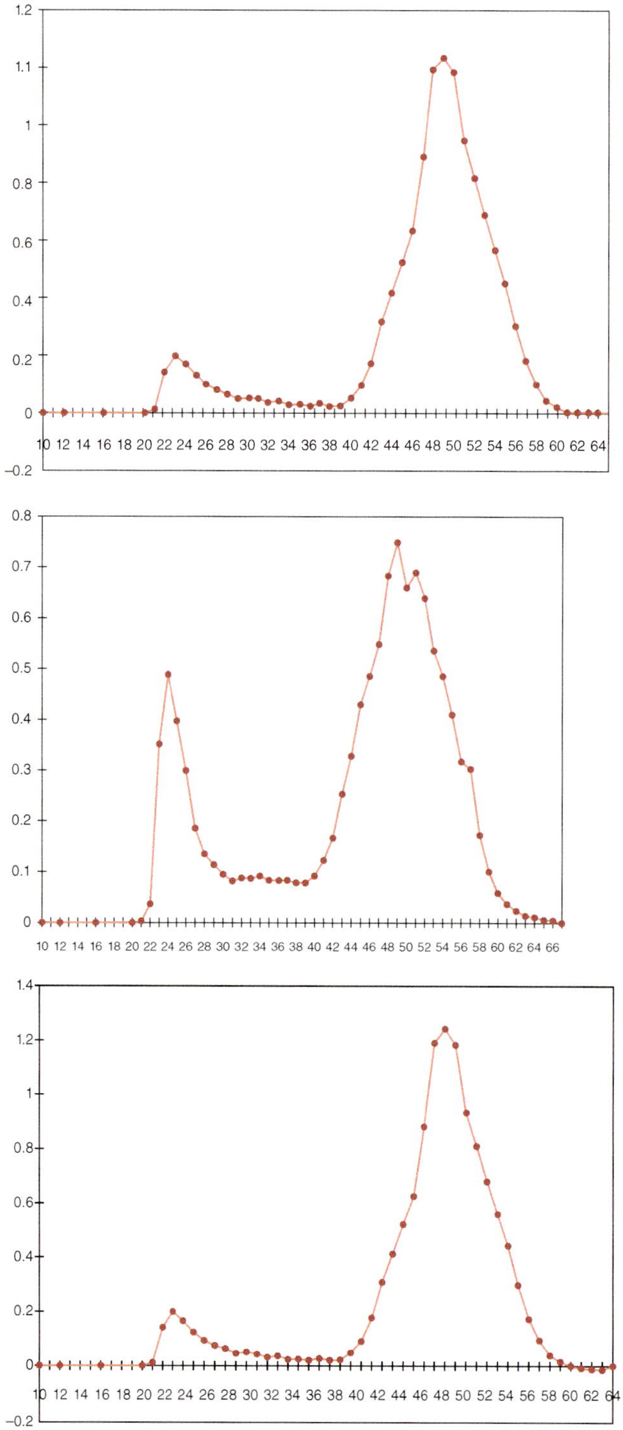

Fig. 5.1. GPC chromatogram of (a) *Amorphophallus* sp. (b) *Xanthosoma* sp. (c) *Colocasia* sp. (From Moorthy, unpublished data)

Fig. 5.2. Scanning electron microscopic images of purified aroid starch granules showing their granule size and morphology. (From Saikia and Konwar, 2012)

X. sagittifolium (Lauzon *et al.*, 1995; Lawal, 2004; Perez-Rea *et al.*, 2013). Valetudie *et al.* (1993) study showed a C_a-type polymorph in *X. sagittifolium*. At 45%, the relative crystallinity of *X. sagittifolium* exceeds that of *Dioscorea alata*, potato and cassava starches (Gunaratne and Hoover, 2002). Starch in *X. caracu* was characterized as an A-type polymorph (Saikia and Konwar, 2012), while *X. yucatanensis* showed as a C_a-type polymorph (Torruco-Uco *et al.*, 2006). The difference in the polymorphism of *Xanthosoma* starches may be attributed to genetic and environmental variations.

5.4.3 Amylose and amylopectin structure

The number-average molecular weight of *C. esculenta* amylose has been found to range from 3.0 to 3.5×10^5. The values are lower than that of cassava starch, but the polydispersity index (~1.2) of these amyloses was similar to cassava. The measurement was made using HPSEC-ELSD with pullulans as standards (Mweta *et al.*, 2010). The weight-average molecular size of amyloses of three *C. esculenta* varieties planted in three different seasons had a degree of polymerization (DP) of 1734–2320, and the size depended on the interactions between planting season and genotype (Lu *et al.*, 2008). Number-average DP of amylose of *X. sagittifolium* was 2775, higher than that of cassava and greater than that of yam (*D. alata*), and lower than that of potato (Gunaratne and Hoover, 2002).

Mweta *et al.* (2010) reported the number-average molecular weight of amylopectins in seven *C. esculenta* varieties to be slightly higher than that in cassava starch, ranging from 1.5×10^6 to 1.6×10^6. The associated polydispersity index was reported to be ~1.0. The weight-average molecular weight and gyration radius of *X. sagittifolium* amylopectin was measured by Tetchi *et al.* (2007)

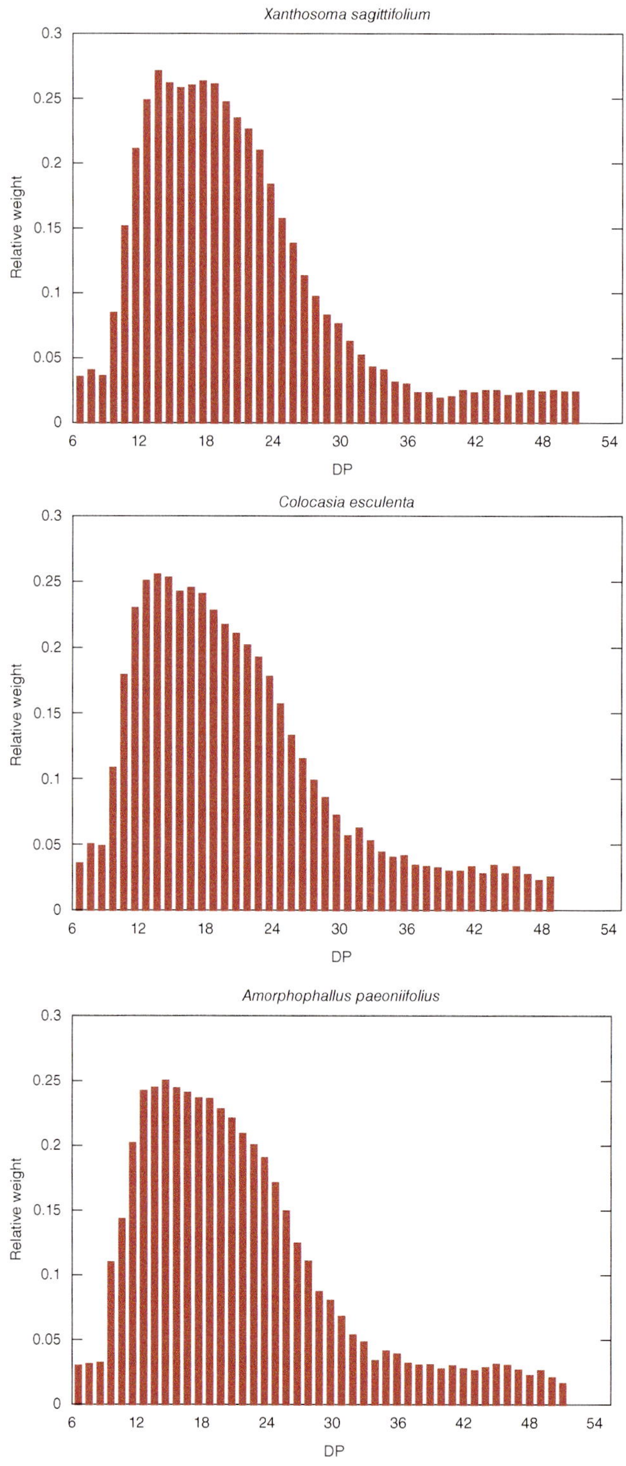

Fig. 5.3. HPSE pattern of *Xanthosoma*, *Colacasia*, and *Amorphophallus* amylopectin.

Table 5.3. Amylose content of different aroid species. (From Moorthy, unpublished data)

Species	SDS	CTAB	Iodine staining	GPC
Xanthosoma saggittifolium	24.9	23.5	19.9	21.4
Colocasia esculenta	7.3	10.6	6.4	9.5
Amorphophallus paeonifolius	25.5	23.8	17.9	22.6

using HPSEC-MALLS (multi-angle laser light scattering) DRI (differential refractometric index) system and reported to be 2×10^8 and 183 nm, respectively. The unit chain-length distribution of amylopectin fraction has also been analysed by using enzymatic debranching (Bertoft, 2004). The amylopectin unit chains' molar compositions in *A. paeoniifolius*, *C. esculenta* and *X. sagittifolium* differed with the degree of polymerization. For a DP of 6–12, the values were found to be 32.4% for *A. paeoniifolius*, 36.3% for *C. esculenta* and 34.7% for *X. sagittifolium*, respectively. The corresponding values for a DP of 13–24 were 60.5%, 57.3% and 58.7%, respectively (Srichuwong *et al.*, 2012). The percentage of short unit chains of amylopectin in two genotypes of *C. esculenta* was higher than that of yams (*D. alata* and *D. batatas*) and normal maize starch (Sugimoto *et al.*, 1986). From the data obtained on three *C. esculenta* varieties planted in three different seasons, the short amylopectin chains were 2.4–2.9 times more abundant than long chains on a weight basis (Lu *et al.*, 2008). The molecular structure of *X. sagittifolium* showed significant differences due to planting season (Lu *et al.*, 2005). The ratio of short to long chains in amylopectin was higher and the chain lengths were shorter for corms that were planted in summer relative to those planted in winter and spring (Taiwan). Gunaratne and Hoover (2002) determined the average unit chain length (CL) of amylopectin in *X. sagittifolium* to be 24.2 glucosyl residues, which is lower that that in potato and *D. alata*. The larger corms of *C. esculenta* were found to have shorter CLs (22 glucosyl residues) than smaller corms (31.7 glucosyl residues) (Tattiyakul *et al.*, 2007). The region of amylopectin between the non-reducing end and the nearest branching point, called the 'external' part can be isolated from the region between the sole reducing end and the outermost branching point, called the 'internal' part, by the action of exo-acting enzymes such as β-amylase or phosphorylase α- and β-amylase. The external part is hydrolysed while the internal part is obtained in the form of β-limit dextrins (β-LDs) or f, β-LDs (Bertoft, 2004). The external chain length (ECL) and internal chain length (ICL) of amylopectin form the difference between amylopectin and β-LDs or f, β-LDs (Bertoft, 2004). The analysis of *C. esculenta* corms of different sizes showed that ECL and ICL were shorter (12.6 and 8.4 glucosyl residues) in larger corms than those of smaller corms (16.9 and 14.2 glucosyl residues) (Tattiyakul *et al.*, 2007). On comparison, the ECL and ICL of *X. sagittifolium* were found to be shorter than those in potato and *D. alata* (Gunaratne and Hoover, 2002). A systematic clustering of amylopectin branches is believed to occur (Pérez and Bertoft, 2010). The cluster structure of aroid amylopectin has not been enumerated.

5.5 Functional Properties

5.5.1 Swelling and solubility

Wide variations in swelling and solubility (at 60°C) of starches from *C. esculenta* and *X. sagittifolium* have been reported.

X. sagittifolium starch showed a lower swelling factor and solubility than potato and cassava starch (Gunaratne and Hoover, 2002) but had higher values than those of sweet potato (Lauzon *et al.*, 1995). *C. esculenta* showed lower values of swelling power and solubility than *X. sagittifolium* as well as sweet potato (Falade and Okafor, 2013; Huang *et al.*, 2010). These values were found to be higher in *C. esculenta* than *D. alata* (Huang *et al.*, 2010). The swelling and solubilization rates in *C. esculenta* were also lower than that in cassava starch, while the reverse was true for the relaxation temperature (Nwokocha *et al.*, 2009). These comparative studies were based on limited numbers (e.g. 1–2) of genotypes, and thus may need confirmation.

Table 5.5. X-ray diffraction data of *Amorphophallus paeoniifolius*. (From John *et al*., 1999)

	Diffraction data		
Sample	'd' spacing (2θ)	Angle Intensity	Intensity (I_0/I_{max})
Starch control	5.86	15.09	86.22
	5.22	16.95	97.42
	4.86	18.25	100.00
	3.89	22.79	66.94
	3.46	25.73	28.35
5% SHMP treated sample	5.9	14.77	72.36
	5.27	16.84	80.00
	4.92	18.00	100.00
	3.88	22.89	73.00
5% NaCl treated sample	5.92	14.93	82.50
	5.29	16.72	93.10
	4.99	17.75	100.00
	3.91	22.70	66.70
5% KMS treated sample	7.90	11.18	24.30
	5.94	14.90	71.04
	5.25	16.85	94.40
	4.92	17.99	100.00
	3.92	22.65	85.11
5% NH_4 OH treated sample	7.89	11.20	20.25
	5.94	14.90	71.04
	5.24	16.90	98.58
	4.94	17.93	100.00
	3.91	22.73	78.45
	3.34	26.62	26.45
	2.35	38.20	20.90
0.1% GMS treated sample	5.89	15.00	80.70
	5.24	16.90	100.00
	4.94	17.90	94.00
	3.90	22.70	83.40
	2.96	30.10	20.80

the pasting viscosity of 5% suspensions of starch obtained from 10 genotypes of *A. paeoniifolius* and *C. esculenta* ranged from 280 to 380 BU and 120–350 BU, respectively (Moorthy *et al*., 1994). At 7%, the pasting viscosity was found to be precipitously higher compared to that at 5% (Moorthy *et al*., 1993, 1994). Studies have compared the pasting properties of aroids with other starches (Huang *et al*., 2010; Pérez *et al*., 2005; Wickramasinghe *et al*., 2009). Pérez *et al*. (2005) showed lower values of PV and BD in *X. sagittifolium* and *C. esculenta* than cassava starch. *C. esculenta* also showed the lowest PV and SB value among diverse tuber and root starches (Wickramasinghe *et al*., 2009). Huang *et al*. (2010) also found that PV, CPV and SB were lower for *C. esculenta* compared to *D. alata* and sweet potato starches. The amylose and amylopectin properties of starches also affected the pasting properties of *C. esculenta* (Lu *et al*., 2008). The amylose content of starch was negatively correlated with PV and BD; the fraction of longer amylopectin unit chains was positively correlated to PV and HPV, while that of shorter unit chains was positively related to BD. These influences have been explained by the entangling action of amylose on amylopectin which hinders swelling and reduces PV. The short amylopectin chains, on the other hand, are too small to establish entangling interactions with other chains (Srichuwong and Jane, 2007). In contrast to the results of Lu *et al*. (2008), Tattiyakul *et al*. (2007) found that a higher PV occurred in *C. esculenta* starch with shorter CL, ECL and ICL of amylopectin. The study also found starch from larger corms showed higher pasting viscosities. In the study of Lu *et al*. (2005), it was found that *X. sagittifolium* planted in winter and

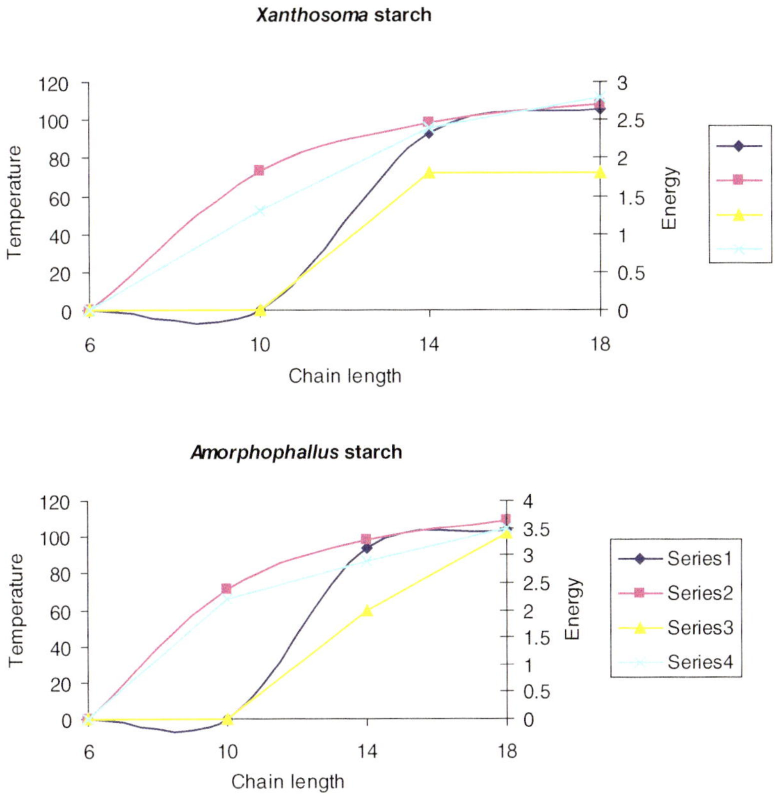

Fig. 5.4. Effect of chain length on starch-lipid complexation. (Source: Moorthy, unpublished data)

Table 5.6. Yield and amylose blue value of starch obtained from *Xanthosoma sagitifolium* roots. (From John *et al*., 1999)

Sample	*Xanthosoma sagitifolium* Yield range (%)	Amylose blue value Apparent	True	Soluble
Starch (control)	43.85	0.0142	0.0259	0.01089
Starch from pretreated roots				
Sodium chloride treated (1–5% w/v)	35.4–37.0	0.0148–0.0154	0.0154–0.0160	0.0087–0.0089
Sodium hexametaphosphate treated (1–5% w/v)	36.7–47.1	0.0131–0.0154	0.0156–0.0162	0.0089–0.0094
Potassium metabisulphite treated (1–5% w/v)	25.5–34.0	0.0140–0.0142	0.0151–0.0160	0.0085–0.0087
Glyceryl monostearate treated (0.025–0.125% w/v)	36.3–45.6	0.0120–0.0130	0.0127–0.0146	0.0071–0.0079
Ammonia treated (1–5% w/v)	15.7–38.3	0.0141–0.0147	0.0151–0.0159	0.0087–0.0092

Table 5.7. Swelling volume and solubility of starch samples prepared from *Amorphophallus paeoniifolius*. (From Rani *et al*., 1998)

Sample	Swelling volume (mL)	Solubility (%)
Starch (control)	27.7	23.0
Starch from pretreated roots		
SHMP Treated		
1% SHMP	24.5	18.5
2% SHMP	26.0	18.5
3% SHMP	25.5	16.5
4% SHMP	24.5	18.5
5% SHMP	24.0	18.8
NaCl treated		
1%	27.5	18.1
2%	25.5	18.6
3%	24.5	ND
4%	27.5	14.5
5%	28.0	8.9
GMS treated		
0.025% GMS	22.5	32.9
0.050% GMS	22.5	19.5
0.075% GMS	19.5	10.25
0.100% GMS	ND	ND
0.125% GMS	20.5	24.0
NH_4OH treated		
1% NH_4OH	23.5	26.8
2% NH_4OH	23.0	32.5
3% NH_4OH	24.0	28.4
4% NH_4OH	25.0	24.4
5% NH_4OH	25.0	24.4
KMS treated		
1% KMS	26.0	22.2
2% KMS	28.0	30.6
3% KMS	27.0	29.2
4% KMS	28.0	28.0
5% KMS	25.0	29.9

Table 5.8. Swelling and solubility characteristics of *Xanthosomas agitifolium* roots starch. (From John *et al*., 1999)

Sample	Swelling volume (mL)	Solubility (%)
Starch (control)	24.0	18.94
Starch from pre-treated roots		
SHMP Treated		
1% SHMP	24.0	11.4
2% SHMP	26.0	14.8
3% SHMP	24.0	15.2
4% SHMP	24.0	15.2
5% SHMP	24.0	15.4
NaCl treated		
1%	26.0	25.0
2%	27.5	21.9
3%	23.5	13.4
4%	27.5	18.3
5%	23.5	19.1
GMS treated		
1% GMS	27.0	12.8
2% GMS	26.5	18.4
3% GMS	26.0	18.5
4% GMS	27.0	10.9
5% GMS	26.5	14.8
NH_3 (aq) treated		
1% NH_3 (aq)	28.0	18.0
2% NH_3 (aq)	27.5	18.2
3% NH_3 (aq)	26.5	18.2
4% NH_3 (aq)	26.5	14.7
5% NH_3 (aq)	27.0	22.2
KMS treated		
1% KMS	24.0	19.0
2% KMS	23.5	22.9
3% KMS	25.0	18.8
4% KMS	23.0	19.3
5% KMS	24.0	19.0

summer produced starch with higher PV and BD values than those planted in spring. Similar results have been reported for *C. esculenta* (Lu *et al.*, 2008). *C. esculenta* starch pastes (0.5–4%) showed lower clarity than cassava starches (Nwokocha *et al.*, 2009; Mweta *et al.*, 2010).

Table 5.9. Brabender viscosity properties of *Colocasia* starch. (From Moorthy *et al*., 1993)

Starch	Concentration (%)	V_p[a](BU)	V_{97}[b](BU)	V_H[c](BU)	Pasting temperature (°C)
C-9	5	350	340	280	81–83
C-9	6	600	640	480	81–83
C-9	7	880	720	640	80–84
C-46	5	160	160	160	85–86
C-46	6	230	230	210	83–84
C-46	7	320	300	280	82–84
C-62	5	220	210	200	81–83
C-62	6	380	360	300	81–83
C-62	7	530	500	460	81–83
C-149	5	130	130	130	83–84
C-149	6	220	200	170	83–84
C-149	7	340	310	270	82–84
C-189	5	250	240	280	82–83
C-189	6	440	390	330	81–83
C-189	7	620	540	500	80–83
C-216	5	280	260	220	83–84
C-216	6	460	410	360	82–83
C-216	7	560	500	490	81–83
C-218	5	170	170	170	83–85
C-218	6	320	310	290	82–83
C-218	7	420	400	380	82–83
C-220	5	120	120	110	84–86
C-220	6	240	230	190	84–86
C-220	7	370	340	280	83–85
C-266	5	180	180	180	81–83
C-266	6	320	310	290	81–83
C-266	7	500	430	380	81–82
C-304	5	140	140	140	84–85
C-304	6	300	290	280	83–84
C-304	7	390	370	320	82–84

5.5.3 Rheological properties

Flow properties of starches from 6 *C. esculenta* varieties fitted in the power law equation (Njintang *et al*., 2008). The consistency index K and flow behaviour index n varied widely among various genotypes (Njintang *et al*., 2008). *C. esculenta* starch paste's (8%) resistance to shearing exceeded that of cassava starch (Nwokocha *et al*., 2009). Moorthy *et al*. (2008) has studied the rheological properties of aroid starches using a Bohlin rheometer at three concentrations. *Colocasia* starch exhibited low values for all the parameters and showed predominant elastic behaviour at all concentrations. For the different aroid starches, elastic behaviour was observed. Meng *et al*. (2015) also found similar results for *Amorphophallus* starch.

Table 5.10. Pasting properties of aroid starches. (From Zhu, 2016)

Species	No.	Equipment	Starch con: (%)	BV	BD	SV	PT (°C)	References
A. peaoniifolius	10	BVA	5	280–380	0–10			Moorthy *et al.*, 1994
A. peaoniifolius			6	450–620	0–20			Moorthy *et al.*, 1994
A. peaoniifolius			7	770–900	18–220			Moorthy *et al.*, 1994
C. esculenta	10	BV	5	120–350				Moorthy *et al.*, 1993
C. esculenta			6	220–600				Moorthy *et al.*, 1993
C. esculenta			7	320–880				Moorthy *et al.*, 1993
C. esculenta	1	BA	8	630			78	Nwokocha *et al.*, 2009
C. esculenta	1	BVA	7	390	10	30		Pérez *et al.* (2005)
C. esculenta	1	BMVA	6	487			77	Zeng *et al.*, 2014
C. esculenta	1	RVA	10.7	370	238	185		Gaosong *et al.*, 1997
C. esculenta	3	RVA	15	191–281	49–142	48–70	84.5–88.8	Falade and Okafor, 2013
C. esculenta	1[a]	RVA	9.2	371–402	193–226	67–68		Tattiyakul *et al.*, 2007
C. esculenta	1	RVA	8	296	143	100	73.1	Huang *et al.*, 2010
C. esculenta	3[b]	RVA	7	755–1321	155–537	309–560		Lu *et al.*, 2008
C. esculenta	1	RVA	6	100	28	38	79.1	Wickramasinghe *et al.*, 2009
X.robustum	1	RHEO	14.3	20870mPas	6987mPas	5573mPas	84	
X. sagittifolium	2	BV	6	230, 300			72.3, 76.3	Lauzon *et al.*, 1995
X.sagittifolium	1	BVA	4	60	–20	70		Pérez *et al.*, 1998

Continued

Table 5.10. Continued

Species	No.	Equipment	Starch con: (%)	BV	BD	SV	PT (°C)	References
X.sagittifolium	1	BVA	7	300	–60	190		Pérez *et al*., 2005
X.sagittifolium	2[b]	RVA	7	883–1974	126–840	526–827		Lu *et al*., 2005
X.sagittifolium	1	RVA	10	2637	1267		81	
X.sagittifolium	2	RVA	15	299–306	102–133	79–130	84.8–85.3	Falade and Okafor, 2013
X.yucatagensis	1	BVA	8	280	–8	172	75	

[a]3 types of corms of large (800–1100 g), medium (400–500 g) and small (300–400 g) sizes;
[b]3 different planting seasons including summer, winter, and spring.
BA, Brabender amylograph; BV, Brabender viscograph; BVA, Brabender Viscoamylograph; BMVA, Brabender Micro Visco-Amylo-Graph; unit for Brabender instruments is BU and for RVA is RVU except specified; PV, peak viscosity; BD, breakdown; SB, setback; PT, pasting temperature (°C); RHEO, rheometer with pasting cell.

Table 5.11. Pasting properties of starch pastes for different ultrasonic treatment conditions. (Sit *et al.*, 2014b)

Treatment	PT (°C)	PV, $_c$P	HV, $_c$P	FV, $_c$P	BV, $_c$P	SV, $_c$P
C1	84.8 ± 0.06	4324 ± 23	2013 ± 19	2939 ± 52	2310 ± 25	926 ± 15
C2	84.7 ± 0.00	4336 ± 51	2017 ± 18	2921 ± 26	2319 ± 48	904 ± 14
C3	84.8 ± 0.06	4295 ± 25	2009 ± 15	2946 ± 27	2286 ± 56	937 ± 11
C4	84.8 ± 0.10	4327 ± 41	2014 ± 29	2939 ± 49	2313 ± 28	925 ± 22
C5	84.9 ± 0.00	4308 ± 42	2179 ± 32	3119 ± 72	2129 ± 34	940 ± 19
C6	85.1 ± 0.05	4297 ± 48	2153 ± 28	3029 ± 29	2144 ± 29	876 ± 14
C7	84.9 ± 0.00	4310 ± 69	2103 ± 14	2973 ± 48	2207 ± 71	870 ± 13
C8	85.2 ± 0.00	4350 ± 53	2054 ± 41	2949 ± 56	2296 ± 28	895 ± 21
C9	85.2 ± 0.06	4395 ± 37	2104 ± 29	2994 ± 49	2291 ± 62	890 ± 26
C10	85.1 ± 0.10	4469 ± 29	2115 ± 25	2968 ± 78	2354 ± 64	853 ± 18
C11	84.9 ± 0.06	4439 ± 27	2067 ± 19	3007 ± 48	2372 ± 43	940 ± 13
C12	85.1 ± 0.00	4365 ± 31	2096 ± 37	3018 ± 34	2269 ± 27	922 ± 24

BV, Breakdown viscosity; FV, Final viscosity; HV, Hold viscosity; PT, Pasting temperature; PV, peak viscosity; SV, Setback viscosity.

A diversity in dynamic rheology properties was found in starches from three *C. esculenta* varieties (Lu *et al.*, 2008). The study also showed that starch from *C. esculenta* planted in summer showed higher G′ value than those planted in winter and spring. Nwokocha *et al.* (2009) showed that 8% starch paste of *C. esculenta* had higher G′ than cassava starch during the frequency sweep, while the G″ values were similar. In a study that compared creep compliance curves of 10% starch gels from *X. sagittifolium* (2 genotypes) and sweet potato (1 genotype) in a range of temperatures (10–40°C), the latter showed more sensitivity to temperature than the former (Shiraishi *et al.*, 1995). Rheologic property of some aroid starches are given in (Figs. 5.6, 5.7 and 5.8). The MDSC of *Amorphophallus*, *Colocasia* and *Xanthosoma* has been recorded and presented in Table 5.5.

5.5.4 Retrogradation

Retrogradation of aroid starches has been studied by analysing thermal properties (Gunaratne and Hoover, 2002), freeze-thaw stability (Nwokocha *et al.*, 2009) and gel texture (Jane *et al.*, 1992) of a starch gel. Using DSC, the degree of retrogradation (DOR,%) is quantified as the ratio of ΔH of retrograded starch to that of gelatinization. Starch from *C. esculenta* has been found to retrograde faster than cassava starch (Mweta *et al.*, 2010). Starch from *X. sagittifolium* showed lower retrogradation than potato and *D. alata* (Gunaratne and Hoover, 2002). The retrogradation properties have been reported in different studies to differ among *C. esculenta* varieties and this behaviour has been attributed to the variations in the structure and content of amylose and amylopectin (Hoover, 1995). The differences across studies may also be due to the use of different experimental conditions and system water content (Hoover, 1995). For example, Jane *et al.* (1992) found DOR of 30.5–53.6% for five *C. esculenta* varieties (conditions: 4°C, 28 days, and moisture content 75%). Agama-Acevedo *et al.* (2011) observed a DOR of 23% for one *C. esculenta* variety (conditions: 4°C, 7 days, and moisture content 70%); and on studying seven varieties of *C. esculenta* at 4°C, for 7 days, at a moisture content of 75%, Mweta *et al.* (2010) found the DOR values of 24.6–45.7%.

5.6 Thermal Properties

5.6.1 DSC

Great diversity in onset (T_o), peak (T_p) and conclusion (T_c) temperatures, and enthalpy

Table 5.12. Rheological properties of *Amorphophallus* starch. (Moorthy *et al.*, 1994)

Concentration (%)	Peak viscosity (Brabender units)			Viscosity at 97°C (Brabender units)			Viscosity after holding at 97°C (Brabender units)			Viscosity breakdown (Brabender units)			Pasting temperature (°C)
	5	6	7	5	6	7	5	6	7	5	6	7	
Am 2	280	450	770	280	450	760	280	450	660	0	0	110	82–85
Am 5	320	480	810	320	480	810	320	480	680	0	0	130	82–85
Am 14	310	480	820	310	470	810	310	470	710	0	10	110	82–85
Am 15	380	620	900	380	600	880	380	600	700	0	20	200	82–85
Am 27	330	575	880	330	570	870	330	570	800	0	5	80	81–84
Am 32	320	570	820	320	570	820	320	570	720	0	0	100	82–85
Am 34	300	580	870	300	570	860	300	570	690	0	10	170	82–84
Am 36	320	600	840	310	600	840	300	600	700	10	0	140	82–85
Am 43	340	560	850	330	550	850	330	550	630	10	10	220	81–84
Am 51	300	500	780	300	500	760	300	500	600	0	0	180	81–84

Table 5.13. Gelatinization and retrogradation properties of isolated starch from three aroid cultivars assessed by differential scanning calorimetry. (From Hoyos-Leyva *et al*., 2017)

Sample	To (°C)	Tp (°C)	Te (°C)	ΔH (J/g)
Gelatinization				
V1	75.4 ± 0.1	80.6 ± 0.3	86.6 ± 0.7	6.6 ± 0.7
V2	80.0 ± 0.1	84.8 ± 0.1	90.0 ± 0.3	7.7 ± 0.4
V3	78.0 ± 0.1	83.1 ± 0.1	90.8 ± 0.1	8.6 ± 0.2
Retrogradation				
V1	50.8 ± 0.7	60.4 ± 0.6	68.8 ± 0.4	2.0 ± 0.1
V2	50.0 ± 0.9	61.7 ± 0.2	68.9 ± 0.4	2.3 ± 0.2
V3	49.7 ± 0.8	61.5 ± 0.1	72.2 ± 0.1	2.8 ± 0.1

Values represent the mean of three determinations ± SE. Means in column not sharing the same letter are significantly different ($p<0.05$).
V1, Colocassia esculenta var. Esculenta;V2, Colocassia esculenta var. Schott;V3, Xanthosoma sagittifolium;To, onset temperature;Tp, peak temperature;Te, end temperature;DH, enthalpy of the phase transition.

change (ΔH) of aroid starch has been observed (Table 5.15).

For example, T_{peak} and ΔH of starches from five genotypes of *C. esculenta* ranged from 72.4 to 79°C and 12.1–15.5 J/g, respectively (Jane *et al*., 1992).

Srichuwong and Jane (2007) attributed these differences to multiple factors like amylose and amylopectin content, granule structure and architecture, and protein and lipid content. Limited analysis has been performed on starch from *X. sagittifolium* and *A. paeoniifolius* due to a limited number of genotypes that have been tested (Wankhede and Sajjan, 1981; Moorthy *et al*., 1994). Gelatinization of aroid starches was compared with other starches (Gunaratne and Hoover, 2002; Huang *et al*., 2010; Pérez *et al*., 2005; Perez-Rea *et al*., 2013; Sugimoto *et al*., 1986). Compared to different tuber and root starches studied by Wickramasinghe *et al*. (2009), *C. esculenta* showed highest T_{onset} and T_{peak} temperatures. *C. esculenta* (2 genotypes) also had higher values of T_{onset}, $T_{conclusion}$, and ΔH than normal maize and yam starches, which was attributed to its lower amylose content (Sugimoto *et al*., 1986). However, another study did not find any correlation between gelatinization parameters and amylose content and amylopectin unit chain composition among *C. esculenta* starches Lu *et al*. (2008). The starch from *X. sagittifolium* showed higher gelatinization temperatures than that from cassava and potato (Gunaratne and Hoover, 2002), and lower ΔH than starch from cassava and canna (Perez-Rea *et al*., 2013). Gelatinization behaviour may also be affected by multiple other factors (Srichuwong and Jane, 2007). Lu *et al*. (2005) found that the planting season of *X. sagittifolium* corms affected their gelatinization behaviour significantly. Starch from corms planted in summer had lower T_{peak}, T_{onset} and ΔH than that from corms planted in winter. This was attributed to the development of structural defects in the granules produced by the former due to a higher amount of shorter unit chain of amylopectin (i.e. a higher ratio of short to long chains) (Srichuwong and Jane, 2007). The effect of *C. esculenta* corm size on starch gelatinization was found to be only minimal (Tattiyakul *et al*., 2007). The DSC gelatinization temperatures of *Amorphophallus* starch were 74.21°C, 80.92°C and 88.49°C for T_{onset}, T_{peak} and T_{end}, and 11.17 J/g for ΔH (Meng *et al*., 2015). Zhu (2016) has reviewed the values for aroids and the values show large variations (see Table 5.15).

5.7 Digestibility

Aroids have good digestibility and this is exemplified by the use of *Colocasia* starch in Poi, a very popular food product (Moorthy *et al*., 2018). Aroid starch is traditionally a major carbohydrate source for many regions such

Table 5.14. Thermal properties of native and retrograded starches; onset (To), peak (Tp), and conclusion temperature (Tc), temperature range (R) and transition energy (ΔH_G). (From Nguimbou *et al.*, 2014)

		Native starches					Retrograded starches				
Source	Genotype	To(°C)	Tp(°C)	Tc(°C)	R(°C)	ΔH_G (J/g)	To(°C)	Tp(°C)	Tc(°C)	ΔH_R (J/g)	Degree of retrogradation (%)
Cocoyam	Chitipa	61.8	69.4	78.6	16.8	14.2	45.9	56.1	62.5	3.5	24.6
	Machinga	71.4	76.7	82.4	11.1	12.1	49.7	58.6	65.2	5.0	41.5
	Mulanje	70.2	75.7	82.6	12.4	14.0	51.2	59.3	66.2	4.4	31.2
	Mzuzu	68.2	73.1	79.2	11.0	8.2	52.0	59.0	65.3	3.8	45.7
	Nkhotakota	75.6	78.7	83.7	8.09	14.7	51.8	59.7	66.0	5.4	37.0
	Thyolo	67.8	73.3	80.9	13.1	14.6	52.5	59.3	65.8	4.2	28.8
	Zomba	71.2	76.4	82.5	11.3	13.8	52.9	60.4	66.1	4.8	35.0
Cassava	Gomani	60.8	65.6	76.9	16.1	14.0	54.3	59.3	63.4	1.1	7.9
	Maunjili	58.2	67.4	77.5	19.3	15.1	54.6	59.7	63.9	1.6	10.6
	Mbundumali	62.1	68.6	76.2	14.1	13.4	54.4	59.7	64.0	1.6	11.9
	Mkondezi	60.2	65.7	76.8	16.6	13.1	54.7	59.7	63.7	1.4	10.6
	Sauti	59.1	65.4	76.4	17.3	14.1	54.6	59.5	63.8	1.1	7.9
Maize	-	66.2	70.9	76.9	10.7	12.8	54.1	60.1	64.6	2.9	22.7

Table 5.15. Gelatinization of aroid starches measured by DSC, KHSM and PPG (Zhu, 2016).

Species	Genotype no.	Starch:water(w:w)	Scanning rate (°C/min)	*To*(°C)	*Tp* (°C)	*Tc* (°C)	ΔH (J/g)	References
A. paeoniifolius	1	KHSM		73	75.5	80.5		Wankhede and Sajjan, 1981
C. esculenta	1	01:02	15	77.2	83.2	89.7	16.7	Pérez *et al.*, 2005
C. esculenta	1	03:07	2	75	77.5		18.1	Wickramasinghe *et al.*, 2009
C. esculenta	1	03:07	10	73.6	80.6	87.6	10.6	Agama-Acevedo *et al.*, 2011
C. esculenta	1	03:07	10	73.6	80.3	88.3		Zeng *et al.*, 2014
C. esculenta	1[a]	07:20	10	72.9–74.8	79–80	89–90.5		Tattiyakul *et al.*, 2007
C. esculenta	5	01:03	10	66.5–74.0	72.4–79	82–93	12.1–15.5	Jane *et al.*, 1992
C. esculenta	7	01:03	10	61.8–75.6	69.4–78.7	78.6–83.7	8.2–14.7	Mweta *et al.*, 2010
C. esculenta	6	01:03	5	48.1–64.4	54.4–74.5	57.9–70.3	11–16.9	Njintang *et al.*, 2008
C. esculenta	3[b]	01:03	1.2	64.5–78.7	74.3–82.6	82.8–88.5	15.8–18.2	Lu *et al.*, 2008
C. esculenta	1	01:04	10	70.1	79.2	88.5	12.9	Huang *et al.*, 2010
C. esculenta	1	01:09	0.5	73	76	80.3	17.2	Nwokocha *et al.*, 2009
X. robustum	1	03:17	7.5	68.3	73.1	77.7	14.1	Londoño-Restrepo *et al.*, 2014
X. sagittifolium	2	PPG, 0.05%[c]		70, 70.5				Lauzon *et al.*, 1995
X. sagittifolium	1	01:02	10	74	78	87		Pérez *et al.*, 1998
X. sagittifolium	1	01:02	15	78	82.6	93.8	14.5	Pérez *et al.*, 2005
X. sagittifolium	1	01:03	10	71.5	77.2	85.4	13.1	Gunaratne and Hoover, 2002
X. sagittifolium	1	01:03	10	62	67.6	73.1	10.8	Perez-Rea *et al.*, 2013
X. sagittifolium	2[b]	01:03	1.2	68.1–77.1	73.8–82.8	86.6–88.7	16.1–18.1	Lu *et al.*, 2005
X. sagittifolium	1	01:05	3	73.5	76.9		17.3	Tetchi *et al.*, 2007
X. yucatanensis	1	03:07	10	72.6	78.5	84.2	15	Torruco-Uco *et al.*, 2006; Torruco-Uco and Betancur-Ancona, 2007

[a] 3 types of corms of large (800e1100 g), medium (400e500 g) and small (300e400 g) sizes;
[b] 3 different planting seasons;
[c] starch concentration.
DSC, differential scanning calorimetry; KHSM, Kofler hot stage microscopy; PPG, photopastegraphy;

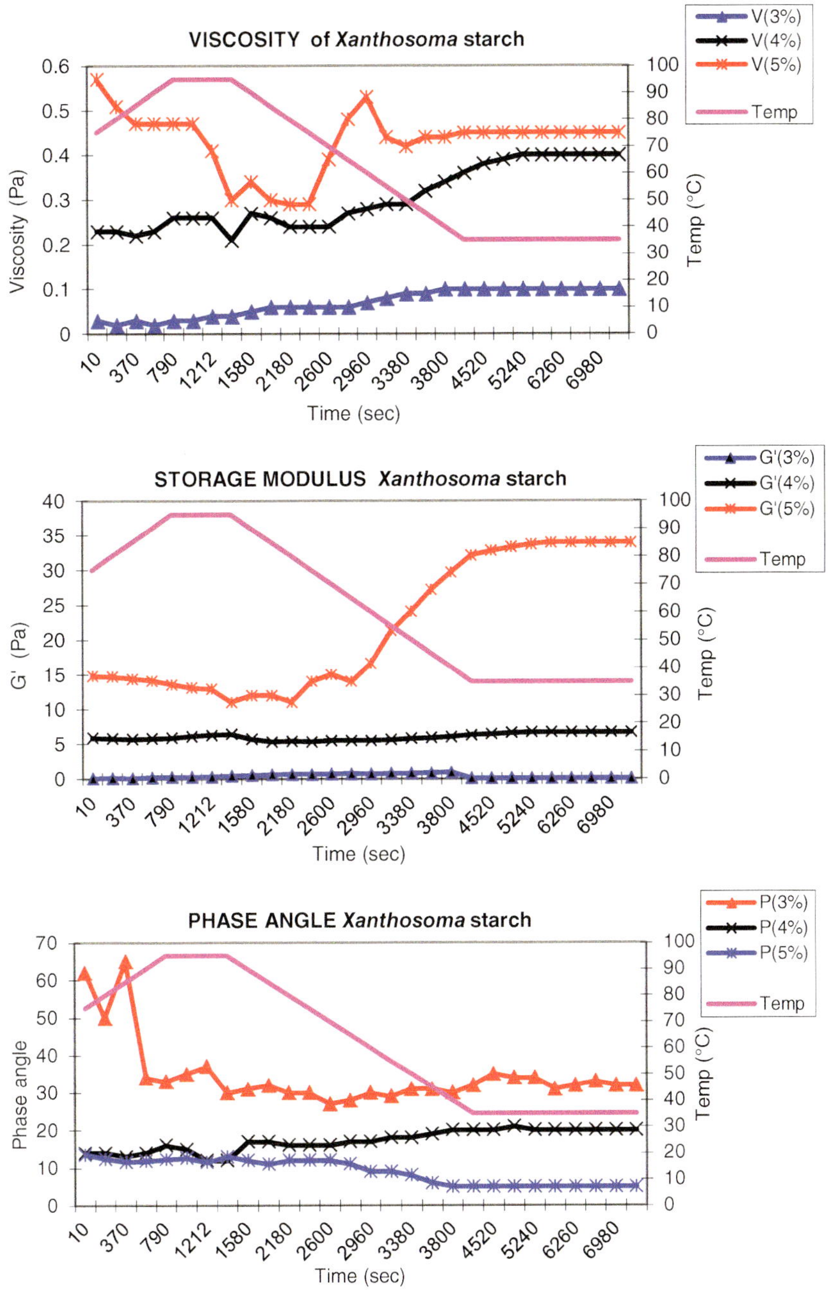

Fig. 5.5. Rheological properties of *Xanthosoma* starch at different concentrations (Moorthy *et al*., 2008).

as the Pacific Islands where now chronic diseases such obesity and diabetes are epidemic (Hughes and Lawrence, 2005). Sugimoto *et al*. (1979) have compared raw starches from diverse botanical sources and found starch from *C. esculanta* has one of the highest enzyme susceptibilities to pancreatin. *In vivo* digestibility studies of raw starch by humans

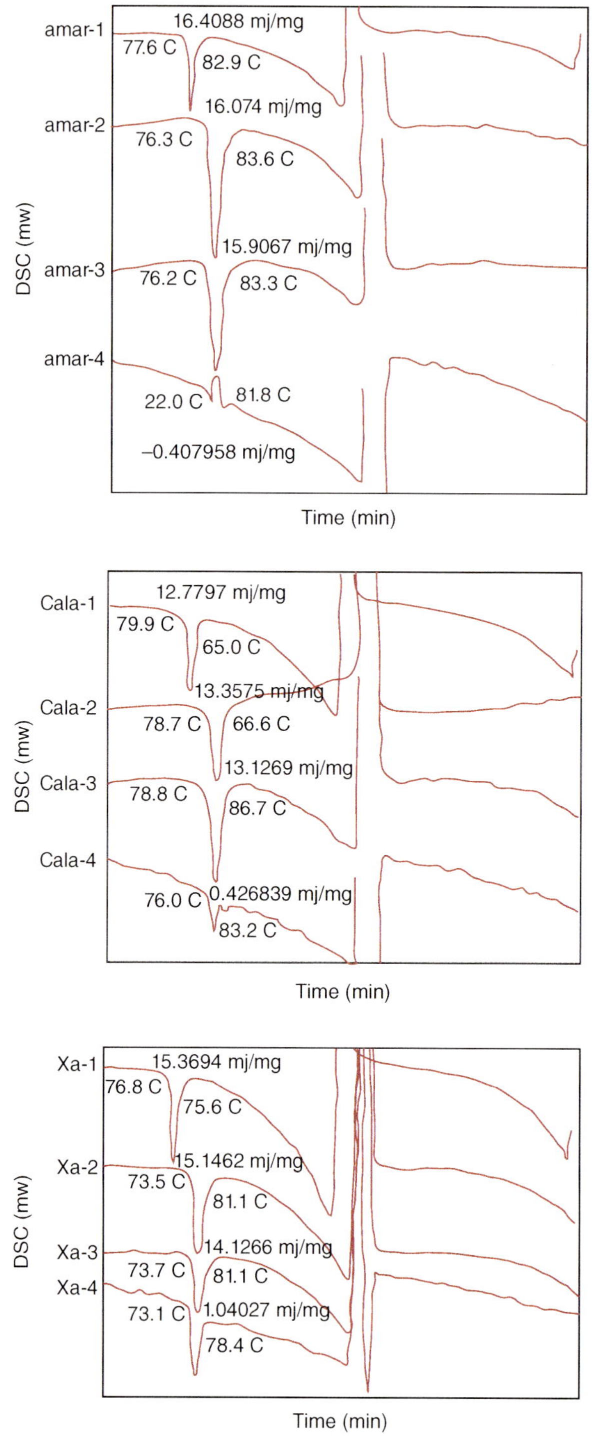

Fig. 5.6. Yield and amylose blue value of starch obtained from *D. alata* and *D. rotundata*. (From John *et al*., 1999)

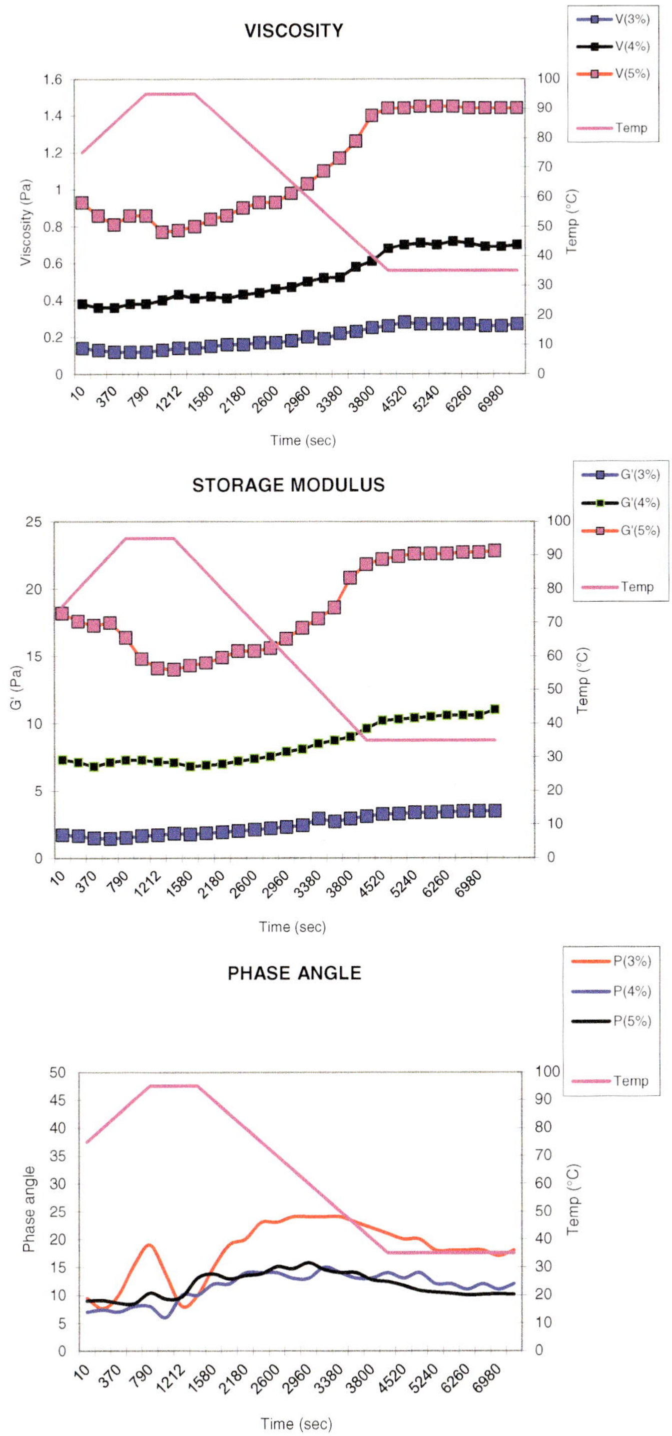

Fig. 5.7. Rheological properties of *Amorphophallus* starch at different concentrations (Moorthy *et al.*, 2008).

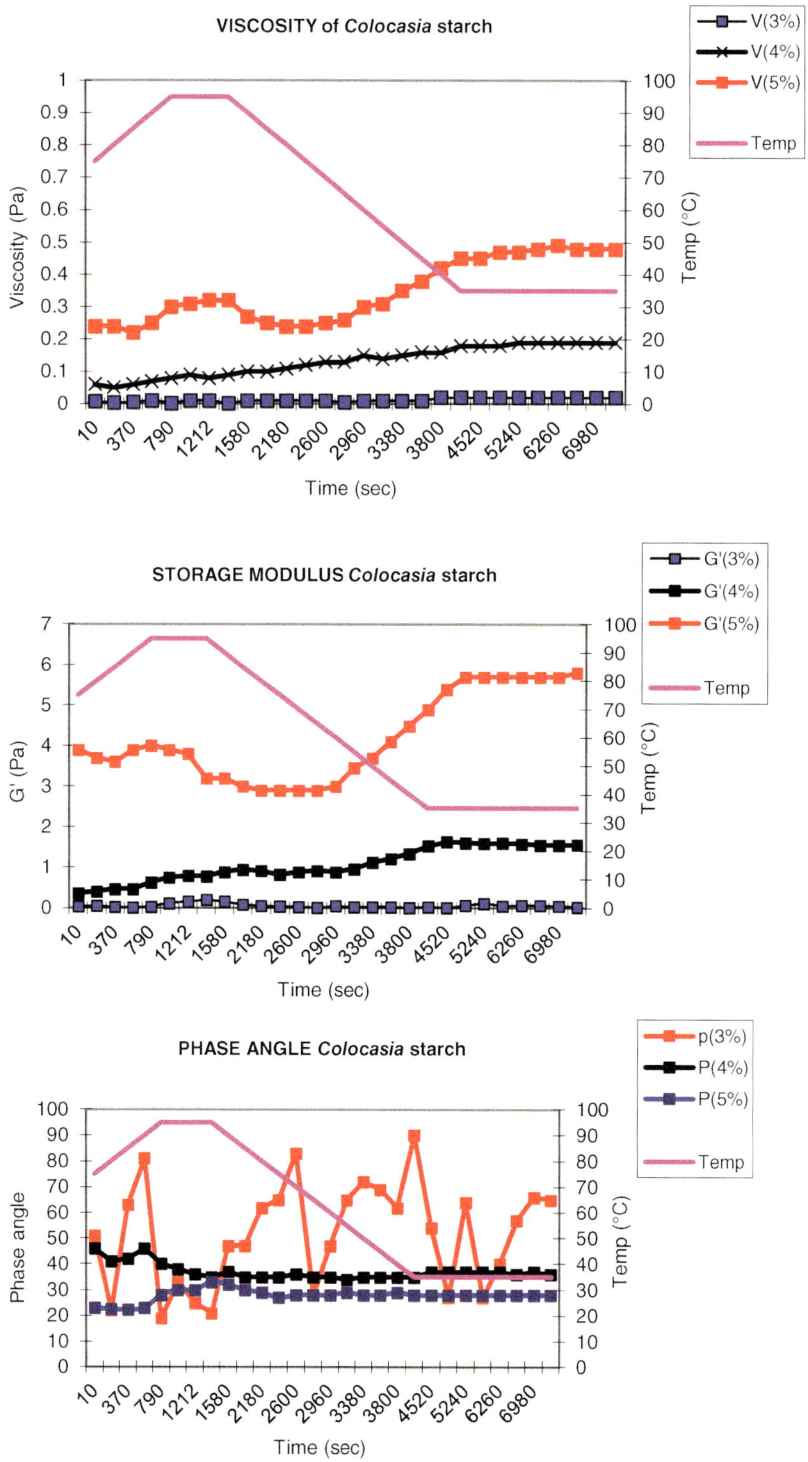

Fig. 5.8. Rheological properties of *Colocasia* starch at different concentrations (Moorthy *et al.*, 2008).

showed that *C. esculenta* is completely digested (Langworthy and Deuel, 1922; Lin *et al.*, 2010). Enzyme sustability of raw starch from *Colocasia esculenta* (2 genotypes) to pancreatin, and glucoamylase from *Rhizopus* sp. was higher compared to yam starch (*D. alata, D. batatus* and *D. opposita*) but similar to maize starch (Sugimoto *et al.*, 1986; Wickramasinghe *et al.*, 2009). When raw starch of *Xanthosoma sagittifolium* was treated with α-amylase of *Aspergillus oryzae* at a temperature range of 30–60°C, two stages of hydrolysis were observed. The intial rate of hydrolysis was fast and observed for the smaller granules while the second stage was slow compared to *Canna* and *Cassava* starches. Besides granule size, amylose content (15–48%) also played an important role in the second stage (Perez-Rea *et al.*, 2013). Similarly, raw *Xanthosoma* starch had higher susceptibility to porcine pancreatic α-amylase compared to *D. alata* starch (Gunaratne and Hoover, 2002); but lower susceptibility compared to cassava and sweet potato starch (Valetudie *et al.*, 1993). Polyhedric granules were more susceptible to hydrolysis compared to spherical granules.

References

Aboubakar, Njintang, Y.N., Scher, J. and Mbofung, C.M.F. (2008) Physicochemical, thermal properties and microstructure of six varieties of taro (*Colocasia esculenta* L. Schott) flours and starches. *Journal of food engineering* 86(2), 294–305. DOI: 10.1016/j.jfoodeng.2007.10.006.

Agama-Acevedo, E., Garcia-Suarez, F.J., Gutierrez-Meraz, F., Sanchez-Rivera, M.M., San Martin, E. *et al.* (2011) Isolation and partial characterization of Mexican taro (*Colocasia esculenta* L.) starch. *Starch/Stärke* 63(3), 139–146. DOI: 10.1002/star.201000113.

Bertoft, E. (2004) On the nature of categories of chains in amylopectin and their connection to the super helix model. *Carbohydrate Polymers* 57(2), 211–224. DOI: 10.1016/j.carbpol.2004.04.015.

Falade, K.O. and Okafor, C.A. (2013) Physicochemical properties of five cocoyam (*Colocasia esculenta* and *Xanthosoma sagittifolium*) starches. *Food Hydrocolloids* 30(1), 173–181. DOI: 10.1016/j.foodhyd.2012.05.006.

FAOSTAT (statistics division of food and agriculture organization of the United Nations) (2015) Available at: http://faostat3.fao.org/browse/Q/QC/E (accessed 18 March 2019).

Gaosong, J., Ramsden, L. and Corke, H. (1997) Effect of water-soluble non-starch polysaccharides from taro on pasting properties of starch. *Starch - Stärke* 49(7-8), 259–261. DOI: 10.1002/star.19970490702.

Gunaratne, A. and Hoover, R. (2002) Effect of heat–moisture treatment on the structure and physicochemical properties of tuber and root starches. *Carbohydrate Polymers* 49(4), 425–437. DOI: 10.1016/S0144-8617(01)00354-X.

Harris, P.J., Ferguson, L.R., Robertson, A.M., Mckenzie, R.J. and White, J.B. (1992) Cell-wall histochemistry and anatomy of taro (*Colocasia esculenta*.). *Australian Journal of Botany* 40(2), 207–222. DOI: 10.1071/BT9920207.

Hoover, R. (1995) Starch retrogradation. *Food Reviews International* 11(2), 331–346. DOI: 10.1080/87559129509541044.

Hoover, R. (2001) Composition, molecular structure, and physicochemical properties of tuber and root starches: a review. *Carbohydrate Polymers* 45(3), 253–267. DOI: 10.1016/S0144-8617(00)00260-5.

Hoyos-Leyva, J.D., Bello-Pérez, L.A., Yee-Madeira, H., Rodriguez-Garcia, M.E. and Aguirre-Cruz, A. (2017) Characterization of the flour and starch of aroid cultivars grown in Mexico. *Starch/Stärke* 69(9–10), 1600370. DOI: 10.1002/star.201600370.

Huang, C.-C., Lai, P., Chen, I.-H., Liu, Y.-F. and Wang, C.-C.R. (2010) Effects of mucilage on the thermal and pasting properties of yam, taro, and sweet potato starches. *Food Science and Technology* 43(6), 849–855. DOI: 10.1016/j.lwt.2009.11.009.

Hughes, R.G. and Lawrence, M.A. (2005) Globalization, food and health in Pacific Island countries. *Asia Pacific Journal of Clinical Nutrition* 14(4), 298–306.

Jane, J., Shen, L., Chen, J., Lim, S., Kasemsuwan, T. *et al.* (1992) Physical and chemical studies of taro starches and flours. *Cereal Chemistry* 69, 528–535.

John, J.K., Rani, V.S., Raja, K.C.M. and Moorthy, S.N. (1999) Physicochemical and enzyme susceptibility characteristics of starch extracted from chemically *pretreated Xanthosoma sagitifolium* roots. *Starch/Stärke* 51(2–3), 86–89. DOI: 10.1002/(SICI)1521-379X(199903)51:2<86::AID-STAR86>3.0.CO;2-P.

Langworthy, C.F. and Deuel, H.J. (1922) Digestibility of raw rice, arrowroot, canna, cassava, taro, tree-fern, and potato starches. *Journal of Biological Chemistry* 52, 251–261.

Lauzon, R.D., Shiraishi, K., Yamazaki, M., Sawayama, S., Sugiyama, N. *et al.* (1995) Physicochemical properties of cocoyam starch. *Food Hydrocolloids* 9(2), 77–81. DOI: 10.1016/S0268-005X(09)80268-3.

Lawal, O. (2004) Composition, physicochemical properties and retrogradation characteristics of native, oxidised, acetylated and acid-thinned new cocoyam (*Xanthosoma sagittifolium*) starch. *Food Chemistry* 87(2), 205–218. DOI: 10.1016/j.foodchem.2003.11.013.

Lebot, V., Champagne, A., Malapa, R. and Shiley, D. (2009) NIR determination of major constituents in tropical root and tuber crop flours. *Journal of Agricultural and Food Chemistry* 57(22), 10539–10547. DOI: 10.1021/jf902675n.

Lebot, V., Malapa, R. and Bourrieau, M. (2011) Rapid estimation of taro (*Colocasia esculenta*) quality by near-infrared reflectance spectroscopy. *Journal of Agricultural and Food Chemistry* 59(17), 9327–9334. DOI: 10.1021/jf202310w.

Lin, M.-H.A., Wu, M.-C., Lu, S. and Lin, J. (2010) Glycemic index, glycemic load and insulinemic index of Chinese starchy foods. *World Journal of Gastroenterology* 16(39), 4973–4979. DOI: 10.3748/wjg.v16.i39.4973.

Londoño-Restrepo, S.M., Rincón-Londoño, N., Contreras-Padilla, M., Acosta-Osorio, A.A., Bello-Pérez, L.A. *et al.* (2014) Physicochemical, morphological, and rheological characterization of *Xanthosoma robustum* Lego-like starch. *International Journal of Biological Macromolecules* 65, 222–228. DOI: 10.1016/j.ijbiomac.2014.01.035.

Lu, T.-J., Chen, J.-C., Lin, C.-L. and Chang, Y.-H. (2005) Properties of starches from cocoyam (*Xanthosoma sagittifolium*) tubers planted in different seasons. *Food Chemistry* 91(1), 69–77. DOI: 10.1016/j.foodchem.2004.05.048.

Lu, T.-J., Lin, J.-H., Chen, J.-C. and Chang, Y.-H. (2008) Characteristics of taro (*Colocasia esculenta*) starches planted in different seasons and their relations to the molecular structure of starch. *Journal of Agricultural and Food Chemistry* 56(6), 2208–2215. DOI: 10.1021/jf0727894.

Meng, F., Liang, Y., Shuai, T., Xiong, J. and Zhong, G. (2015) Physicochemical properties of *Amorphophallus paeoniifolius* (Dennst): Nicolson starch and its blend with konjac glucomannan. *Starch* 67, 820–828.

Moorthy, S.N., Larsson, H. and Eliasson, A.-C. (2008) Rheological characteristics of different tropical root starches. *Starch/Stärke* 60(5), 233–247. DOI: 10.1002/star.200700708.

Moorthy, S.N., Sajeev, M.S. and Anish, R.J. (2018) Functionality of tuber starches. In: *Starch in Food: Structure, Function and Applications*, 2nd edn. Woodhead Publishing, Cambridge, Massachusetts, pp. 421–508.

Moorthy, S.N., Thankamma Pillai, P.K. and Unnikrishnan, M. (1993) Variability in starch extracted from taro. *Carbohydrate Polymers* 20(3), 169–173. DOI: 10.1016/0144-8617(93)90147-V.

Moorthy, S.N., Unnikrishnan, M. and Lakshmi, K.R. (1994) The physicochemical properties of starch of some accessions of *Amorphophallus paeoniifolius*. *Tropical Science* 34, 371–376.

Mweta, D.E., Labuschagne, M.T., Bonnet, S., Swarts, J. and Saka, J.D.K. (2010) Isolation and physicochemical characterisation of starch from cocoyam (*Colocasia esculenta*) grown in Malawi. *Journal of the Science of Food and Agriculture* 90(11), 1886–1896. DOI: 10.1002/jsfa.4029.

Nguimbou, R.M., Boudjeko, T., Njintang, N.Y., Himeda, M., Scher, J. *et al.* (2014) Mucilage chemical profile and antioxidant properties of giant swamp taro tubers. *Journal of Food Science and Technology* 51(12), 3559–3567. DOI: 10.1007/s13197-012-0906-6.

Nwokocha, L.M., Aviara, N.A., Senan, C. and Williams, P.A. (2009) A comparative study of some properties of cassava (*Manihot esculenta*, Crantz) and cocoyam (*Colocasia esculenta*, Linn) starches. *Carbohydrate Polymers* 76(3), 362–367. DOI: 10.1016/j.carbpol.2008.10.034.

Opara, L.U. (2003) Edible aroids: post-harvest operations. In: Mejia, D. (ed.) *AGST/FAO*. Food and Agriculture Organization of the United Nations (Technical), Rome.

Pérez, E.E., Breene, W.M. and Bahnassey, Y.A. (1998) Variations in the gelatinization profiles of cassava, Sagu and Arrowroot native starches as measured with different thermal and mechanical methods. *Starch/Stärke* 50(2–3), 70–72. DOI: 10.1002/(SICI)1521-379X(199803)50:2/3<70::AID-STAR70>3.0.CO;2-U.

Pérez, E., Schultz, F.S. and de Delahaye, E.P. (2005) Characterization of some properties of starches isolated from *Xanthosoma sagittifolium* (tannia) and Colocassia esculenta (taro). *Carbohydrate Polymers* 60(2), 139–145. DOI: 10.1016/j.carbpol.2004.11.033.

Pérez, S. and Bertoft, E. (2010) The molecular structures of starch components and their contribution to the architecture of starch granules: a comprehensive review. *Starch/Stärke* 62(8), 389–420. DOI: 10.1002/star.201000013.

Perez-Rea, D., Rojas, C., Carballo, S., Aguilar, W., Bergenståhl, B. *et al.* (2013) Enzymatic hydrolysis of *Canna indica*, *Manihot esculenta* and *Xanthosoma sagittifolium* native starches below the gelatinization temperature. *Starch/Stärke* 65(1–2), 151–161. DOI: 10.1002/star.201200103.

Rani, V.S., John, J.K., Moorthy, S.N. and Raja, K.C.M. (1998) Effect of pretreatment of fresh *Amorphophallus paeoniifolius* on physicochemical properties of starch. *Starch/Stärke* 50(2–3), 72–77. DOI: 10.1002/(SICI)1521-379X(199803)50:2/3<72::AID-STAR72>3.0.CO;2-I.

Saikia, J.P. and Konwar, B.K. (2012) Physicochemical properties of starch from Aroids of North East India. *International Journal of Food Properties* 15(6), 1247–1261. DOI: 10.1080/10942912.2010.491929.

Salwa, M., Abo- El-, F., Hanan, M., Sayed, A.L. and Nesserin, M.N. (2010) Physiocochemical properties of starch extracted from different sources and their applications in puddings and white souces. *World Journal of Dairy & Food Science* 5(2), 173–182.

Shiraishi, K., Lauzon, R.D., Yamazaki, M., Sawayama, S., Sugiyama, N. *et al.* (1995) Rheological properties of cocoyam starch paste and gel. *Food Hydrocolloids* 9(2), 69–75. DOI: 10.1016/S0268-005X(09)80267-1.

Sit, N., Misra, S. and Deka, S.C. (2014a) Yield and functional properties of taro starch as affected by ultrasound. *Food and Bioprocess Technology* 7(7), 1950–1958. DOI: 10.1007/s11947-013-1192-7.

Sit, N., Deka, S.C. and Misra, S. (2014b) Combined effect of ultrasound and enzymatic pretreatment on yield and functional properties of taro (*Colocasia esculenta*) starch. *Starch/Stärke* 66(11–12), 959–967. DOI: 10.1002/star.201400085.

Soni, P.L., Sharma, H.W., Dobhal, N.P., Bisen, S.S., Srivastava, H.C. *et al.* (1985) The starches *of Dioscorea ballophylla and Amorphophallus campanulatus*. Comparison with tapioca starch. *Starch/Stärke* 37(1), 6–9. DOI: 10.1002/star.19850370103.

Srichuwong, S. and Jane, J.L. (2007) Physicochemical properties of starch affected by molecular composition and structures: A review. *Food Science and Biotechnology* 16, 663–674.

Srichuwong, S., Isono, N., Jiang, H., Mishima, T. and Hisamatsu, M. (2012) Freeze–thaw stability of starches from different botanical sources: correlation with structural features. *Carbohydrate Polymers* 87(2), 1275–1279. DOI: 10.1016/j.carbpol.2011.09.004.

Sugimoto, Y., Ohnishi, K., Takaya, T. and Fuwa, H. (1979) Comparative susceptibility to pancreatin of starch granules from different plant species. *Journal of the Japanese Society of Starch Science* 26(3), 182–190. DOI: 10.5458/jag1972.26.182.

Sugimoto, Y., Nishihara, K. and Fuwa, H. (1986) Some properties of taro (ishikawa-wase and takenoko-imo) and yam (iseimo and nagaimo) starches. *Journal of the Japanese Society of Starch Science* 33(3), 169–176. DOI: 10.5458/jag1972.33.169.

Tattiyakul, J., Pradipasena, P. and Asavasaksakul, S. (2007) Taro *Colocasia esculenta* (L.) Schott amylopectin structure and its effect on starch functional properties. *Starch/Stärke* 59(7), 342–347. DOI: 10.1002/star.200700620.

Tetchi, F.A., Rolland-Sabaté, A., Amani, G.N. and Colonna, P. (2007) Molecular and physicochemical characterisation of starches from yam, cocoyam, cassava, sweet potato and ginger produced in the Ivory Coast. *Journal of the Science of Food and Agriculture* 87(10), 1906–1916. DOI: 10.1002/jsfa.2928.

Torruco-Uco, J. and Betancur-Ancona, D. (2007) Physicochemical and functional properties of makal (*Xanthosoma yucatanensis*) starch. *Food Chemistry* 101(4), 1319–1326. DOI: 10.1016/j.foodchem.2006.03.047.

Torruco-Uco, J.G., Chel-Guerrero, L.A. and Betancur-Ancona, D. (2006) Isolation and molecular characterization of Makal (*Xanthosoma yucatanensis*) starch. *Starch/Stärke* 58(6), 300–307. DOI: 10.1002/star.200500451.

Valetudie, J.-C., Colonna, P., Bouchet, B. and Gallant, D.J. (1993) Hydrolysis of tropical tuber starches by bacterial and pancreatic α-amylases. *Starch/Stärke* 45(8), 270–276. DOI: 10.1002/star.19930450805.

Wankhede, D.B. and Sajjan, S.U. (1981) Isolation and physico-chemical properties of starch extracted from yam, elephant (*Amorphophallus campanulatus*). *Starch/Stärke* 33(5), 153–157. DOI: 10.1002/star.19810330503.

Wickramasinghe, H.A.M., Takigawa, S., Matsuura-Endo, C., Yamauchi, H. and Noda, T. (2009) Comparative analysis of starch properties of different root and tuber crops of Sri Lanka. *Food Chemistry* 112(1), 98–103. DOI: 10.1016/j.foodchem.2008.05.046.

Williams, J.T. and Haq, N. (2002) *Global Research on Underutilized Crops: An Assessment of Current Activities and Proposals for Enhanced Cooperation*. ICUC, Southampton, UK.
Zeng, F.K., Liu, H. and Liu, G. (2014) Physicochemical properties of starch extracted from *Colocasia esculenta* (L.) Schott (Bun-long taro) grown in Hunan, China. *Starch/Stärke* 66(1–2), 142–148. DOI: 10.1002/star.201300039.
Zhu, F. (2015) Impact of ultrasound on structure, physicochemical properties, modifications, and applications of starch. *Trends in Food Science & Technology* 43(1), 1–17. DOI: 10.1016/j.tifs.2014.12.008.
Zhu, F. (2016) Structure, properties, and applications of aroid starch. *Food Hydrocolloids* 52, 378–392. DOI: 10.1016/j.foodhyd.2015.06.023.

6

Yams

6.1 Introduction

Though yams do not constitute a promising raw material for the commercial manufacture of starch on economic grounds, their starches possess some special properties, and recently there have been detailed studies on starch from different yam species. The starches of a number of Indian species have been examined by Rao and Beri (1952a, b, 1953a, b, 1955), while those of some material, both indigenous and exotically grown in Nigeria, have been described in a report by Greenwood-Barton (1961). A very comprehensive review of *Dioscorea* starches has been published by Seidemann (1964). Some fairly detailed studies have been made by Hollo (1964) on the starch of some species which he describes as *D. batatas*. Other west African material has been investigated and figured by Miege (1948, 1957). Zhu (2016) has comprehensively reviewed the available information on yam starches.

6.2 Extraction

Though the starch content of yams is quite high, extraction of starch from yams has not been commercially practised due to the difficulty in extraction and getting pure starch.

The basic steps of starch isolation from yam tubers are similar to other tubers. The tubers are washed, peeled, sliced into smaller pieces and homogenized in a blender with water, sodium bisulphite solution (0.075%) or alkaline solution (Alves *et al.*, 1999; Riley *et al.*, 2004; Otegbayo *et al.*, 2014). To remove impurities, before sedimentation, the slurry is then filtered and sieved. The supernatant is then discarded and sediment re-suspended in water. To clean starch of any impurities, this washing procedure has to be repeated several times. Further, the starch thus separated is treated with sodium hydroxide solution (about 0.1%) and washed with additional water before drying (Alves *et al.*, 1999; Otegbayo *et al.*, 2014). Emiola and Delarosa (1981) reported the use of mercury (II) chloride (0.01 M) to suppress any microbial growth during homogenization.

Extraction of starch from *D. dumetorum* is more difficult because it contains smaller granules compared to those from *D. cayenensis*, *D. bulbifera*, *D. rotundata* and *D. alata* (Otegbayo *et al.*, 2014), mainly due to two reasons: (i) starch with small granules settles more slowly during extraction compared to starch with large granules; (ii) the small-sized granules get trapped in the fibrous matrix more easily. The presence of non-starch polysaccharides (NSP) in some species, such as *D. alata*, considerably impact starch extraction by giving rise to a viscous suspension after homogenization, trapping small-sized granules and carrying them over to the waste portion.

Various chemicals, such as NaOH (0.03 M), pectinase and oxalate (10%), have been tested to reduce the effect of NSP during extraction. Within these chemicals, oxalate was found to be effective in reducing the slurry viscosity and increasing the starch recovery to 63% (Daiuto *et al.*, 2005). Use of dilute ammonia has been found to enhance the starch yield from different yam species without affecting the physicochemical and functional properties (Moorthy, 1991).

Employing increased ratio of water to solid during extraction and use of diverse enzymes to degrade NSP and protein also benefit the starch yield and purity from an NSP/protein-rich slurry (Vasanthan, 2001). To negate the challenges due to the presence of NSP and proteins, a continuous bubble separation process has been found to be effective. This pilot scale system creates a gas–liquid interface within the system to extract the starch and mucilage from *D. pseudojaponica* tubers (Fu *et al.*, 2005). The separation system consists of bubbling (Fig. 6.1A) and defoaming (Fig. 6.1B) components. The slurry suspension from the reservoir tank is pumped into foam column I. Bubbles are produced by passing air through the slurry, resulting in foam, which is collected in column II. The bubbles are eventually collected and taken to the de-foaming system. The starch recovery is as high as 91.2% and has a purity of 99% with a trace amount of fat (0.12%), protein (0.25%) and ash (0.08%). The bubble separation process, being suitable to separate starch from raw materials rich in viscous mucilage, is simple and economically feasible (Fu *et al.*, 2005).

One patent claimed that the starch can be extracted directly from the tubers without peeling (Woodward, 1982). In this method, the whole unskinned roots are pulverized and then pressed to obtain a mix of starch and juice. After the separation of starch from the juice, the liquid is then pumped back into the initial mixture for recycling. The major advantage of this process is that it does not require additional liquid in the process and the separation is completed with the natural juice present in the tubers (Woodward, 1982). Though the purity of starch separated using this process is not studied, this extraction process appears to be feasible for industries. Most of the above-mentioned reports are on the laboratory scale and need to be scaled up for viability.

The yield of starch ranged from 10% to 25% (wet basis) for *D. rotundata* (Moorthy and Nair, 1989), *D. opposita* (Zhou *et al.*, 2012), *D. esculenta* and *D. alata* (Jayakody *et al.*, 2007) and from 85% to 88% (dry basis) for *D. alata, D. cayenensis, D. dumetorum* and *D. esculenta* (Emiola and Delarosa, 1981). The starch yield and recovery from the tubers depend on various factors including the starch content in the original tubers, structure and composition of the tubers, post-harvest storage conditions, and the efficiency of the isolation method (for example, starch content in dried flour of *D. dumetorum* ranged from 40.7% to 63.3%) (Akinoso and Abiodun, 2013; Cereda *et al.*, 2005).

The effects of blanching, sulphite treatment and cooking pre-treatments on colour, functional and pasting characteristics of flours of *Dioscorea rotundata* (*Abuja, Amula, Gbangi*) and *Dioscorea cayenensis* (*Alo*) cultivars were investigated. The L^* (71.45–79.98), a^* (2.36–4.72), b^* (19.96–25.71), ΔE (0–6.68), ΔC (0–5.75) and hue angle (78.99–84.77) varied significantly with pre-treatment and yam variety (Falade *et al.* 2015).

6.3 Physiochemical Properties

6.3.1 Biochemical property

Great variation in the composition of minor components in yam starches has been observed (Table 6.1; Zhu, 2015). Protein content of yam starches varied greatly among different reports, mostly being <0.2%. The lower protein content may be due to a more efficient method for starch extraction and purification, and the nature of protein components associated with the granule that can be obtained only by extensive 'washing' steps in yam starches has not be investigated (Baldwin, 2001). The lipid content of yam starches in most reports was much less than 1%, similar to other root and tuber starches but lower than cereal starches. Lipid content was reported to be more than 1% for *D. abyssinica* (Gebre-Mariam and Schmidt, 1998) and *D. ballophylla* (Soni *et al.*, 1985) starches, but needs confirmation, as only one genotype was tested

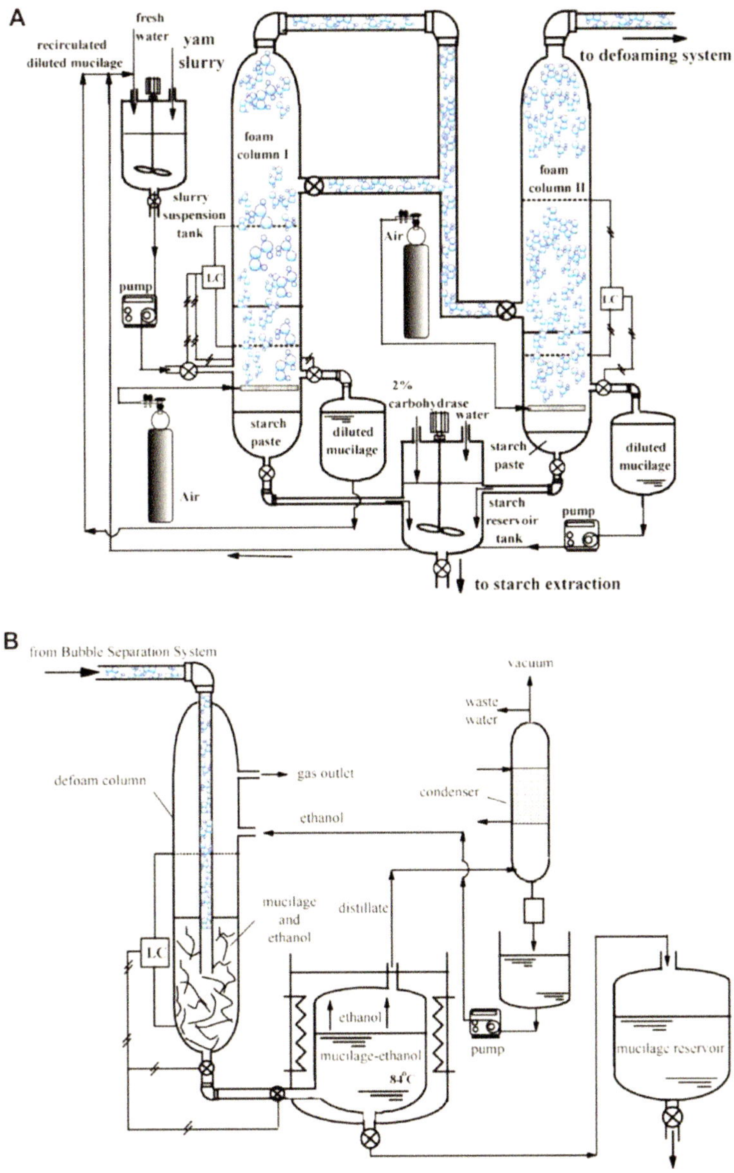

Fig. 6.1. Schematic diagram of the continuous bubble separation system (A) for separation of starch and viscous mucilage from yam tubers; (B) diagram of the ethanol-defoaming system coupled to (A); the resulting starch from *D. pseudojaponica* had a recovery of 91.2% and a purity of 99%. (From Fu *et al.*, 2005)

in each study. Elmi Sharlina *et al.* (2017) have given the values for different components in *D. pyrifolia* starch as protein 1.34%, ash 0.88% and no lipid or fibre. *D. opposita* starch had 0.23% protein, 0.11% lipid and 0.03% ash (Zhou *et al.*, 2012). For *D. zingiberensis*, the values were lipid

Table 6.1. Chemical composition of yam starches. (Source: Zhu, 2016)

Species	No.	Amylase content (%)	Amylase measuring method	Minor components				References
				Protein (%)	Ash (%)	Lipids (%)	Phosphorus (%)	
D.abyssinica	1	29.7	Iodine –SP	0.5	0.1	1		Gebre-Mariam and Schmidt, 1998
D. alata	9	21.7 (average)	Iodine –Sp					Otegbayo *et al.*, 2014
	2	21–21.1	Iodine –APT	0.13–0.15	0.20–0.23			Rašper and Coursey, 1967
	1	22.2	Iodine –Sp	0.06	0.13	0.03		Emiola and Delarosa, 1981
	1	25	Iodine –SP					Farhat *et al.*, 1999
	3	25–27[a], 34–41[b]	A, DSC; b, HPSEC-MALLS-DRI					Rolland-Sabaté *et al.*, 2003
	1	36	Iodine –SP	0.1	0.09			Freitas *et al.*, 2004
	1	21	Iodine –SP					Riley *et al.*, 2004
	5	25.3–27.4	DSC	0.19–0.35	0.08–0.38	0.09–0.11		Amani *et al.*, 2004
	2,A	22.3–25.4	Iodine –APT					Brunnschweiler *et al.*, 2005
	4,B	23.5–40.1	Iodine –SP,[e]					Huang *et al.*, 2006
	2	24.7–31.0	Iodine –SP,[f]		013–0.17	0.28–0.3,[f]	0.04–0.05	Jayakody *et al.*, 2007
	8	20.1–23	Iodine –SP		0.19–0.51	0.50–0.73	0.0144–0.0281	Huang *et al.*, 2006
	1	18.1	Iodine –SP		0.87			Jiang *et al.*, 2012b
D. ballophylla	1	24.1	n.a	2.6	0.27	1.1	0.005	Soni *et al.*, 1985
D. bulbifera	5	19 (average)	Iodine –SP					Otegbayo *et al.*, 2014
	3	12.2–14.3	Iodine –SP	0.01	1.72–2.01			Jiang *et al.*, 2012a

Continued

Table 6.1. Continued

Species	No.	Amylase content (%)	Amylase measuring method	Minor components			References
	1	17.6	Iodine –SP		1.71		Jiang *et al*., 2012b
D. cayenensis	1	22.1 (average)	Iodine –SP				Otegbayo *et al*., 2014
	1	21.6	Iodine –SP	0.05	0.08	0.02	Emiola and Delarosa, 1981
	1	23.6	Iodine –SP				Farhat *et al*., 1999
	1	26.5	Iodine –SP				Riley *et al*., 2004
D. cayenensis-rotundata	5	25–29a, 40–47b	a, DSC; b, HPSEC-MALLS-DRI				Rolland-Sabaté *et al*., 2003
	11	25.2–28.8	DSC	0.11–0.26	0.04–0.26	0.11–0.26	Amani *et al*., 2004
	1	25.6	Iodine –APT				Brunnschweiler *et al*., 2005
D. dumetorum	1	15.3 (average)	Iodine –SP				Otegbayo *et al*., 2014
	1	15	Iodine –APT	1.49	0.29		Rašper and Coursey, 1967
	1	25.4	Iodine –SP	0.08	0.09	0.04	Emiola and Delarosa, 1981
	2	11.1–12.4	Iodine –SP	0.8–1.51	1.13–0.22		Nkala *et al*., 1994
	1	12.3	Iodine –SP				Farhat *et al*., 1999
	1	17[a]; 50[b]	a, DSC; b, HPSEC-MALLS-DRI				Rolland-Sabaté *et al*., 2003
	1	16.6	DSC	0.68	0.02	0.08	Amani *et al*., 2004
	2C	14.7–17.4	Iodine –SP				Akinoso and Abiodun, 2013

Continued

Table 6.1. Continued

Species	No.	Amylase content (%)	Amylase measuring method	Minor components				References
	1	14.2	Iodine –APT	0.08	0.44			Rašper and Coursey, 1967
	1	16[a]; 30[b]	a, DSC; b, HPSEC-MALLS-DRI					Rolland-Sabaté *et al.*, 2003
	1	11.1	Iodine –SP					Riley *et al.*, 2004
	4	14.1–17.1	DSC	0.07–0.39	0.21–0.23	0.06–0.09		Amani *et al.*, 2004
	3	15.6–24	Iodine –SP,F		0.17–0.32	0.36–0.47,F	0.05–0.10	Jayakody *et al.*, 2007
	1	17.9	GPC (CL-6B) after debranching					Bertoft *et al.*, 2008
D. hispida	1	39.3	Iodine –SP	0.1	<0.05	0.1		Tattiyakul *et al.*, 2006
D. nipponica	1	26.3	Iodine –SP					Yuan *et al.*, 2007
	1	17.7	Iodine –SP		1			Jiang *et al.*, 2012b
	3	10.8–17.2	Iodine –SP	0.24–0.3				Jiang *et al.*, 2013b
D. opposita	8	19.4–25.9	Iodine –SP					Wang *et al.*, 2006a, 2006d, 2008a
	3	33.3–36.7	Iodine –SP	0.21–0.24	0.03–0.12	0.11–0.12		Zhou *et al.*, 2012
	9	9.9–23.4	AOAC 1990	0.01–0.03				Jiang *et al.*, 2013b
D. polygonoides	1	11.9	Iodine –SP					Riley *et al.*, 2004
D. persimilis	1	23.9	Iodine –SP	0.0142				Jiang *et al.*, 2013b
D. rotundata	27	22.1 (average)	Iodine –SP					Otegbayo *et al.*, 2014
	1	22.4	Iodine –SP	0.03	0.22	0.04		Emiola and Delarosa, 1981
	5	21.1–23	Iodine –APT	0.05–0.15	0.15–0.28			Rašper and Coursey, 1967

Continued

Table 6.1. Continued

Species	No.	Amylase content (%)	Amylase measuring method	Minor components				References
	1	23.8	Iodine –SP					Farhat *et al.*, 1999
	1	22.1	Iodine –SP					Riley *et al.*, 2004
	6	20.9–24.6	Iodine –SP				0.011–0.015	Moorthy and Nair, 1989
D. septemloba	1	13.6	Iodine –SP		1.92			Jiang *et al.*, 2012b
D. trifida	4D	1.37–12.0	DSC, Iodine –SP, Iodine –APT	0.02–0.14	0.02–0.08	0.05–0.1	0.03–0.07	Pérez *et al.*, 2011
D. zingiberensis	1	23.6	Iodine –SP	0.54	4.2	0.6		Zhou *et al.*, 2013

No. = number of genotypes tested in the study; n.a. = no data available; for the amylase content measurement, DSC = differential scanning calorimetry-based; HPSEC–MALLS–DRI = high-performance size-exclusion chromatography-multi-angle laser light scattering-differential refractometric index dection-based method; GPC(CL-6B) = gel-permeation chromatography of debranched starch on Sepharose CL6B; iodine-APT and iodine-SP represent iodine-binding amperometric-titration-based and iodine-binding spectrophotometric/colorimetric methods, respectively. In the report by Rolland-Sabate and others (2003), the letters a and b (following the amylase content) denote DSC and HPSEC–MALLS–DRI-based methods, respectively.

[a] different parts of tuber and 3 different storage conditions for 1 genotype;

[b] 4 varieties harvested at 4 different dates (155, 190, 225 and 260 days) after planting;

[c] 2 varieties harvested at 5 different months (7,8, 9, 10, 11) after planting;

[d] 3 varieties harvested at 2 years;

[e] measured before/after the removal of bound lipids;

[f] measured before/after the removal of bound lipids and 2 types of solvent used for bound and unbound lipid. In the report by Rolland-Satat'e *et al.*, 2003.

0.6% and fibre 0.07% (Zhou *et al.*, 2013). The composition of *D. esculenta* and *D. alata* starches were lipids 0.36–0.47% and 0.28–0.30%, respectively, and phosphorus 0.071 and 0.045%, respectively. The lipid content of starch also depends on the solvent used. For example, the bound lipid in starch can be extracted using an n-propanol-water mixture, whereas the unbound lipids can be extracted using a chloroform–methanol mixture (Jayakody *et al.*, 2007). Ash content varied widely among various reports, most of them having a value of <0.3%. Higher ash content (>0.5%) may indicate a less efficient method for purifying starch (Jiang *et al.*, 2012b; Zhou *et al.*, 2013). A typical well-washed root or tuber starch contains about 0.05% protein, 0.05−0.1% lipid (Baldwin, 2001) and a small amount of ash (<0.2%), and higher values may indicate incomplete extraction. The measuring methods used may also influence the values (spectrophotometry compared with atomic absorption spectroscopy) (Pérez *et al.*, 2011). In general, phosphorus, in starch from tubers, is in the form of phospholipids, phosphate monoester or inorganic phosphates (Kasemsuwan and Jane, 1996). It has been reported that the phosphorus content of starch from yam is relatively high compared with the other tuber starches; but there is a lack of systematic studies that analyse the composition of phosphorus from diverse tuber species, and there is no published study on the comparative analysis between yams and other tuber and root starches.

6.3.2 Amylose and amylopectin content

Various methods have been used to quantify the amylose content of yam starches. These include iodine binding spectrophotometry/colorimetry (Otegbayo *et al.*, 2014), iodine binding amperometric titration (Brunnschweiler *et al.*, 2005), enthalpy changes of amylose lysophospholipid inclusion complex transition measured by DSC (Amani *et al.*, 2004; Pérez *et al.*, 2011) and CTAB complex (Moorthy *et al.*, 2006), size-exclusion chromatography of whole starch by high-performance size-exclusion chromatography coupled to multi-angle laser light scattering and differential refractometric index (HPSEC–MALLS–DRI) (Rolland-Sabaté *et al.*, 2003), and size exclusion chromatography of debranched starch (Bertoft *et al.*, 2008). However, the different methods can give widely different values for the amylose content for the same samples.

Different methods of quantification can give rather different values of amylose content for the same sample as revealed by comparative analyses (Rolland-Sabaté *et al.*, 2003; Pérez *et al.*, 2011). For instance, values for amylose content (50%) obtained from the chromatography were significantly higher than the value (17%) by HPSEC–MALLS–DRI (based on the DRI signal of amylose fraction in proportion to the complete DRI signal) which was considerably higher than that obtained from DSC analysis (Rolland-Sabaté *et al.*, 2003). Pérez *et al.* (2011) found that iodine binding amperometric titration, iodine binding spectrophotometry/colorimetry, and DSC-based method gave rather different values of amylose content for the same sample.

Another important factor affecting the observed amylose content is the endogenous lipid matter which can form inclusion complexes with amylose, thus lowering the values. Removal of the bound lipids can give rise to higher amylose content even up to 4% in the resulting starch (Jayakody *et al.*, 2007). The percentage of amylose bound to lipid was estimated as 15–22% for *D. esculenta* (3 genotypes) and 5.6–8.3% for *D. alata* (2 genotypes) (Jayakody *et al.*, 2007). In iodine binding methods, the amylopectin fraction can also bind with iodine resulting in colour change, contributing to a possible artefact in the amylose content measurement. This can be avoided by using the concanavalin A precipitation-based method. Concanavalin A specifically binds with amylopectin molecules but not with the linear amylose chains, thus excluding the amylopectin component on the measured amylose content of starch (Matheson, 1990).

Ketiku and Oyenuga (1973) observed only small variations in the relative amylose and amylopectin content with age. The levels in the *D. rotundata* tuber were 16.7–18.6% amylose and 80.2–86.0% amylopectin. Delpeuch *et al.* (1978) found that the amylose content was related to the granular morphology. Small granules have the lowest amylose content. *Dioscoria libreschiana*, which has the largest granules, contains as much as 34.5%. The amylose content of three edible species has been determined by GPC and the values are given in Table 6.2.

Table 6.2. Amylose content (%) in yams. (Moorthy *et al.*, 2008)

Species	SDS	CTAB	Iodine staining	GPC
Dioscorea rotundata	23	24.2	22.5	23.6
Dioscorea esculenta	19.4	12.0	14.4	14.0
Dioscorea alata	20	21	23.7	21.2

In the case of *D. hispida* and *D. alata* (Rao and Beri, 1952a,b, 1955) the contents of amylose were found to be 10.24% and 24.77%, respectively, while Hollo (1964), in his west African material, found 20–22%. Farhat *et al.* (1999) have obtained the following values for amylose content in starches of different *Dioscorea* species: *D. alata*, 25%; *D. rotundata* and *D. cayensis*, 23.8%; *D. dumetorum*, 12.6%. A value of 29.7% was reported for yam starch from Ethiopia, while Soni *et al.* (1985) reported 24.1% for *D. ballophylla* starch. A minor variation in the amylose content was observed during the growth period for *D. esculenta*, *D. alata* and *D. rotundata* starches. *D. hispida* tends to have higher amylose content (39.3%) than the other species (Tattiyakul *et al.*, 2006). Amylose content of genotypes from various yam species ranged from 1.4% (Perez *et al.*, 2011) to 50% (Rolland-Sabaté *et al.*, 2003). Zhou *et al.* (2013) reported 23.62% amylose for *D. zingiberensis* starch. Nwokoha and Williams obtained 22.8% for *D. alata* starch. Yuan *et al.* (2007) found the amylose content of *D. nipponica* to be 26.3%. Zhang *et al.* (2014) compared sulphur fumigation and freeze-drying and the amylose content was only 8.6% compared to 16% for native and sulphur fumigated sample. *D. opposita* had a high amylose content of 33% (Zhou *et al.*, 2012). Variation in amylose content with species and method was observed by Pérez *et al.* (2013). According to Sharlina *et al.* (2017) *D. pyrifolia* had very high amylose content of 44%.

Otegbayo *et al.* (2014) found the amylose content to vary considerably among yam species and varieties: 23.4% for *D. bulbifera*, 22.05% for *D. rotundata*, 22.06% for *D. cyanensis*, 15.27% for *D. dumetorium* and 21.73% for *D. alata*. Apparent amylose content from 43 genotypes in five species ranged from 15.1% to 27.1% (Otegbayo *et al.*, 2014). Among the most studied species, *D. dumetorum* and *D. esculenta* tend to have lower amounts of amylose (<20%) compared to *D. alata*, *D. rotundata*, *D. cayenensis-rotundata* and *D. cayenensis* (>20%) (Amani *et al.*, 2004; Otegbayo *et al.*, 2014).

Amylose content in distal and proximal regions of *D. alata* tubers was rather similar (Brunnschweiler *et al.*, 2005) and indicated an even distribution of amylose throughout the whole tuber. Amylose content was considerably influenced by environmental factors and agronomic practices (Huang *et al.*, 2006; Pérez *et al.*, 2011; Akinoso and Abiodun, 2013; Otegbayo *et al.*, 2014). Drought lowered the amylose content of starch from *D. dumetorum* (Otegbayo *et al.*, 2014). The harvesting time influenced the amylose content of *D. trifida* starch (Pérez *et al.*, 2011). Harvesting the crop 7–11 months after planting had only a minor effect on the amylose content of starch from *D. dumetorum* (Akinoso and Abiodun, 2013). On the other hand, starch from *D. alata* harvested on the 260th day of planting presented a higher amylose content than from the crop harvested in advance (Huang *et al.*, 2006). The factors that could have influenced this discrepancy are the harvesting period, yam genotype and the environmental conditions. The amylose content of starch from *D. alata* showed a decrease during a three-month post-harvest storage period (Brunnschweiler *et al.*, 2005). The influence of environmental and human factors indicate possible manipulation of the amylose starch content and further artificially diversify the variation. Freeze-drying decreased amylose content from 16.68% to 8.59%.

Waxy genotypes of *D. trifida* with amylose content as low as 1–2%, measured by differential scanning calorimetry (DSC) based on iodine spectrophotometry and iodine amperometric titration methods, have been discovered from the Venezuelan Amazon in South America (Pérez *et al.*, 2011). These special genotypes could become excellent genetic resources for further diversifying the amylose content of yam starches.

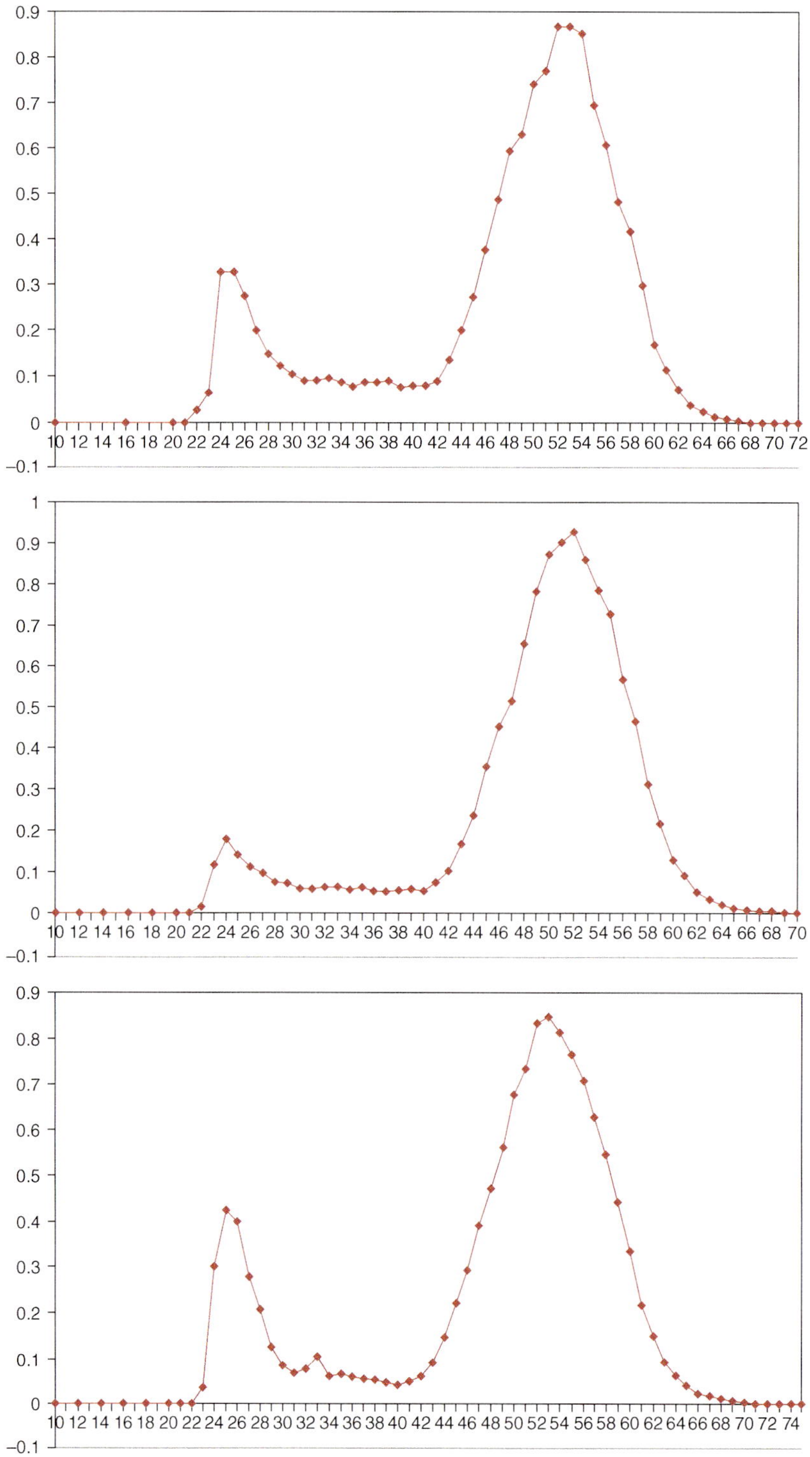

Fig. 6.2. GPC of *D. esculenta, D. alata* and *D. rotundata* starches (Moorthy, unpublished data).

Table 6.3. Shape, diameter and spectrum of diffraction of starch granules. (Faboya and Asagbra, 1990)

Species	Shape	Diameter	Spectrum	References
D. alata	Oval, ovo-triangular	5–50	B	Allen, 1951; Coursey, 1967; Rašper and Coursey, 1967; Delpeuch *et al*., 1978; Rašper and Coursey, 1969
D. bulbifera	Ovo-triangular	5–45	B	Coursey, 1967; Delpeuch *et al*., 1978
D. cayenesis-cayenensis	Oval, Ovo-triangular	3–40	B	Coursey, 1967; Rašper and Coursey, 1967; Delpeuch *et al*., 1978; Rašper and Coursey, 1969; Szylit *et al.,* 1977
D. cayenensis-rotundata	Oval	5–45	B	Coursey, 1967; Rašper and Coursey, 1967; Delpeuch *et al*., 1978; Rašper and Coursey, 1969
D. liberschiana	Oval	15–70	B	Delpeuch *et al*., 1978
D. luzonensis	Ovo-circular	12–48	–	Allen, 1951
D. opposita	–	5–60	–	Coursey, 1967
D. pentaphylla	Oval ± deformed	30–40	–	Allen, 1951
D. schimperiana	Oval	10–35	B	Delpeuch *et al*., 1978
D. trifida	Oval	10–65	?	Coursey, 1967
D. dumetorum	Polyhedral ± aggregate	1–5	Ca-A	Coursey, 1967; Rašper and Coursey, 1967; Delpeuch *et al*., 1978; Rašper and Coursey, 1969; Szylit *et al.,* 1977
D. esculenta	Polyhedral ± aggregate	1–15	Cb	Allen, 1951; Coursey, 1967; Rašper and Coursey, 1967; Delpeuch *et al*., 1978; Rašper and Coursey, 1969

6.4 Structural Properties

6.4.1 Granular morphology

Granule size and shape of starches from diverse yam species vary considerably, even within the same species, as revealed by various techniques, including light microscopy (LM), scanning electron microscopy (SEM), and laser light diffraction (LLD)-based particle size analysis. It is important to consider the variation in the particle size measurement methods because the reported particle size is limited to a specific method of measurement (Lindeboom *et al*., 2004). Yam starches are simple granules, with mono-modal size distribution, possessing small fissures on the surface. Yam starches have a large variability in shape: round, triangular, oval and elliptical (Tables 6.3–6.5) (Faboya and Asagbra, 1990). Otegbayo *et al.* (2014) studied different yam species and found oval, round and triangular shapes and sizes varying from 7.5 µm for *D. dumetorum* to 49.5 µm for *D. bulbiferra*. Pérez *et al.* (2013) compared the size of five wild yam species and obtained values ranging from 24.5 µm to 60.0 µm with average of 45 µm.

Table 6.4. Granular size and gelatinization temperature of *Dioscorea* species (Rao and Beri, 1952).

Species	Granular size um	Gelatinization temperatures (°C)
D. alata	5–50	69.0–78.5
D. rotundata	5–45	64.5–75.5
D. cayenensis	3–25	71.0–78.0
D. opposita	5–60	65.5–75.5
D. bulbifera	5–45	72.0–80.0
D. esculenta	1–15	69.5–80.5
D. hispida	1–5	75.5–83.0
D. dumetorum	1–4	77.0–85.5
D. trifida	10–65	-

Table 6.5. Granular morphology of yam starches. (Zhu, 2015)

Species	No.	Granular size (μm)	Method of measuring	Shape description	References
D. abyssinica	1	29.2 (average)	LLD, SEM	Round	Gebre-Mariam and Schmidt, 1998
D. alata	9	28–40.25 (average)	LM	Round, triangular, oval	Otegbayo *et al.*, 2014
	2	10–70	LM		Rašper and Coursey, 1967
	1	37 (max. diameter mean)		Lenticular	Valetudie *et al.*, 1995
	1	10–35 × 25–40 (E), 31 (average)	LM, LLD	Ellipsoid, rounded	Farhat *et al.*, 1999
	1	8–38	SEM		Riley *et al.*, 2004
	5	18.6–29.3 (average)	LLD, SEM		Amani *et al.*, 2004
	1,A	1.94–92.22 (Individual)			Cereda *et al.*, 2005
		23.73–36.97 (average)			
	4,B	10–40	LLD, SEM	Oval, angular	Huang *et al.*, 2006
	2	30–45	SEM, LM	Truncated spade/oval	Jayakody *et al.*, 2007
	8	22.89–28.01	SEM, LM	Ellipsoid, polyhedral, triangular, rod-like	Riley *et al.*, 2006b
D. bulbifera	5	33–49 (average)	LM	Triangular	
	3	10–60	LM	Irregular flat cake	Otegbayo *et al.*, 2014
D. ballophylla	1	7–50	SEM	Elliptical	Jiang *et al.*, 2012a
D. cayenesis	1	41 (average)	LM	Oval, triangular, ellipsoidal	Soni *et al.*, 1985
	1	10–30 × 20–50 (E), 35 (average)	LM, LLD	Polyhedral, ellipsoid	Otegbayo *et al.*, 2014
	1	16–42	SEM		Farhat *et al.*, 1999
D. cayenensis-rotundata	11	20.4–30.9 (average)	LLD, SEM		Riley *et al.*, 2004
D. dumetorum	1	7 (average)	LM	Polygonal	Amani *et al.*, 2004

Continued

Table 6.5. Continued

Species	No.	Granular size (μm)	Method of measuring	Shape description	References
	1	1–5	LM		Otegbayo *et al*., 2014
	2	1.5 and 2 (average)	LM		Rašper and Coursey, 1967
	1	3–5, F	LM,LLD	Rounded	Nkala *et al*., 1994
	1	2.2 (average)	SEM	Polygonal	Farhat *et al*., 1999
	2, C	2.2–3.6 (average)	SEM	Polygonal	Amani *et al*., 2004
D. esculenta	1	1–5	LM		Akinoso and Abiodun, 2013
	1	1–3	SEM		Rašper and Coursey, 1967
	4	5.8–6.2	LLD, SEM	Polygonal	Riley *et al*., 2004
	3	4–10	SEM, LM	Polygonal	Amani *et al*., 2004
D. opposita	10	1–90, 23.39–40.3 (average)	LLD, SEM	Round or oval, irregular or cuboidal, elliptic shaped	Jayakody *et al*., 2007
	3	20.17–26.34	LLD, SEM	Round, oval or irregular	Wang *et al*., 2006b, Wang *et al*., 2006c; Wang *et al*., 2008a
D. nipponica	1	3–22 (range), 9.5 (average)	LLD, SEM	Oval, dissimilar	Zhou *et al*., 2012
D. polygonoides	1	4–17	SEM		Yuan *et al*., 2007
D. rotundata	27	28–47 (average)	LM	Round, elongated oval, oval, irregular, triangular, Polygonal	Riley *et al*., 2004
	5	10–70	LM	Polyhedral	Otegbayo *et al*., 2014
	1	10–30 × 20–50 (E), 32 (average)	LM, LLD	Polyhedral	Rašper and Coursey, 1967
	1	16–32	SEM		Farhat *et al*., 1999
	6	7.5–75 (range)	n.a.	Round to oval	Riley *et al*., 2004
		27.2–34.3 (average)			Moorthy and Nair, 1989

Continued

Table 6.5. Continued

Species	No.	Granular size (μm)	Method of measuring	Shape description	References
D. septemloba	1		SEM	Elongated or oval shaped	Jiang *et al*., 2012b
D. trifida	4,D	24.5–35.5 (average)	LLD, SEM, LM	Oval or shell-shaped, trancuted and with an obtuse end, smooth surface.	Jiang *et al*., 2011
D. zingiberensis	1	0.40–56.37, 21.19 (average)	LLD, SEM	Flattened, ovoid	Zhou *et al*., 2013

No. = number of genotypes tested in the study; n.a. = not available; LM = light microscopy; LLD = laser light diffraction-based method; SEM = scanning electron microscopy.
[a] 4 extraction methods;
[b] 4 varieties harvested at four different dates (155, 190, 225, and 260 days) after planting;
[c] 2 varieties harvested at five different months (7, 8, 9, 10, 11) after planting;
[d] 3 of the four varieties harvested at 2 years;
[e] short-axis range × long axis range;
[f] bimodal distribution centered at 4.5 μm and 9 μm.

Table 6.3 gives an overview of the different possibilities (Faboya and Asagbra, 1990). The 13 species have been classified into two groups depending upon the sizes and shapes of the granule. The larger group includes those with oval granules that are generally at least 30 microns in diameter, while the smaller group entails *D. dumetorum*, *D. esculenta* and *D. hispida*, whose polyhedral granules do not exceed 15 µm.

In most cases, the sizes of the granules are higher from the proximal to the medio-distal sections, then decrease beyond that. In the median region, they decrease from the periphery to the centre. These gradients exhibit topographical irregularities that may be linked to eventual digitations and ecological developments.

According to Seidemann (1964), the granules can be classified into four groups, as follows:

1. Starch containing many fairly large granules, oval or egg-shaped, elongated rounded squares or mussel-shell shaped, sometimes with one side flattened, e.g. *D. alata*, *D. rotundata*, *D. opposita*.
2. Starch containing many fairly large granules, of rounded triangular form, sometimes elongated, rarely of trapezoidal form, eg. *D. bulbifera*, *D. cayenensis*.
3. All granules small, rounded or polyhedral, sometimes complex, as though built up from many smaller granules, e.g. *D. esculenta*, *D. hispida*, *D. dumetorum*.
4. All granules small, rounded in form, often joined together at one or more surfaces, e.g. *D. digitata*, *D. sinuata*, *D. belizensis*. Starches of this group are very similar to cassava or sweet potato starches.

Most yam starches contain, in addition to the main granules, a large number of very small grains. The distribution of particle size has only been studied in detail in the case of *D. esculenta* (Miege, 1948). Although the small grains are far more numerous than the larger ones, they only account for a few per cent of the total mass of the starch, on account of their small size.

The shapes of granules differ from each other from diverse species but also among the same species. The granules of diverse yam species have been described, for example, as round to oval (Moorthy and Nair, 1989) or polyhedral (*D. rotundata*) (Farhat *et al.*, 1999), and polygonal (*D. esculenta*) (Amani *et al.*, 2004). The position of the hilum in starch granules of *D. opposita* (C-type polymorph) was eccentric and more terminally situated with a different disruption pattern of the crystalline part during gelatinization (Cai and Wei, 2013).

The sizes of some yam species from Sri Lanka have been compared. In fact, the species exhibit dramatic variability in granular size ranging from very tiny granules of *D. esculenta* (1–5 µm) to the very large ones observed for *D. alata* (16–110 µm). Farhat *et al.* (1999) have examined the distribution in size of different yam species. *D. rotundata* and *D. cayensis* starches showed a narrow symmetrical distribution centred at 32 and 35 µm, respectively, while *D. alata* starch exhibited a non-symmetrical wider particle distribution (centred around 31 µm). *D. dumetorum* starch had an uneven distribution. Polarized light microscopy confirmed the findings and corroborated the results of Rašper and Coursey (1967).

In addition, the size can also be influenced by the shape of granules. The reported size of individual yam starch granules ranged from around 1 µm (*D. esculenta*) (Rašper and Coursey, 1967) to 90 µm (*D. alata*) (Cereda *et al.*, 2005). The average size of granules from nine genotypes of *D. alata* ranged from 28 to 40.25 µm as measured by light microscopy (Otegbayo *et al.*, 2014). *D. rotundata* and *D. cayenensis* had a narrower distribution compared to *D. alata*, as indicated by light-scattering techniques (Farhat *et al.*, 1999). *D. dumetorum* and *D. esculenta* starches possess smaller granules than other species. Granules from *D. dumetorum* were as small as 0.8 µm (Nkala *et al.*, 1994). It is significant that *D. dumetorum* and *D. esculenta* species contain lower amylose content than the other species. This indicates that the amylose component is correlated with the size of the granules. Furthermore, amylose component is mainly concentrated in the periphery of granules and the larger granules tend to have a higher amylose content (Jane *et al.*, 2003). The effect of some environmental conditions on granule size has been brought out. The granule size of starch from *D. alata* showed a steady increase with growth time up to 260 days after planting (Huang *et al.*, 2006). Drought conditions decreased the starch granule size in *D. dumetorum* (Nkala *et al.*, 1994).

Table 6.6. Polymorph and crystallinity of yam starches. (From Falade and Omiwale, 2015)

Species	No.	Polymorph type	Crystalline composition (%)	Percentage of crystallinity (%)	References
D. abyssinica	1	B			Gebre-Mariam and Schmidt, 1998
D. alata	1	C			Farhat *et al*., 1999
	1	B			Riley *et al*., 2004
	5	C and B	65–100	31–41	Amani *et al*., 2004
	2	C			Brunnschweiler *et al*., 2005
	4,a	B			Huang *et al*., 2006
	2	C and B		43	Jayakody *et al*., 2007
	8	B			Huang *et al*., 2006
D. bulbifera	3	C		27.1–53	Jiang *et al*., 2012a
D. cayenensis	1	B			Farhat *et al*., 1999
	1	B			Riley *et al*., 2004
D. cayenensis-rotundata	11	C and B	65–100	26–35	Amani *et al*., 2004
	1	C			Brunnschweiler *et al*., 2005
D. dumetorum	1	A			Farhat *et al*., 1999
	1	A	0	37	Amani *et al*., 2004
D. esculenta	1	C			Riley *et al*., 2004
	4	C and B	65–100	26–35	Amani *et al*., 2004
	3	B		49–53	Jayakody *et al*., 2007
	1	B		25.1	Bertoft *et al*., 2008
D. hispida	1	B			Tattiyakul *et al*., 2006
D. nipponica	1	C		48.5	Yuan *et al*., 2007
D. opposita	3	C		31.0–34.9	Zhou *et al*., 2012
	10	C_B and C		31.5–50.5	Huang *et al*., 2006; Wang *et al*., 2006d, 2008, a, b
D. persimillis	1	C		28.1	Jiang *et al*., 2013a
D. polygonoides	1	A			Riley *et al*., 2004
D. rotundata	1	C			Farhat *et al*., 1999
	1	B			Riley *et al*., 2004
D. septemloba	1	C		53.4	Jiang *et al*., 2012b
D. trifida	4,b	B		24–33	Jiang *et al*., 2011
D. zingiberensis	1	A			Zhou *et al*., 2012

No. = number of genotypes tested in the study; crystalline composition (%B) denotes the percentage of B-type polymorph part in the crystallinity; CB denotes the B-type polymorph dominant in the granules; a = 4 varieties harvested at four different dates (155, 190, 225, and 260 days) after planting; b = 3 of the 4 varieties harvested at 2 years.

For several varieties of *D.alata* and *D. cayenensis* grown in the Ivory Coast, Miege (1957) has shown that there are considerable variations in the properties of the starch between one part of the tuber and another. Though there are differences across specimens, in general, the largest granules are found in the central portion of the tubers, those from the ends, especially the 'head' ends, being substantially smaller, and those from the peripheral layers slightly smaller than those on the axis.

6.4.2 XRD and starch crystallinity

Wide variation in the degree of crystallinity (%) has been observed not only between the various species but also within the same species (Amani *et al.*, 2004; Jayakody *et al.*, 2007; Pérez *et al.*, 2011). For example, five genotypes from *D. alata* and *D. esculenta* possessed crystallinity ranging from 43% to 53% (Jayakody *et al.*, 2007) (Table 6.7). The observed differences in the degree of crystallinity (%) and polymorphic composition in the same study may be due to variation in the size of crystals, number of crystals that are arranged in a crystalline array, and the moisture content (Hoover, 2010). *D. opposita* starch granules having C-pattern were subjected to acid hydrolysis and the results reveal that amorphous and less ordered crystalline components are located around the hilum. A-type polymorph is located within the periphery while B-type polymorph, more susceptible to acid hydrolysis, is concentrated in the central region (Wang *et al.*, 2009). The A-type polymorph amounted to 50% in *D. cayenensis-rotundata*, *D. esculenta* and *D. alata*, which have C-type starches (Amani *et al.*, 2004). The polymorphic composition of C-type starches from many other yam species has not been studied. It was found that the XRD pattern of extracted starch is the same throughout the growth period of *D. rotundata*.

Starches from diverse yam species possess A-, B- or C-polymorphism. The edible *Dioscorea* starches (*D. alata*, *D. colocasiata*, *D. esculenta*, *D. rotundata*, *D. abysinica* and *D. cayenensis*) starch exhibited 'B' XRD patterns. Among the most studied species are *D. alata* (B- or C-type) (Amani *et al.*, 2004), *D. cayenensis-rotundata* (B- or C-type) (Amani *et al.*, 2004), and *D. opposita*

Table 6.7. X-ray diffraction patterns and crystallinity (%) of native *Dioscorea* starches. (From Jayakody *et al.*, 2007)

Starch sources	Moisture content	Crystallinity (%)	Crystalline type
D. esculenta			
Kukulala	34	49	B-type
Java-ala	35	50	B-type
Nattala	33	53	B-type
D. alata			
Hingurala	24	43	C-type
Raja-ala	28	43	B-type

All data reported on dry basis and represent the mean of at least four replicates. Values followed by the same superscript in each column are not significantly different ($P<0.05$) by Tukey's HSD test.

(C-type) (Huang *et al.*, 2006; Wang *et al.*, 2006a, 2006b, 2006c, Wang *et al.*, 2008b). *D. zingiberensis* (Zhou *et al.*, 2013), *D. polygonoides* (Riley *et al.*, 2004) and *D. dumetorum* (Farhat *et al.*, 1999; Amani *et al.*, 2004) exhibited A-type polymorphism. *D. cayenensis*, *D. rotundata* and the *D. cayenensis-rotundata* complex had either B- or C-type polymorphic pattern, and hence suggested that *D. cayenensis* and *D. rotundata* should be classified as the *D. cayenensis-rotundata* complex (Hamon and Tour, 1990). However, both 'A' and 'C' patterns have been reported for *D. dumetorum* starch. Yuan *et al.* (2007) obtained C pattern for starch from *D. nipponica*, while Hoover (2000) reported A pattern for *D. zinziberensis* starch. Pérez *et al.* (2013) obtained B patterns for five wild yam varieties and crystallinity of 35–40%. Jayakody *et al.* (2007) reported B pattern for *D. esculenta* with crystallinity of 43%, and B/C pattern for *D. alata* with crystallinity of 49%. An isolated study on the effect of environmental conditions on starch crystalline structure showed that the polymorph pattern (B-type) of starch from *D. alata* changed a little during the growth up to 260 days after planting (Huang *et al.*, 2006).

6.4.3 Amylose and amylopectin structure

Chemical structural features of amyloses from several yam species have been studied in detail.

Table 6.8. Branch chain length distribution and average chain length (CL) of native *Dioscorea* starches. (Source: Jayakody *et al.*, 2007)

Source	% distribution (dp_n)				
	6–12	13–24	25–36	37–54	CL
D. esculenta					
Kukulala	29.06 ± 0.13	53.78 ± 0.55	13.07 ± 0.29	4.09 ± 0.14	17.95 ± 0.07
Java-ala	28.95 ± 1.85	55.46 ± 2.04	11.75 ± 0.22	3.84 ± 0.41	17.65 ± 0.00
Nattala	27.85 ± 2.94	56.46 ± 2.43	11.93 ± 0.80	3.75 ± 0.29	17.75 ± 0.18
D. alata					
Hingurala	23.09 ± 0.93	57.34 ± 0.71	15.33 ± 0.28	4.24 ± 0.06	18.81 ± 0.10
Raja-ala	21.06 ± 1.21	59.26 ± 0.80	15.16 ± 0.41	4.52 ± 0.00	19.03 ± 0.14

All data reported on dry basis and represent the mean of triplicates. Values followed by the same superscript in each column are not significantly different ($P<0.05$) by Tukey's HSD test.

The degree of polymerization of the amylose fraction corresponded to 660 glucose units, and that of the amylopectin to 25. Iodine-binding capacity (IBC) of amylose that is correlated to the ability of amylose to form amylase inclusion complexes with iodine solution, was approximately 20% for *D. rotundata*, *D. dumetorum*, *D. alata* and *D. cayenensis* (Emiola and Delarosa, 1981). IBC (%) of amylose was 2–3 times that of its amylopectin fraction (Emiola and Delarosa, 1981), the percentage of β-amylolysis of amylose from *D. rotundata*, *D. dumetorum*, *D. alata* and *D. cayenensis* (Emiola and Delarosa, 1981). The average molecular weight of amyloses from *D. rotundata*, *D. dumetorum*, *D. alata* and *D. cayenensis*, measured by osmotic pressure of the acetylated starch solution-based method, ranged from 2.1 to 2.65×10^5 (Emiola and Delarosa, 1981). Rolland-Sabaté *et al.* (2003) studied the weight-average molecular weight of amyloses from *D. esculenta*, *D. dumetorum*, *D. cayenensis-rotundata* and *D. alata*, using HPSEC–MALLS–DRI detection, and obtained the value range 6.5–12.0 × 10^6. Radius of gyration (RG), indicating the hydrodynamic size of amylose ranged from 60 nm to 142 nm for *D. esculenta* and *D. alata*, respectively. The amylose structure did not indicate any species-specific pattern for *D. cayenensis-rotundata* and *D. alata* or for other species. Some other structural parameters of yam amyloses, such as the number of chains per molecule and average chain length, have not been reported.

The considerable diversity in chemical structural features of amylopectins from different yam species has been reported. IBC (%) and blue value, indicating the ability of amylopectin to bind iodine in the form of helix, ranged from 0.64% to 0.76% (Emiola and Delarosa, 1981) and 0.118–0.132 (Wang *et al.*, 2006d), respectively. The weighted average molecular weight and the z-average radius of gyration (RG) of amylopectins from *D. alata*, *D. cayenensis-rotundata*, *D. dumetorum*, *D. esculenta* and *D. trifida* (Rolland-Sabaté *et al.*, 2003; Pérez *et al.*, 2011) was in the range 0.9–6 × 10^8 and 147–247 nm, respectively. The two techniques used in the study are HPSEC-MALLS-DRI and asymmetrical field-flow fractionation (A4F) coupled with MALLS-DRI (A4F-MALLS-DRI), which gave different results for the same sample indicating the method for the molecular size determination is very relevant (Pérez *et al.*, 2011). However, the molecular weight of amylopectins from *D. alata*, *D. cayenensis-rotundata*, *D. dumetorum* and *D. esculenta* was higher than the amylopectin from the maize, potato and cassava starches. Comparing the carbohydrate content to the amount of reducing sugars revealed that the degree of polymerization (DP) of amylopectins from *D. alata* and *D. batatas* was 3469–4474 (Wang *et al.*, 2006d). The size of amylopectins does not seen to be species-specific for *D. alata* and *D. cayenensis-rotundata* (Rolland-Sabaté *et al.*, 2003), and for *D. alata* and *D. batatas* (Wang *et al.*, 2006d).

The amylopectin Mw, Rg and Mw/Mn values of amylopectin fractions for five wild yam species ranged from 174 × 106 g/mole to 237 ×

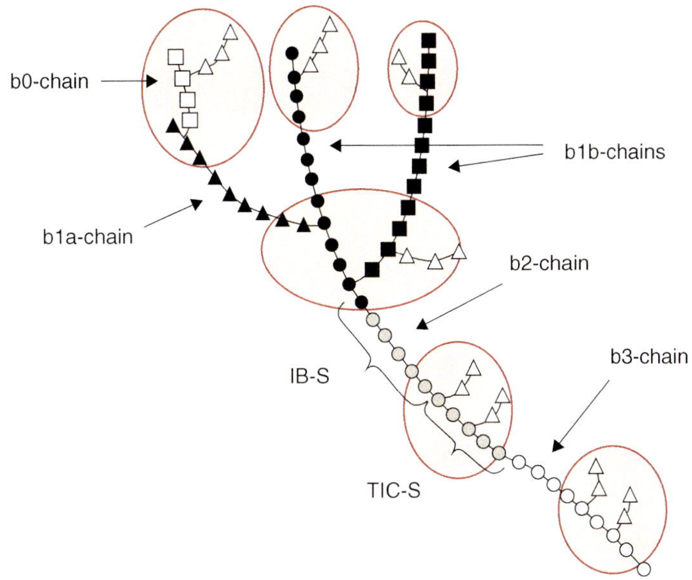

Fig. 6.3. Schematic illustration of how the individual chains and building blocks can be built up into a cluster. IB-S = interblock segment; TIC-S = total internal chain segments; b0-chains (contain no IB-S); b1a-chains (contain 1 IB-S); b1b-chains (contain 1 IB-S and TIC-S); b2-chains (contain 2 IB-S and TIC-S); b3-chains (contain 3 IB-S and TIC-S) (Bertoft *et al.*, 2012a).

10^6 g/mole, 201–233 nm, 2.0–2.2, respectively (Perez *et al.*, 2013).

By enzyme debranching, the amylopectin with enzymes uses isoamylase and/or pullulanase aids in obtaining the average unit chain length (CL) using high-performance anion-exchange chromatography with pulsed amperometric detection (HPAEC-PAD) technique. This helps with comparing the carbohydrate content to the amount of reducing sugars (Wang *et al.*, 2006d; Jayakody *et al.*, 2007; Bertoft *et al.*, 2008). Because of the polydispersity in granule size and amylopectin unit chains, CL of *D. alata*, *D. batatas* and *D. esculenta* amylopectins ranged from 18 to 28 glucosyl residues. By comparing the molecular size and CL of amylopectin, the number of chains per molecule (NC) for *D. alata* and *D. batatas* were 121.8–169.1 and 232.8–237.2, respectively (two genotypes for each) (Wang *et al.*, 2006d).

The unit chain length distribution of amylopectin from *D. esculenta* obtained by HPEAC-PAD after debranching with pullulanase and isoamylase is presented in Fig. 6.3. Two distinctive peaks (weight-based) of the unit chain distribution from amylopectins of *D. esculenta* centred at DP15 and DP48 (1 genotype) (analysed by HPEAC-PAD) (Bertoft *et al.*, 2008),and of *D. batatas* at DP 23 and DP 54 (1 genotype) (analysed by an HPSEC differential refractometer) (Hanashiro *et al.*, 2002). The unit chain length of amylopectin has been categorized into several groups based on chain length, thereby linked to the starch physical structure and properties of starch. A wider heterogeneity has been observed within the same species and also between different species such as *D. alata* and *D. esculenta* (Jayakody *et al.*, 2007). The area under curve ranging from DP 6 to 12 is 21.1–29.1%, DP 3–24 is 53.8–59.3%, DP 25–36 is 11.8–15.3% and DP 37– 54 is 3.8–4.5%, respectively. Compared with *D. esculenta* starch, *D. alata* apparently had a lower amount of short chains with DP 6–12 and a higher amount of chains with DP >12, but no relation with the polymorph type of the starch (Jayakody *et al.*, 2007). The unit chains with DP >54, which can play an important role in starch structure (Bertoft *et al.*, 2008), were not detected by Jayakody *et al.* (2007). However, extra-long unit chains (DP >100) of amylopectins (four genotypes) from *D. alata* and *D. batatas*, amounting to 2.3–8.2% (weight percentage), have been observed by gel-permeation chromatography of debranched samples (Wang *et al.*, 2006d). The molar percentage of the A-chain of amylopectin from *D. esculenta*, estimated from φ, β-limit dextrins (φ, β-LDs) of which all the A-chains appear as a maltose stub, was 55.9% (1 genotype) (Bertoft *et al.*, 2008). The C-chain length distribution of *D. batatas* amylopectin, analysed using size-exclusion chromatography after fluorescent labelling with 2-aminopyridine (Hanashiro *et al.*, 2002), followed a normal distribution centred at DP 49 with a shoulder extending at DP 27. The distribution pattern shows close similarity with that of normal starches from cereals and roots and tubers.

Removal of external portions of amylopectin starch from the non-reducing end to the nearest branches by treatment with β-amylase and/or phosphorylase led to β-limit dextrins (β-LDs) or XXI,β-LDs, which represent the internal part of amylopectin (Bertoft *et al.*, 2008). The β-limit value (or the φ,β-limit value) (%), representing the removed portion of diverse amylopectins, ranged from 50.5 to 66.7% (Emiola and Delarosa, 1981; Wang *et al.*, 2006d; Bertoft *et al.*, 2008). By comparing the average unit chain length of amylopectin with their β-LDs or φ,βLDs, the average lengths of external chains (ECL) and internal chains (ICL) could be calculated. Amylopectins from *D. alata*, *D. batatas*, and *D. esculenta* had ECL of 11.8 to 16.4 glucosyl residues and ICL of 6 to 11.1 glucosyl residues (Wang *et al.*, 2006d; Bertoft *et al.*, 2008).The ECL and ICL of amylopectin from *D. esculenta* (B-type starch polymorph) were longer than A-type polymorph starches (Bertoft *et al.*, 2008). The unit chain length distribution of the internal part of amylopectin from *D.esculenta*, as measured by HPAEC after debranching φ, β-LDs with pullulanase and isoamylase, is presented in Fig. 6.7. Two distinctive peaks (weight-based) centred at DP 10 and DP 34, with a shoulder at DP 6, have been observed (Bertoft *et al.*, 2008). Based on their internal unit chain composition, amylopectins from varied sources can be categorized into four groups (Bertoft *et al.*, 2008). Group one contains the highest amount of short internal unit chains and the lowest number of long internal unit chains. Whereas, the trend in group four was the opposite. Groups two and three values within the range observed for group one and four. The amylopectin of *D. esculenta* could be categorized as group 4 (Bertoft *et al.*, 2008). This revealed an apparent correlation between amylopectin's internal unit chain composition and unit chain length distribution of amylopectin, and the polymorph type of starch (Bertoft *et al.*, 2008).

The cluster structure of amylopectin from one genotype of *D. esculenta* having B-type polymorph has been enumerated (Bertoft *et al.*, 2012a, b). The size of clusters (DP), number of chains per cluster (NC), and number of building blocks per cluster (NBbl) were found to be 59.7, 8.5, and 3.9, respectively, the values being lower than those of starches with A-type polymorph (Bertoft *et al.*, 2012a). The interblock chain-length (IB-CL), reflecting the average distance of blocks between each other within a cluster, was 7.4 glucosyl residues, which is longer than that for amylopectins from starches with A-type polymorph. The building blocks of clusters have been classified into six structural groups (Bertoft *et al.*, 2012b). Groups one to six correspond to building blocks with the number of chains from one to six per block, respectively. The relative molar amounts of building blocks of groups 2 to 6 in the clusters from *D.esculenta* amylopectin were 51.3%, 26.4%, 11.4%, 8.2%, and 2.7%, respectively (Bertoft *et al.*, 2012a). Large building blocks in yam clusters showed a higher value of the A- to B-chain ratio compared to the cereal grains which have A-type polymorphs, indicating the higher ratio of Staudinger conformation than the Haworth conformation. Amylopectin composition and structure of clusters can be correlated to the type of polymorph, the physical structure, and the starch properties (Bertoft *et al.*, 2012a, 2012b; Vamadevan *et al.*, 2013). But, the structure of amylopectins from other yam species may be different compared with *D. esculenta*, since this study examined only one genotype from one species in detail.

6.5 Functional Properties

6.5.1 Swelling pattern and solubility

Among the tuber starches, yam starch has been found to have higher swelling characters.

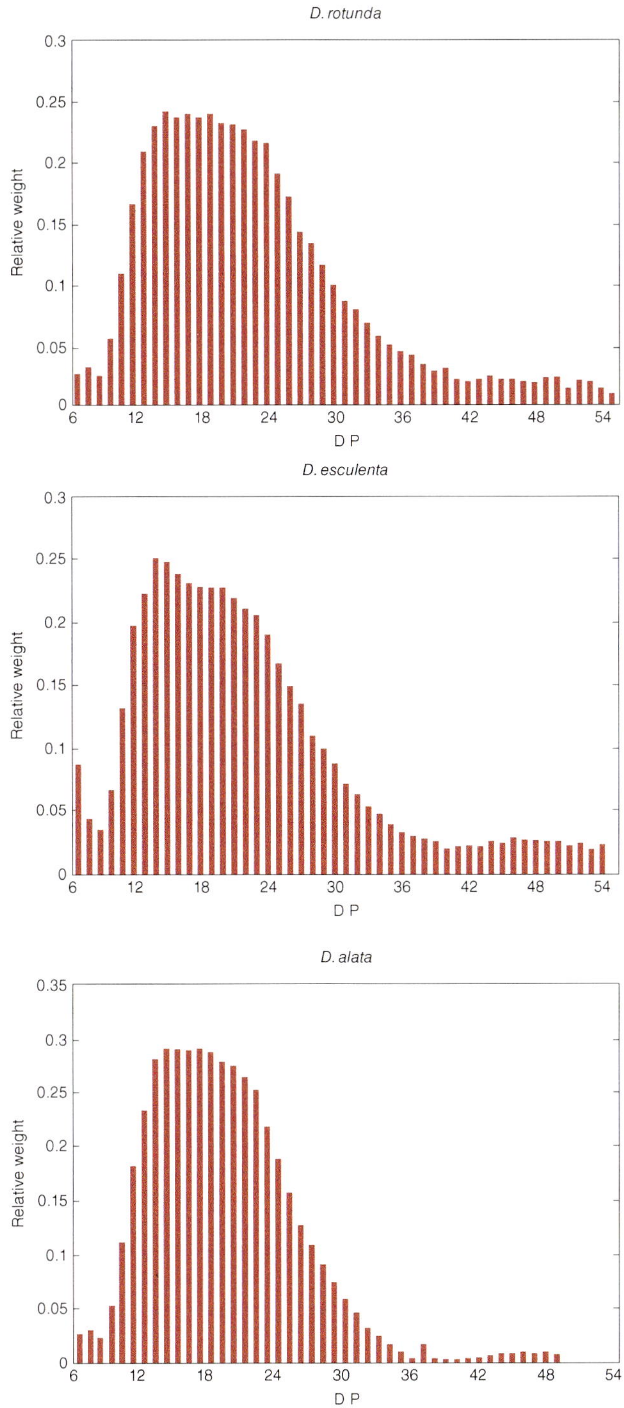

Fig. 6.4. HPEAC-PAD pattern of yam species. (From Moorthy *et al*., unpublished data)

Compared with potato starch, yam starch appeared to have lower swelling and solubility values (Gunaratne and Hoover, 2002). Great diversity has been reported in swelling (SP) and solubilization (S) properties among diverse species and also within the same species of yam (Emiola and Delarosa, 1981; Amani *et al.*, 2004; Pérez *et al.*, 2011; Jiang *et al.*, 2013b). For example, *D. cayenensis-rotundata* (11 genotypes) had SP values of 11.8 to 16.8 g/g and S values of 8.6–23.2% at 50°C, respectively (Amani *et al.*, 2004). Swelling volume values of starch of six clonal selections of *D. rotundata* showed only slight differences among them. The swelling volume of *D. abyssinica* starch increased from 10 ml/g to 23 ml/g as temperature was increased from 65°C to 85°C. The relatively lower swelling of *Dioscorea* starch compared to potato starch has been attributed to the higher intrinsic lipid content in the starch and also stronger inter-associative forces. Zhu has reviewed the swelling and solubilization of yam starches in Table 6.9.

Otegbayo *et al.* (2014) obtained swelling power between 4.5 and 10.75 among different yam species. Jiang *et al.* (2013b) studied swelling and solubility of *D. niponica* starch and recorded values of 10–2.5 for swelling power in Table 6.10. The values of SP and S at 95°C for *D. alata*, *D. cayenensis*, *D. dumetorum* and *D. rotundata* (one genotype for each) were 16.9– 21.5 g/g and 7.8–16.8%, respectively, and diverse species had different effects on SP and S to a range of temperatures (60–95°C) (Emiola and Delarosa, 1981). A comparative study revealed no difference in SP and S at 50°C and 90°C among *D. cayenensis-rotundata*, *D. alata*, *D. dumetorum* and *D. esculenta* (Amani *et al.*, 2004), even though *D. dumetorum* and *D. esculenta* have smaller granule size and lower amylose contents. This suggests that factors other than granule size and amylose content play important roles in swelling and solubilization. Environmental factors can influence the swelling and solubilization of granules (for example, starch granules from the crop harvested in 2005 were different compared with 2009) (Pérez *et al.*, 2011). The structure of the leached components from the granules of six genotypes from three species (*D. alata*, *D. cayenensis-rotundata* and *D. esculenta*) at 90°C was studied by the HPSEC–MALLS–DRI system (Rolland-Sabaté *et al.*, 2003). The leached fractions had a weight-average molecular weight from 10.47 to 1.14×10^6 and RG 55–96 nm, which are much smaller than the amylopectin and amylose fractions (Rolland-Sabaté *et al.*, 2003). Since the molecules are smaller there is less possibility of entangling with other molecules to get fixed, thus they have a greater tendency to leach out during swelling.

The solubility of *D. abyssinica* starch was found to be enhanced with temperature and resembled that of maize starch. The solubility of starch of yams pre-treated with various chemicals was affected to different extents by the chemicals used and also the concentration, and the values were between 18% and 32%. Both guar gum and locust bean gum affected the swelling and solubility of yam starches (Table 6.11).

6.5.2 Viscosity

All the yam starches showed a characteristic pattern of slow rise in viscosity and even after 95°C an increase in viscosity was noticed. This implies that all the granules do not gelatinize at 95°C and some of them gelatinize only during the holding period. In this respect, the yam starches resemble potato starch (Moorthy *et al.*, 2018).

The gelatinization characteristics of yam starch show wide variation among different species (Coursey, 1967; Rašper and Coursey, 1969; Treche, 1989; Faboya and Asagbra, 1990). High levels of maximal viscosity (470–980) and swelling (24.9) reached by *D. cayenesis-rotundata* were observed compared to *D. dumetorum* (185 and 13.9). The viscosity of several yam starches produced in west Africa have been studied by Coursey (1967). The results obtained are given in Table 6.12.

It was observed that, upon heating, the viscosity of the starch solution did not decrease significantly. Similar observations have been made in the case of *D. hispida* starch by Rao and Beri, 1952a, b) and by the same authors (1955) for *D. alata*, *D. sativa* and *D. anguina*. Their viscosities are comparatively low, except for the value recorded for *D. rotundata*.

Studies of starch from a west African yam of uncertain identification have been made using the Brabender amylograph by Hollo (1964).

Table 6.9. Swelling and solubilization of yam starches. (From Zhu, 2016)

Species	No.	Parameters	Temperatures (°C)					References
			60	70	80	90	95	
D. abyssinica	1	SP	10(65)	17 (75)	23 (85)			Gebre-Mariam and Schmidt, 1998
	1	S	3.5 (65)	5.5 (75)	11 (85)			
D. alata	1	SP	5.01	7.05	11.50	15.45	20.5	Emiola and Delarosa, 1981
		S	2.70	3.5	4.55	6.98	7.8	
	5	SP	15.4–16.8 (50)			13.8–16.0		Amani *et al*., 2004
		S	8.1–21.4 (50)			7.3–13.5		
D. bulbifera	3	SP	5.4–6	6.1–6.5	7.9–10.6	11.7–18.6		Jiang *et al*., 2012a
		S	3.1–4.1	5.2–5.6	11.3–13.8	12.5–18.2		
D. cayenensis	1	SP	8.25	9.88	11.97	14.83	16.9	Emiola and Delarosa, 1981
		S	1.3	2.4	4.25	7.9	13.8	
D. cayenensis-rotundata	11	SP	11.8–16.8 (50)			10.8–16.4		Amani *et al*., 2004
	1	S	8.6–23.2 (50)			6.1–11.4		
D. dumetorum	1	SP	7.38	8.96	12.15	15.52	18.65	Emiola and Delarosa (1981)
		S	2.64	4.35	7.02	13.05	8.42	
	1	SP	14.9 (50)			13.9		Amani *et al*. (2004)
		S	14.1 (50)			8.7		
D. esculenta	4	SP	14.8–16.1 (50)			13.9–14.8		Amani *et al*. (2004)
		S	7.1–16.5 (50)			7.3–12.5		
D. hispida	1	SP				15.6		Tattiyakul *et al*. (2006)
		AML				15.8		
D. nipponica	1	SP	6.4 (65)	6.9 (75)	7.4 (85)	8.9 (95)		Jiang *et al*. (2012b)
		S	3.6 (65)	5.2 (75)	5.4 (85)	11.1 (95)		

Continued

Table 6.9. Continued

Species	No.	Parameters	Temperatures (°C)					References
	3	SP	3.9–6.6 (65)	5.9–8.7 (75)	8.7–11.6 (85)	9.8–12.6 (95)		Jiang *et al*. (2013b)
		S	2.0–4.5 (65)	6.8–9.0 (75)	12.2–17.4 (85)	16.6–21.6 (95)		
D. opposita	4	SP				9.6–12.4		Wang *et al*. (2008a)
		S				8.7–10.4		
	3	SP				10.1–11.3		Zhou *et al*. (2012)
		S				10.1–11		
	9	SP	2.74–4.33 (65)	4.52–7.58 (75)	10.01–13.98 (85)	10.83–16.45 (95)		Jiang *et al*. (2013b)
		S	1.76–3.59 (65)	3.71–9.09 (75)	7.96–18.03(85)	12.21–19.24 (95)		
D. persimilis	1	SP	5.77 (65)	4.52–7.58 (75)	10.01–13.98 (85)	10.83–16.45 (95)		Jiang *et al*. (2013b)
		S	1.76–3.59 (65)	3.71–9.09 (75)	7.96–18.03 (85)	12.21–19.24 (95)		
D. septemloba	1	SP	7.7 (65)	8.4 (75)	11.9 (85)	12.1 (95)		Jiang *et al*. (2012b)
		S	3.3 (65)	3.6 (75)	11.0 (85)	14.4 (95)		
D. rotundata	1	SP	3.1	7.5	12.55	16.5	21.5	Emiola and Delarosa (1981)
		S	1.8	4.21	6.05	8.42	11.9	
D. trifida	4 A	SP		8.6–26.5 (75)		20.5–32.2		Jiang *et al*. (2011)
		S		2.1–4.3 (75)		2.1–4.3		

No. = number of genotypes tested in the study; SP = swelling power; S = solubility index; AML = amylase leaching; figures in parentheses denote the actual experimental temperature used; A = 3 of the four varieties harvested at 2 years; Amani *et al*. (2004) used 23-min. incubation time at 50°C and 12 min. at 90°C.

Table 6.10. Swelling power (SP) and solubility (SOL) of starches separated from *D. nipponica makino* of different origins at different temperatures. (From Jiang *et al.*, 2013b)

	Swelling power (g/g)			Solubility (%)		
Samples	Funiu	Qingyuan	Yichun	Funiu	Qingyuan	Yichun
65 (°C)	3.91 ± 0.82	6.40 ± 0.71	6.58 ± 0.13	2.03 ± 0.53	3.48 ± 0.27	4.49 ± 0.80
75 (°C)	5.94 ± 0.27	7.46 ± 0.51	8.67 ± 0.07	6.81 ± 1.82	7.97 ± 1.06	8.96 ± 0.36
85 (°C)	10.35 ± 0.19	11.61 ± 1.21	8.68 ± 0.50	17.41 ± 1.12	13.70 ± 1.41	12.16 ± 1.03
95 (°C)	11.61 ± 0.55	12.56 ± 0.52	9.77 ± 0.05	21.55 ± 1.33	20.93 ± 2.14	16.63 ± 0.23

All the values and mean ± SD of the three replicate analyses. Means within columns with different letters are significantly different ($P<0.05$).

The amylograph curves obtained by him were generally similar to those for potato starch. The transition temperature was 77.5°C and the maximum viscosity 290 BU. As in the case of the samples examined by the other values, little change in viscosity occurred on heating of the starch solution.

It has been shown (Rašper and Coursey, 1967) that out of a number of west African yam starches, only that of *D. rotundata* gave a high viscosity maximum of 470–690 Brabender units (BU), compared with *D. alata*, 100–200 BU, *D. esculenta*, 55 BU and *D. dumetorum*, 25 BU. Both *D. rotundata* and *D. alata* starches, however, were found to have high gel strength compared with the other species.

The viscosities of starch of four varieties of *D. esculenta* extracted using water and ammonia solution showed only minor differences between the varieties (800–950 BU), but the viscosities of starch extracted using ammonia solution were much higher than those obtained by water extraction. For most yam starches, the viscosity breakdown was quite low in spite of the high viscosity levels. The yam starches contain three to four times more phosphorus than cassava and aroid starches. It has been reported that the phosphate linkages in potato starch are responsible for its high viscosity and such effects may also be important in the yam starches. Zhu (2016) has reviewed various reports on yam starches (see Table 6.13).

Gebre-Mariam and Schmidt (1998) have obtained 781, 756, 1282 BU for peak viscosity, hot paste viscosity and cold viscosity respectively for *D. cayensis* starch indicating low breakdown, but high setback for this starch. Pasting properties of yam tuber flour of 65 accessions in Southern Ethiopia have been investigated and are presented in Table 6.13 by (Tamiru *et al.*, 2008).

Effects of guar gum (GG) and locust bean gum (LBG) on the pasting properties of yam

Table 6.11. Effects of guar gum (GG) and locust bean gum (LBG) on the swelling power and solubility of root and tuber starches. (From Huang, 2009)

Starch + gum	Swelling power (g/g starch)	Solubility (%)
Yam	10.23	7.36
Yam + GG	9.65	6.92
Yam + LBG	9.73	7.15
Yam bean	14.92	11.28
Yam bean + GG	14.02	10.91
Yam bean + LBG	13.78	11.17
Potato	42.92	15.28

Table 6.12. Viscosity characteristics of yam starches. (Source: Coursey, 1967)

Species	Viscosity (BU)	Gelatinization temperatures (°C)
D. alata	150	88.0
D. rotundata	720	77.5
D. cayenensis	220	90.5
D. bulbifera	40	89.0
D. esculenta	65	79.0
D. dumetorum	120	85.0

Table 6.13. Pasting properties. (Source: Zhu, 2015)

Species	No.	Instrument	Starch conc. (%)	PV	BD	SB	PT (°C)	References
D. abyssinica	1	BV	6	781	25	501	72.9	Gebre-Mariam and Schmidt (1998)
D. alata	9	RVA		207.2–417.7	28.4–104.3	79–230	83.5–85.8	Otegbayo *et al*. (2014)
	2	BV	4.3	110–200			81–85	Rašper and Coursey (1967)
	1	RVA	9.1	3230 mPas	196 mPas	2782 mPas	79.8	Farhat *et al*. (1999)
	1 A	RVA	9.1	152–204	8–39	87–148	75.5–78.4	Cereda *et al*. (2005)
	4 B	RVA	8	217–708	0–370	63–337	75.1–82.2	Huang *et al*. (2006)
D. bulbifera	5	RVA		251–355	34–69	39–283	84.0–87.7	Otegbayo *et al*. (2014)
D. cayenensis	1	RVA		304	136	67.9	88.4	Otegbayo *et al*. (2014)
	1	RVA	9.1	3893 mPas	820 mPas	892 mPas	75	Farhat *et al*. (1999)
D. dumetorum	1	RVA		462	94	188	79.5	Otegbayo *et al*. (2014)
	1	BV	4.3	25			82	Rašper and Coursey (1967)
	1	RVA	9.1	2028 mPas	438 mPas	688 mPas	83.1	Farhat *et al*. (1999)
	2 C	RVA		200–374	97–217	50–112		Akinoso and Abiodun (2013)
D. esculenta	1	BV	4.3	55			82	Rašper and Coursey (1967)
D. hispida	1	RVA	9.2	361	76	92	78.3	Tattiyakul *et al*. (2006)
D. nipponica	1	RVA	6	362	179	222	73	Yuan *et al*. (2007)
D. opposita	3	RVA	8.9	291–415	80–190	123–148	83.5–87.3	Zhou *et al*. (2012)
D. rotundata	27	RVA		177.4–528.9	6.7–229	40–339	80–86.5	Otegbayo *et al*. (2014)
	5	BV	4.3	470–650			76–80	Rašper and Coursey (1967)
	1	RVA	9.1	3237 mPas	458 mPas	1448 mPas	78.2	Farhat *et al*. (1999)
	6	BV	5 or 6	325–550 (5%) 680–920 (6%)				Moorthy and Nair (1989)

Continued

Table 6.13. Continued

Species	No.	Instrument	Starch conc. (%)	PV	BD	SB	PT (°C)	References
D. septemloba	1	RVA	11.	1148	1025	38	51	Jiang *et al*. (2012b)
D. trifida	4D	RVA		1428–2252 cP	220–966 cP	−175–1071 cP	75.5–80	Jiang *et al*. (2011)
D. zingiberensis	1	BV	8	309	46	161	73.4	Zhou *et al*. (2013)

No. = number of genotypes tested in the study; BV = Brabender viscograph, unit as BU; RVA = rapid viscoanalyzer, unit as RVA except where specified; PV = peak viscosity; A = 4 extraction methods; B = 4 varieties harvested at four different dates (155, 190, 225 and 260 days) after planting; C = 2 varieties harvested at five different months (7, 8, 9, 10, 11) after planting; D = 3 of the four varieties harvested at 2 years.

Table 6.14. Summary statistics and mean comparison of pasting properties of yam tuber flour (*Dioscorea* sp.) among 65 accessions collected in southern Ethiopia based on type, maturity time and geographic origin. (Source: Tamiru *et al.*, 2008)

Pasting properties								
Summary	N	Peak viscosity ($_c$P)	Trough viscosity ($_c$P)	Breakdown viscosity ($_c$P)	Final viscosity ($_c$P)	Setback ($_c$P)	Peak time (min)	Pasting temperatures (°C)
Mean	65	5329.9	2321.6	3008.3	4995.4	2678.8	4.9	73.9
SD	65	784.1	607.6	793.0	575.5	766.1	0.35	2.3
CV (%)	65	14.7	26.2	26.4	11.5	28.6	7.2	3.1
Range	65	3725.5–7336.0	1288.0–3968.0	86.5–4206.0	3424.5–6092.5	871.3–4085.0	4.0–6.7	66.7–78.7
Grouping criterion								
Type of tuber								
Aerial yam	5	4444.1^{a}	3272.4^{b}	1771.7^{a}	4536.7^{a}	1264.3^{a}	5.0^{a}	67.9^{a}
Underground	60	5403.8^{b}	2242.4^{a}	3161.8^{b}	5033.6^{a}	2791.2^{b}	4.8^{a}	74.4^{b}
Maturity time								
Early	32	5555.0^{b}	2307.6^{b}	3247.4^{a}	5070.0^{a}	2762.4^{a}	4.9^{b}	74.3^{a}
Late	15	4884.3^{a}	1901.8^{a}	2982.5^{a}	4853.2^{a}	2951.4^{a}	4.5^{a}	73.6^{a}
Collection area								
Areka	13	5630.8^{a}	2474.8^{a}	3156.0^{a}	5152.2^{a}	2677.4^{ab}	4.9^{a}	74.8^{a}
Gamo-gofa	4	5670.3^{a}	2524.0^{a}	3146.3^{a}	4941.0^{a}	2417.0^{a}	4.7^{a}	73.9^{a}
Gedeo	7	5400.2^{a}	2011.2^{a}	3389.3^{a}	5262.4^{a}	3251.2^{b}	4.8^{a}	75.0^{a}
Sidama	11	5611.5^{a}	2469.2^{a}	3142.3^{a}	5306.4^{a}	2837.2^{ab}	4.9^{a}	74.5^{a}
Wolayita	25	5152.7^{a}	2041.4^{a}	3111.3^{a}	4802.7^{a}	2761.2^{ab}	4.7^{a}	74.1^{a}

Values in the same column and group followed by different letters are significantly different by Duncan test ($P<0.05$).
[a] Excluding accessions from Areka (because of unknown maturity time) and aerial yam accessions;
[b] Excluding aerial yam accessions;
[c] Peak viscosity–trough viscosity;
[d] final viscosity–trough viscosity;
[e] Time at which peak viscosity is reached.

Table 6.15. Effects of guar gum (GG) and locust bean gum (LBG) on the pasting properties of yam starches. (From Huang, 2009)

Parameters of pasting behaviours (RVU)						
Tuber starch + gum	Pasting temperature (°C)	Peak viscosity (P)	95°C viscosity (H)[2]	Final viscosity (F)[3]	Breakdown (P-H)	Setback (F-H)
Yam	77.3	268	256	367	12	111
Yam + GG	74.6	504	480	892	24	412
Yam + LBG	75.9	428	403	880	24	477
Yam bean	69.3	206	148	238	58	90
Yam bean + GG	66.8	376	221	337	155	116
Yam bean + LBG	68.2	354	233	345	121	112
Potato	63.5	458	146	214	312	68

starches has been examined by (Huang, 2009) (Table 6.15).

Farhat *et al.* (1999) have studied the viscosity properties of four yam species. The peak viscosities ranged from 2028 mPas for *D. dumetorum* to 3893 mPas for *D. cayenensis*, compared to 8900 mPas for potato and 3134 mPas for cassava starches. The breakdown in viscosity was much lower in comparison to cassava and potato starches and in conformity with the results obtained with the Brabender viscograph. The final viscosity was slightly higher than peak viscosity indicating a tendency to retrograde. Among these species, *D. dumetorum* starch had all the viscometric parameters lower than the other three and in support of earlier reports. Nkala *et al.* (1994) did not observe much difference in the viscosity parameters of *D. dumetorum* starch during the dry and wet seasons. The RVA profiles of starch of some varieties of *D. alata*, *D. esculenta* and *D. rotundata* harvested at different maturities have revealed that maturity did not affect the rheological properties to any major extent. Pre-treatment prior to air drying on pasting properties of flours of different varieties of yam have been compared by Falade *et al.* (2015) in Table 6.16.

Species had a major influence on the pasting properties of starch (Rašper and Coursey, 1967; Farhat *et al.*, 1999; Jiang *et al.*, 2012b; Otegbayo *et al.*, 2014). In addition, differences were also observed within the same species (Rašper and Coursey, 1967; Moorthy and Nair, 1989; Huang *et al.*, 2006; Yuan *et al.*, 2007; Pérez *et al.*, 2011; Zhou *et al.*, 2012). Moorthy and Nair (1989) studied the viscosity of six varieties of *D. rotundata*. For example, PV of starches from *D. alata*, *D. rotundata*, *D. cayenensis* and *D. dumetorum* (one genotype for each) ranged from 2028 to 3893 mPas (Farhat *et al.*, 1999). PV of starches from 27 genotypes of *D. rotundata* ranged from 177 to 529 RVU (Otegbayo *et al.*, 2014). Using cluster analysis, the pasting properties of 43 genotypes from five species can be categorized into three clusters. Starch from *D. rotundata* fell under cluster I with high PV, low SB and PT. Cluster II included starch from *D. alata* and *D. bulbifera* and a few starches from *D. rotundata*, which had high PT, SB and low PV. Cluster III included starches from *D. rotundata* and a few from *D. alata* and *D. bulbifera*, which had high CPV and SB (Otegbayo *et al.*, 2014). This clustering analysis indicated that the pasting properties of starches from yam are non-species-specific, though this study had included only one genotype from *D. dumetorum* and *D. cayenensis*. Starch from *D. alata* exhibited high viscosity under heat treatment and mechanical shearing in neutral to slightly acidic conditions as revealed by pasting analysis (Mali *et al.*, 2003), suggesting its potential use in some acidic food products which also require thermal processing. Starch from *D. zingiberensis* had a very stable viscosity with little shear-thinning when kept at 93°C for 5 minutes (Zhou *et al.*, 2013). Some reports showed that *D. dumetorum* and *D. esculenta* tended to have lower viscosity during pasting than other

Table 6.16. Effect of pre-treatments prior to air drying on pasting properties of flours of yam (*Dioscorea rotundata* and *D. cayenensis*) varieties. (From Falade and Omiwale, 2015)

Cultivar	Pre-treatment	Peak viscosity (RVU)	Trough viscosity (RVU)	Breakdown viscosity (RVU)	Final viscosity (RVU)	Setback viscosity (RVU)	Peak time (min)	Pasting temp (°C)
Abuja	Control	233.75 ± 13.68	225.21 ± 15.15	8.54 ± 1.47	335.41 ± 26.64	110.20 ± 11.49	6.05 ± 1.22	49.60 ± 0.07
Abuja	Blanched	174.96 ± 17.73	166.67 ± 16.97	8.29 ± 0.76	226.38 ± 27.75	59.71 ± 10.78	6.91 ± 0.00	49.48 ± 0.11
Abuja	Sulfited	46.96 ± 0.88	42.54 ± 1.71	4.42 ± 2.59	58.71 ± 0.06	16.17 ± 1.77	6.67 ± 0.16	49.53 ± 0.11
Abuja	Cooked	19.58 ± 0.35	18.55 ± 0.18	1.05 ± 0.18	29.66 ± 0.57	11.13 ± 0.42	6.75 ± 0.09	49.63 ± 0.11
Alo	Control	388.25 ± 49.26	289.58 ± 44.90	98.67 ± 4.36	459.46 ± 29.76	169.88 ± 15.14	5.05 ± 0.04	49.43 ± 0.11
Alo	Blanched	184.83 ± 30.88	159.12 ± 20.79	25.71 ± 10.08	237.45 ± 26.69	78.34 ± 5.89	6.97 ± 0.00	49.38 ± 0.11
Alo	Sulfited	163.92 ± 6.36	143.12 ± 5.01	20.79 ± 1.36	225.54 ± 5.71	82.42 ± 0.07	6.97 ± 0.00	49.38 ± 0.04
Alo	Cooked	61.88 ± 10.89	51.71 ± 10.55	10.17 ± 0.35	96.46 ± 10.55	44.75 ± 0.00	6.91 ± 0.04	49.53 ± 0.11
Amula	Control	264.50 ± 4.24	229.71 ± 3.95	34.79 ± 0.29	283.34 ± 4.36	53.63 ± 0.42	5.25 ± 0.05	49.45 ± 0.00
Amula	Blanched	66.96 ± 1.00	54.29 ± 1.00	12.67 ± 0.00	125.00 ± 3.29	70.71 ± 2.29	6.96 ± 0.02	49.73 ± 0.32
Amula	Sulfited	131.84 ± 6.95	127.67 ± 4.59	4.17 ± 2.36	209.92 ± 0.71	82.25 ± 3.89	6.50 ± 0.11	49.48 ± 0.04
Amula	Cooked	35.33 ± 3.89	27.25 ± 3.07	8.09 ± 0.83	70.25 ± 5.66	43.00 ± 2.59	6.97 ± 0.00	49.35 ± 0.00
Gbangi	Control	296.91 ± 13.20	255.67 ± 53.51	41.25 ± 40.31	448.00 ± 3.65	192.33 ± 49.85	5.00 ± 0.11	49.35 ± 0.14
Gbangi	Blanched	298.67 ± 5.66	278.91 ± 0.47	19.75 ± 6.12	418.21 ± 11.26	139.29 ± 11.72	6.91 ± 0.04	49.43 ± 0.11
Gbangi	Sulfited	79.00 ± 7.18	68.33 ± 6.01	10.67 ± 1.18	113.59 ± 9.07	45.25 ± 3.07	6.96 ± 0.02	49.30 ± 0.14
Gbangi	Cooked	92.12 ± 6.89	88.00 ± 6.48	4.12 ± 0.42	122.71 ± 8.66	34.71 ± 2.18	6.91 ± 0.04	49.55 ± 0.07

Moorthy and Nair (1989) Studies on *Dioscorea rotundata* starch properties. *Starch/Stärke* 41(3), 81–83.
The viscosity properties of six varieties of *D. rotundata* have been studied by Moorthy and Nair (1989).

common species (*D. alata*, *D. rotundata*, *D. cayenensis*) (Rašper and Coursey, 1967 ; Farhat *et al.*, 1999; Amani *et al.*, 2004), and can be attributed to the smaller granules and lower amylose contents of the former two. Wang *et al.* (2006d) correlated the composition of starch and molecular structure of amylopectin to the pasting properties (Wang *et al.*, 2006d). Rašper and Coursey (1967) showed that amylopectin had an influence on the pasting properties and gel strength for starches containing similar amylose contents. Analysis of four genotypes from two species (*D. alata* and *D. batatas*) indicated that the PV was positively related to CL, ICL, ECL and phosphate content, and negatively related to NC (Wang *et al.*, 2006d). Comparative studies have been conducted on yam starches and starches from other sources (Farhat *et al.*, 1999; Brunnschweiler *et al.*, 2005; Pérez *et al.*, 2011). Waxy starches from *D. trifida* had lower viscosity compared to normal and waxy potato starches, but had higher viscosity than starches from normal and waxy maize, rice and cassava (Pérez *et al.*, 2011). Starches from *D. alata*, *D. rotundata* and *D. cayenensis* had lower PV and higher CPV content compared with potato and cassava starches (Farhat *et al.*, 1999). On the other hand, the viscosity of starches from these yams had a slight increase upon heating that stabilized with further thermal treatment (Brunnschweiler *et al.*, 2005). This dissimilar behaviour was partially attributed to the difference in the changes in granular structure upon heating (Brunnschweiler *et al.*, 2005; N'daKouamé *et al.*, 2011). Agronomic practices exerted a great influence on the pasting properties of starch. Harvesting time has been linked to the pasting of yam starch (Huang *et al.*, 2006; Pérez *et al.*, 2011; Akinoso and Abiodun, 2013). Harvesting time (7–11 months after planting) and genetic origin drastically affect the pasting properties of starch from *D. dumetorum*, either increasing or decreasing the PV, SB and BD of starch (Akinoso and Abiodun, 2013). PV of starches from four varieties of *D. alata* increased, and pasting time (PT) decreased gradually with growth period up to 260 days after planting, while BD and SB either increased or decreased, depending on the variety and harvesting time. The increasing PV of starch has been traced to the increasing granule size (Huang *et al.*, 2006). Another study showed that starches of *D. trifida* harvested at different years, 2005 and 2009, had different pasting behaviours (Pérez *et al.*, 2011).

Average peak viscosity of cooked and sulphated starch from four varieties of yam, Abuja, Amula, Gbangi and Alo, was 19.58 and 16.96, 35.33 and 131.84, 92.18 and 79.00, and 61.88 and 163.92 RVU, respectively. Peak, trough, breakdown and setback viscosities of pre-treated yams were significantly lower than the untreated control. Flours of pre-treated Alo and Gbangi varieties showed higher values of peak, trough, breakdown and setback viscosities than Amula and Abuja varieties (Falade and Omiwale, 2015).

6.5.3 Rheological properties

Rheological tests other than pasting, such as flow and dynamic oscillatory analysis, have been conducted on yam starches to gain

Table 6.17. Viscosity properties and pasting temperature of *Dioscorea rotundata* starch. (Source: Moorthy and Nair, 1989)

Variety	2% viscosity (s)	Paste viscosity (BU) 1)pv (5%)	Pv (6%)	2)V_{97} (5%)	V_{97} (6%)	3)V_H (5%)	V_H (6%)	4)V_C (5%)	V_C (6%)	Pasting temp. (°C)
T3-1	37.5	470	730	450	720	480	770	620	1000	83–97
I-146	42.0	435	700	420	660	435	740	500	1000	81–97
U-195 (2)	46.0	550	920	570	900	550	1000	750	1000	80–97
I-212	45.0	500	770	490	760	500	800	770	1000	79–97
183	37.0	370	680	360	650	370	610	480	720	83–97
184	38.5	325	680	320	630	350	675	450	800	82–97

Table 6.18. Thermal properties of different yam starches was reviewed by Zhu (2016).

Species	Number of genotypes	To (°C)	Tp (°C)	Tc (°C)	ΔH (J/g)	References
D. abyssinica	1	64.2	68.2	74.8	19.2	Gebre-Mariam and Schmidt (1998)
D. alata	1	65	69	71.5		Emiola and Delarosa (1981)
	1	76.5	78.8	81.9	19.8	Farhat *et al.* (1999)
	1	70.2	74.4	80.9	20.9	Valetudie *et al.* (1995)
	1	69–69.5	77–79	72.5–73	11.5–12	Freitas *et al.* (2004)
	5	74.7–75.8			14.4–18.1	Amani *et al.* (2004)
	2 A		77.8		19.9–24.0	Brunnschweiler *et al.* (2005)
	4 B	68.8–75.8	73.5–82.4	79.9–87.8	11.8–18	Huang *et al.* (2006)
	2	75.5–78.2	78.4–85.1	85.7–92.9	18.6–19	Jayakody *et al.* (2007)
	8	71.4–74.2	74.3–76.9	77.1–80.2	4.8–12.8	Riley *et al.* (2006)
D. bulbifera	3	71.5–75.3	75.7–78.0	82.5–93.1		Jiang *et al.* (2012a)
D. cayenensis	1	68	72	74.5		Emiola and Delarosa (1981)
	1	69.4	72.9	76.7	19.8	Farhat *et al.* (1999)
D. cayenensis rotundata	11	69.9–76.9			13.7–16.7	Amani *et al.* (2004)
	3	67.6–74.2	72.9–77.9		10.6–14.6	N'daKouamé *et al.* (2011)
D. dumetorum	1	65.5	68	72.5		Emiola and Delarosa (1981)
	1	78.1	81.3	86.4	18.6	Farhat *et al.* (1999)
	1	81.7			16.7	Amani *et al.* (2004)
D. esculenta	4	72.3–73.6			18.0–20.3	Amani *et al.* (2004)
	3	72.3–72.6	75–75.7	81.7–85.4	17.3–18.1	Jayakody *et al.* (2007)
	1	74.8	77.5	81.3	15.7	Vamadevan *et al.* (2013)
D. nipponica	1	67.4	76.0	81.0	18.6	Yuan *et al.* (2007)
	3	68.4–71.9	73.8–77.8	78.9–82.5		Jiang *et al.* (2013b)
D. opposita	14	70.2–75.8	77.5–81.5	82.1–87.2	6.5–12.5	Wang *et al.* (2008a)

Continued

Table 6.18. Continued

Species	Number of genotypes	To (°C)	Tp (°C)	Tc (°C)	ΔH (J/g)	References
	9	63.9–69.5	70.4–79.2	84.4–91.0		Jiang *et al.* (2013b)
D. persimilis	1	64.9	68.5	72.4		Jiang *et al.* (2013b)
D. rotundata	1	63.5	66	71		Emiola and Delarosa (1981)
	1	71.5	74.8	80.5	19.8	Farhat *et al.* (1999)
D. septemloba	1	75.1	80.7	84.7		Jiang *et al.* (2012b)
D. trifida	4 C	69.1–73.4	73.0–77.3	79.8–83.6	22.4–25.3	Pérez *et al.* (2011)
D. zingiberensis	1	70.5	78.5	86	12.73	Zhou *et al.* (2013)

The gelatinization temperatures (T_{onset} T_{peak} and T_{end}) were averages of four tests with different starch-to-water ratios (1:2 to 1:6), and ΔH was obtained as water-to-starch ratio of 6:1.

information on starch gelatinization and retrogradation. Differences in rheological properties were noted among starches from yams compared to other tubers and roots (Freitas *et al.*, 2004; Brunnschweiler *et al.*, 2005; Moorthy *et al.*, 2008; N'da Kouamé *et al.*, 2011) in Table 6.19 and Fig. 6.5. The instrumentation and experimental conditions varied greatly among various studies, making direct comparison of data often difficult. For example, starch rheology is highly influenced by the water content (Yeh *et al.*, 2009). An increase in water content from 60% to 90% reduced the maximum G′ of starch from *D. alata* and increased with temperature where G′ reached the maximum during temperature sweeping in the dynamic rheological test (Yeh *et al.*, 2009). Nevertheless, the same authors reported a variation in rheological properties (when tested using temperature sweep, time dependence and strain sweep methodologies) among yam starches from *D. esculenta*, *D. rotundata* and *D. alata*. Comparative analysis placed yam starch uniquely among starches from other botanical origins (Moorthy *et al.*, 2008). Dynamic oscillatory analysis, including temperature (15–85°C) and frequency (0.01–10 Hz) sweeps, showed that yam starch (*D. alata*) had a higher gelatinization temperature, higher G′ and G″, and stronger gel compared to cassava starch (Freitas *et al.*, 2004). This could be due to the higher amylose content of the former. Flow analysis (change in viscosity under a range of shear stress) and the pasting data showed that the gels from *D. alata* and *D. cayenensis-rotundata* had high viscosity at low stress and strong shear-thinning behaviour compared with cassava potato starch gels (Brunnschweiler *et al.*, 2005; N'da Kouamé *et al.*, 2011). Studies on flow curves of yam starches in comparison with potato and cassava starches showed the higher viscosity of yam starches at the beginning of shearing. This was attributed to better preservation of granular structure upon heating and the subsequent destruction of the granule remnant by stronger shearing during the rheological analysis, as evidenced by microscopic analysis (Brunnschweiler *et al.*, 2005; N'da Kouamé *et al.*, 2011). But, there is a lack of studies correlating the rheological properties with granular and molecular structure of starch.

Table 6.19. Rheological properties of Yam starches. (Source: Moorthy *et al.*, 2008)

Species	G (3%)	G (4%)	G (5%)
D. alata	6–12	36–53	54–92
D. esculenta	0.02–0.1	0.1–0.28	0.6–5
D. rotundata	12–35	23–39	48–66

6.5.4 Retrogradation

The retrogradation process involves two steps, namely initial gel firming due to amylose matrix gel followed by gradual increase in gel firmness attributed to amylopectin recrystallization. This phenomenon involves increase in crystallinity, gel firmness, syneresis and formation of B-type polymorph during storage (Miles *et al.*, 1985). Retrogradation has been investigated by using several techniques such as the gelation behaviour, WAXS, gel texture analysis, DSC, syneresis and gel turbidity. Wide variability has been observed in gelling capacity and gel properties of starches from different species and also genotypes within the same species (Amani *et al.*, 2005; Otegbayo *et al.*, 2014). The least gelation concentrations (LGC) of starches of 43 genotypes from five yam species were found to be 4–16% (Otegbayo *et al.*, 2014). Gels of starches isolated from 21 genotypes of four species were subjected to various processes, including thermal treatment, long-term freezing, refrigeration, high-speed shearing, and acidic treatment. There was wide variability in the stability of the gels during these processes and it was noticed that some genotypes had stability comparable to modified starches, suggesting their potential in various food-processing systems as replacements for modified starches (Amani *et al.*, 2005).

Firming of gel from *D. alata* and *D. cayenensis-rotundata* starches measured by uni-axial compression was more prominent compared to potato starch during the first days following gelatinization, suggesting the role of amylose in the initial gelling of starch (Brunnschweiler *et al.*, 2005). The results from rheological and gelling analysis suggested that starches of *D. alata* and *D. cayenensis-rotundata* can be suitable as thickening and gelling agents for food (Brunnschweiler *et al.*, 2005; N'da Kouamé *et al.*, 2011). The retrogradation kinetics of heated flour dough of *D. rotundata* was monitored by

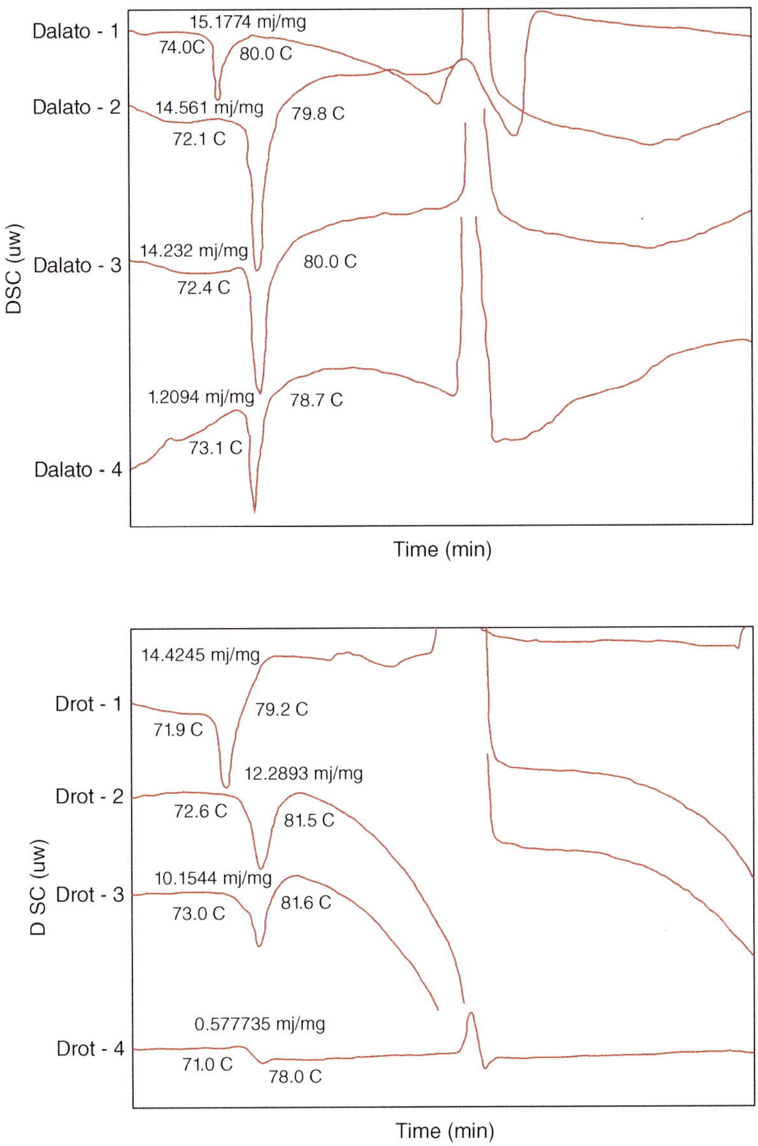

Fig. 6.5. MDSC pattern of starches obtained from *D. alata* and *D. rotundata*. (Source: unpublished data S.N. Moorthy).

textural analysis (compression) (Brennan and Sodah-Ayernor, 1973). The rate of retrogradation decreased with increase in temperature from 2°C to 50°C. The retrogradation rate of yam dough was much faster compared to that of wheat dough. Apart from the starch, other factors and chemical components present in each system may contribute to retrogradation characteristics. The retrogradation kinetics of heated flour dough of *D. rotundata* has been examined by WAXS besides textural analysis (Brennan and Sodah-Ayernor, 1973). Both techniques (WAXS and textural measurements) complemented each other in the quantification of retrogradation (Brennan and Sodah-Ayernor, 1973).

Table 6.20. Effects of guar gum (GG) and locust bean gum (LBG) on the thermal properties of root and tuber starches. (Source: Huang, 2009)

Parameters of endothermic properties					
Starch + gum	To (°C)	Tp (°C)	Tc (°C)	Gel range	ΔH (J/g)
Yam	73.1	81.4	89.6	16.5	13.1
Yam + GG	73.8	81.9	91.2	17.4	12.6
Yam + LBG	73.2	82.6	91.3	18.1	12.4
Yam bean	65.2	73.3	80.5	15.3	13.2
Yam bean + GG	65.9	74.1	81.6	15.7	13.0
Yam bean + LBG	66.1	76.5	83.7	17.6	13.3
Potato	60.8	66.5	72.7	11.9	14.9

To (°C): onset temperature. Tp (°C): peak temperature, Tc (°C): end temperature, ΔH: enthalpy gelatinization range.

Retrogradation properties of diverse yam starches measured by DSC also show large diversity (Brunnschweiler *et al.*, 2005; Jayakody *et al.*, 2007). Data on ΔH for melting of retrograded starches of five genotypes from *D. esculenta* and *D. alata* up to seven days after gelatinization indicated considerable variability in retrogradation behaviours of these starches and the results have been attributed to differences in the phosphorus content, amylose content and amylopectin structure (Jayakody *et al.*, 2007). Compared to potato starch, DSC melting parameters of recrystallized amylopectin (7–9 days after gelatinization) were similar, and firming of starch gel (measured by uniaxial compression) was prominent during the first few days, confirming the role of amylose in the initial gelling of *D. alata* and *D. cayenensis-rotundata* starches (Brunnschweiler *et al.*, 2005). The results of DSC and gel textural measurements also suggested that DSC primarily measures the recrystallization of amylopectin (Brunnschweiler *et al.*, 2005). The influence of amylopectin structure and the presence of minor components in the starch have not been elucidated. Factors such as the section within the tuber (distal compared with proximal) where the starch was extracted and the post-harvest storage conditions (gibberellic acid treated, desprouted biweekly, and control) had little influence on the retrogradation (similar T_{peak} and ΔH) (Brunnschweiler *et al.*, 2005).

Syneresis involves the exudation of water from gels during starch retrogradation, and is influenced by factors such as starch concentration, thermal history, freeze-thawing cycles and centrifugation conditions. The syneresis of gelatinized starch (2% solids concentration) from 43 genotypes from five species (4°C, up to 96 hours) varied considerably during storage. In addition, the values were high in yam starch as compared with some waxy starches (Otegbayo *et al.*, 2014). It would be worthwhile if waxy genotypes of diverse yam species are developed in addition to genotypes with a higher amount of short amylopectin unit chains (DP<12) with decreasing syneresis during various processing treatments (Bertoft *et al.*, 2012a). The syneresis of *D. alata* starch could also be decreased or prevented by adding certain hydrocolloids such as xanthan gum (Mali *et al.*, 2003). During the freeze-thawing cycle, the retrogradation led to syneresis and also the formation of resistant starch (Zhang *et al.*, 2014). Continuous freeze-thawing treatment (−18°C for 24 hours and 30°C for 2 hours) of starch gel from *D. cayenensis* up to five times increased the melting parameters and resistant starch content of retrograded starch that had a B-type polymorph (Zhang *et al.*, 2014).

Changes in the gel turbidity (transmittance, clarity) during storage reflect the retrogradation of starch and can be monitored by a spectrophotometer at fixed wavelength (620 nm). There is great variation in gel turbidity after gelatinization of starch, as well as the changing pattern among diverse species and within the same species during storage (Amani *et al.*, 2005; Otegbayo *et al.*, 2014). *D. rotundata* apparently had lower gel turbidity compared to *D. bulbifera* and *D. dumetorum* (Otegbayo et al., 2014). Yam starches with diverse species had higher gel turbidity than potato and cassava starches (Amani *et al.*, 2005). Gel turbidity is determined by the presence of granule remnants after gelatinization and the molecular association of starch chains in the gel (Moorthy and Nair, 1989). Microscopic analysis showed that the granules of yam starches remained more intact after gelatinization than did potato and cassava starches. In addition, there was a difference in the integrity and morphology of granule remnants among yam genotypes (Brunnschweiler *et al.*, 2005).

6.6 Thermal Properties

6.6.1 DSC

DSC data on starch depends on the experimental conditions employed, most important being water-to-starch ratio and heating rate (Freitas *et al.*, 2004; Yeh *et al.*, 2009). Decreasing water content from 90% to 40% had only minor influence on T_{onset} and T_{peak} but increased T_{end} and lowered the ΔH of starch from *D. alata* (Yeh *et al.*, 2009). A higher heating rate resulted in higher gelatinization temperatures of starch from *D. alata* (Freitas *et al.*, 2004). Even maintaining the same experimental conditions, a great diversity in gelatinization parameters has been observed within the same species (Amani *et al.*, 2004; Huang *et al.*, 2006; Jayakody *et al.*, 2007; Wang *et al.*, 2008a; Pérez *et al.*, 2011; Jiang *et al.*, 2013a, 2013b) and among diverse species (Emiola and Delarosa, 1981; Farhat *et al.*, 1999; Amani *et al.*, 2004; Jayakody *et al.*, 2007, N'da Kouamé *et al.*, 2011). Wang *et al.* (2006b) and Wang *et al.* (2008a) found that under the same DSC experimental conditions, starch of *D. opposite* from 14 genotypes has values for T_{onset} T_{peak} T_{end} and ΔH, which are 70.2–75.8°C, 77.5–81.5°C, 82.1–87.2°C and 6.5–12.5 J/g, respectively. Starch from 11 genotypes of *D. cayenensis-rotundata* had T_{onset} of 69.9–76.9°C and ΔH of 13.7–16.7 J/g (Amani *et al.*, 2004). Starches from *D. alata*, *D. rotundata*, *D. cayenensis* and *D. dumetorum* (one genotype for each) gave values for T_{onset} of 69.4–78.1°C, T_{peak} of 72.9–81.3°C, T_{end} of 76.7–86.4°C and ΔH of 18.6–19.8 J/g, respectively (Farhat *et al.*, 1999). Starch from distal and proximal regions of tubers from *D. alata* had similar T_{peak} and ΔH of gelatinization (Brunnschweiler *et al.*, 2005). Comparative analysis showed that *D. esculenta* tended to have higher ΔH than starches from *D. alata*, *D. cayenensis-rotundata* and *D. dumetorum*, while *D. dumetorum* starch was found to have higher gelatinization temperatures than the other yam starches (Farhat *et al.*, 1999; Amani *et al.*, 2004). The T_{onset} T_{peak} T_{end} and ΔH for *D. pyrofolia* starch were 71.5, 75.05, 78.25°C and 3.86 J/g (Sharlina *et al.*, 2017). The T_{onset} T_{peak} and ΔH of five wild yam varieties were 69.7–72.4°C, 78.0– 82.2°C and 20–21.4 J/g, respectively (Akinoso and Abiodun, 2013). Wickramasinghe *et al.* (2009) studied starch from three Sri Lankan yams and obtained values of 64–72°C, 67–75°C and 20.2 J/g for T_{onset} T_{peak} and ΔH, respectively. *D. niponica* starch had T_{onset} T_{end} and ΔH as 68.4–71.96°C, 78.9–82.5°C and 0.55 J/g (Akinoso and Abiodun, 2013). Yuan *et al.* (2007) obtained 67.4°C, 81.0°C and 18.6 J/g for T_{onset} T_{end} and ΔH, respectively, for starch from *D niponica*. Zhang *et al.* (2011) found that sulphur fumigation and freezing increased the DSC parameters considerably while lowering the ΔH. Among several yam species, waxy starches from *D. trifida* apparently had the highest ΔH (Pérez *et al.*, 2011). Compared with other species, *D. dumetorum*, having A-type polymorph, lower amylose content than *D. trifida* and fewer structural defects in the granules, had therefore higher gelatinization parameters (Srichuwong and Jane, 2007). The relationship between molecular structure and DSC gelatinization parameter has been explored. Based on research from four genotypes from two species (*D. alata* and *D. batatas*), T_{peak} was negatively correlated to CL, ICL and ECL, and positively correlated to NC (Wang *et al.*, 2006d). The gelatinization temperature of yam starches (T_{onset} T_{peak} and T_{end}) were higher than the other tubers and root starches such as potato, sweet potato and cassava starches (Valetudie *et al.*, 1995; Farhat *et al.*, 1999; Amani *et al.*, 2004). Starch from *D. alata* species exhibited a higher energetic gelatinization with a slower process compared with cassava starch (Freitas *et al.*, 2004). The gelatinization of starch from *D. alata* was followed by scanning electron microscopy (SEM) (Valetudie *et al.*, 1995). One notable feature was that during the expansion stage, amylose escaped through pinholes formed across the yam starch granules, whereas that of sweet potato leached from micropores (Valetudie *et al.*, 1995). Thermal properties of root and tuber starches in the presence of guar gam and locust bean gum were studied by Huang (2009) (Table 6.20).

A comparative study of 17 starches from diverse botanical origins, *D. esculenta*, explored the amylopectin cluster structure and the gelatinization behaviour (Vamadevan *et al.*, 2013). Starch of *D. esculenta* possesses highest T_{onset} T_{peak} and T_{end} as well as a relatively high ΔH among the 17 starches (Vamadevan *et al.*, 2013). The amylopectin internal and cluster structure of these starches was elaborated and *D. esculenta* was characterized as having a longer IB-CL and ECL, as well as the lowest molar amount of fingerprint A-chain (Afp) and NBbl among the 17 starches (Bertoft *et al.*, 2008, Bertoft *et al.*, 2012a, b). ECL was positively correlated to ΔH, and T_{onset} positively correlated to ECL and negatively correlated to NBbl (Vamadevan *et al.*, 2013). The authors proposed a backbone model to explain the thermal behaviour of *D. esculenta* starch that was in contrast to the other starches. Basically, small clusters with low NBbl and long IB-CL in the amylopectin of *D. esculenta* favour the parallel packing of adjacent double helices, and a less branched backbone renders amylopectins better organized in the granules (Vamadevan *et al.*, 2013). Agronomic practices can influence the DSC gelatinization parameters (Brunnschweiler *et al.*, 2005; Huang *et al.*, 2006; Pérez *et al.*, 2011). Starch of *D. trifida* harvested in different years (2005 compared with 2009) exhibited different gelatinization parameters (Pérez *et al.*, 2011). Gelatinization temperatures (T_{onset} T_{peak} and T_{end}) of starches from four varieties of *D. alata* increased gradually with growth time up to 260 days after planting, while ΔH either increased or decreased, depending on the variety and the harvesting time (Huang *et al.*, 2006). T_{peak} and ΔH of gelatinization of starch from *D. alata* were not affected to any major extent by post-harvest storage conditions (gibberellic acid treated, desprouted biweekly, and control) during a three-month period (Brunnschweiler *et al.*, 2005).

6.7 Digestibility

Great diversity in enzyme susceptibility of starches from several yam species and within the same species has been observed (Table 6.21). The digestibility of raw yam starches is lower compared to other tuber starches both *in vivo* and *in vitro* (Moorthy and Padmaja, 1991). Pérez *et al.* (2013) found only 4.6–7.8% digestibility for starch of five wild yam species. Similarily, Wikramsinghe *et al.* (2009) found <4% digestibility of starch from Sri Lankan yam varieties. The resistant starch content of *D. niponica* starch was around 66% (Jiang *et al.*, 2013a).

Compared to *D. alata*, *D. esculenta* has lower amylose content, smaller granules and a higher amount of short unit chains of amylopectin, and thus a higher susceptibility to porcine pancreatic α-amylase (Jayakody *et al.*, 2007). *D. esculenta* (C-type polymorph) and *D. polygonoides* (A-type polymorph) starches were low in amylose content with smaller granules and higher specific surface areas, and thus were more susceptible to porcine pancreatic α-amylase hydrolysis (Riley *et al.*, 2004). Systematic studies have revealed that starches with A-type polymorph and less/little amylose content tend to be more susceptible to enzyme hydrolysis compared to starches with B-type polymorph and higher amylose content (Jane *et al.*, 2003). Branches in A-type starch seem to scatter into the crystals in the granules, while those in B-type starch tend to concentrate in the amorphous regions (Jane *et al.*, 2003). Amylopectin with a higher amount of short unit chains (DP 6–12) may also contribute to structural defects in the granules (Vamadevan *et al.*, 2013), thereby enhancing the enzyme penetration and action.

Considering the susceptibility to α-amylase and glucoamylase enzymes, starch can be classified as rapidly digestible starch (RDS), slowly digestible starch (SDS) and resistant starch (RS) (Englyst *et al.*, 1992). Significant variation in the composition of starches from diverse yam genotypes among and within the species has

Table 6.21. Enzyme susceptibility of yam starches. (Source: Zhu, 2015)

Species	No.	Form	Enzyme	Reaction condition	Hydrolysis	References
D. alata	1	Raw	Porcine pancreatic α-amylase	40°C, 24 hour, pH 5.5	15.3	Riley *et al.* (2004)
	2	Raw	Porcine pancreatic α-amylase	37°C, 72 hour, pH 6.9	53.6–56.1	Jayakody *et al.* (2007)
	8	Raw	Porcine pancreatic α-amylase	40°C, 24 hour, pH 5.5	15.3–15.5	Riley *et al.* (2004)
D. cayenensis	1	Raw	Porcine pancreatic α-amylase	40°C, 24 hour, pH 5.5	13.7	Riley *et al.* (2004)
D. esculenta	1	Raw	Porcine pancreatic α-amylase	40°C, 24 hour, pH 5.5	21.3	Riley *et al.* (2004)
	3	Raw	Porcine pancreatic α-amylase	37°C, 72 hour, pH 6.9	66.1–76.6	Jayakody *et al.* (2007)
D. opposita	1	Raw	Bacterial α-amylase or glucoamylase	50°C, up to 20 hour, pH 6.9 for α-amylase, pH 4.5 for glucoamylase	17.5 (α-amylase); 22.4 (glucoamylase)	Li *et al.* (2011)
	1	Raw	Glucoamylase	45°C, up to 24 hour, pH 4		Jiang *et al.* (2011)
D. polygonoides	1	Raw	Porcine pancreatic α-amylase	40°C, 24 hour, pH 5.5	18.1	Riley *et al.* (2004)
	1	Raw	Porcine pancreatic α-amylase	40°C, 24 hour, pH 5.5	15.0	Riley *et al.* (2004)

No. = number of genotypes tested in the study.

been observed (including for *D. nipponica*, *D. opposita* and *D. persimilis*) (Jiang *et al.*, 2013a, b). For example, RDS, SDS and RS from three genotypes of *D. nipponica* ranged from 3.4% to 10.2%, 23.5–29.3% and 66.3–67.3%, respectively (Jiang *et al.*, 2013b). Higher amylose content was related to the higher RS content of starch (Kouadio *et al.*, 2013). The mealy (as opposed to hard) texture of cooked tubers was correlated to a higher total starch and a lower RS content in starch (Kouadio *et al.*, 2013).

In vivo digestion and *in vitro* enzyme susceptibility of yam starch have been found to be related to each other. The glycaemic index (GI) of cooked tubers of *D. cayenensis*, *D. alata*, *D. rotundata* and *D. esculenta* consumed by humans ranged from 68 to 97, and was positively correlated to enzyme susceptibility to porcine pancreatic α-amylase (13.7–21.3%). High GI, like the enzyme susceptibility, could be related to lower amylose content, the presence of A-type polymorph and smaller granules with higher specific surface area.

Starch of *D. opposita* was subjected to hydrolysis by bacterial α-amylase or glucoamylase (Li *et al.*, 2011a; Akinoso and Abiodun, 2013; Jiang *et al.*, 2011). The enzyme treatment changed the polymorph of granules from C- to A-type. The hydrolysis by both enzymes increased the T_{peak} and decreased the ΔH of gelatinization. Presence of bacterial α-amylase created a cavity at the centre of the granule, while glucoamylase peeled off the outer layer of the granules (Li *et al.*, 2011b). It has also been shown that the glucoamylase treatment decreased the T_{onset} T_{peak} and ΔH of gelatinization (Li *et al.*, 2011a; Jiang *et al.*, 2011). This could be due to the difference in the measurement conditions such as the enzyme activity.

References

Akinoso, R. and Abiodun, O.A. (2013) Effect of harvesting periods on the morphology and physico-chemical properties of trifoliate yam starches. *Starch/Stärke* 65(9–10), 753–761. DOI: 10.1002/star.201200171.

Alves, R.M.L., Grossmann, M.V.E. and Silva, R.S.S.F. (1999) Gelling properties of extruded yam (*Dioscorea alata*) starch. *Food Chemistry* 67(2), 123–127. DOI: 10.1016/S0308-8146(99)00064-3.

Amani, N.G.G., Buléon, A., Kamenan, A. and Colonna, P. (2004) Variability in starch physicochemical and functional properties of yam (*Dioscorea* sp) cultivated in Ivory Coast. *Journal of the Science of Food and Agriculture* 84(15), 2085–2096. DOI: 10.1002/jsfa.1834.

Amani, N.G.G., Kamenan, A., Rolland-Sabate, A. and Colonna, P. *et al*. (2005) Stability of yam starch gels during processing. *African Journal of Biotechnology* 4, 94–101.

Baldwin, P.M. (2001) Starch granule-associated proteins and polypeptides: a review. *Starch - Stärke* 53(10), 475–503. DOI: 10.1002/1521-379X(200110)53:10<475::AID-STAR475>3.0.CO;2-E.

Bertoft, E., Piyachomkwan, K., Chatakanonda, P. and Sriroth, K. (2008) Internal unit chain composition in amylopectins. *Carbohydrate Polymers* 74(3), 527–543. DOI: 10.1016/j.carbpol.2008.04.011.

Bertoft, E., Koch, K. and Åman, P. (2012a) Building block organisation of clusters in amylopectin from different structural types. *International Journal of Biological Macromolecules* 50(5), 1212–1223. DOI: 10.1016/j.ijbiomac.2012.03.004.

Bertoft, E., Koch, K. and Åman, P. (2012b) Structure of building blocks in amylopectins. *Carbohydrate Research* 361, 105–113. DOI: 10.1016/j.carres.2012.08.012.

Brennan, J.G. and Sodah-Ayernor, G. (1973) A study of the kinetics of retrogradation in a starch-based dough made from dehydrated yam (*Dioscorea rotundata* L. Poir). *Starch/Stärke* 25(8), 276–280. DOI: 10.1002/star.19730250807.

Brunnschweiler, J., Luethi, D., Handschin, S., Farah, Z., Escher, F. *et al*. (2005) Isolation, physicochemical characterization and application of yam (*Dioscorea* spp.) starch as thickening and gelling agent. *Starch/Stärke* 57(3–4), 107–117. DOI: 10.1002/star.200400327.

Cai, C. and Wei, C. (2013) In situ observation of crystallinity disruption patterns during starch gelatinization. *Carbohydrate Polymers* 92(1), 469–478. DOI: 10.1016/j.carbpol.2012.09.073.

Coursey, G. (1967) *Yams*. Longman, London.

Daiuto, E., Cereda, M., Sarmento, S. and Vilpoux, O. (2005) Effects of extraction methods on yam (*Dioscorea alata*) starch characteristics. *Starch/Stärke* 57, 153–60.

Delpeuch, F., Favier, J.C. and Charbonniere, R. *et al*. (1978) Characteristics of starches from tropical food plants. *Annales de Technologie Agricole* 27, 809–826.

Emiola, L.O. and Delarosa, L.C. (1981) Physicochemical characteristics of yam starches. *Journal of Food Biochemistry* 5(2), 115–130. DOI: 10.1111/j.1745-4514.1981.tb00665.x.

Englyst, H.N., Kingman, S.M. and Cummings, J.H. (1992) Classification and measurement of nutritionally important starch fractions. *European Journal of Clinical Nutrition* 46(Suppl 2), S33–S50.

Faboya, O.O.P. and Asagbra, A.A. (1990) The physico-chemical properties of starches from some Nigerian cultivars of white yam (*Dioscorea rotundata* Poir). *Tropical Science* 30, 51–57.

Falade, K.O. and Omiwale, O.O. (2015) Effect of pretreatments on color, functional and pasting properties of white (*Dioscorea rotundata*) and yellow yam (*Dioscorea cayenensis*) varieties. *Journal of Food Processing and Preservation* 39(6), 1542–1554. DOI: 10.1111/jfpp.12380.

Farhat, I.A., Oguntona, T. and Neale, R.J. (1999) Characterisation of starches from West African yams. *Journal of the Science of Food and Agriculture* 79(15), 2105–2112. DOI: 10.1002/(SICI)1097-0010(199912)79:15<2105::AID-JSFA510>3.0.CO;2-R.

Freitas, R.A., Paula, R.C., Feitosa, J.P.A., Rocha, S. and Sierakowski, M.-R. (2004) Amylose contents, rheological properties and gelatinization kinetics of yam (*Dioscorea alata*) and cassava (*Manihot utilissima*) starches. *Carbohydrate Polymers* 55(1), 3–8. DOI: 10.1016/S0144-8617(03)00142-5.

Fu, Y.-C., Huang, P.-Y. and Chu, C.-J. (2005) Use of continuous bubble separation process for separating and recovering starch and mucilage from yam (*Dioscorea pseudojaponica yamamoto*). *LWT - Food Science and Technology* 38(7), 735–744. DOI: 10.1016/j.lwt.2004.09.008.

Gebre-Mariam, T. and Schmidt, P.C. (1998) Some physico-chemical properties of *Dioscorea* starch from Ethiopia. *Starch/Stärke* 50(6), 241–246. DOI: 10.1002/(SICI)1521-379X(199806)50:6<241::AID-STAR241>3.0.CO;2-R.

Greenwood-Barton, L.H. (1961) *Yam Starches from Nigeria*. Report No.51 of the Tropical Products Institute. London.
Gunaratne, A. and Hoover, R. (2002) Effect of heat–moisture treatment on the structure and physicochemical properties of tuber and root starches. *Carbohydrate Polymers* 49(4), 425–437. DOI: 10.1016/S0144-8617(01)00354-X.
Hamon, P. and Tour, B. (1990) The classification of the cultivated yams (*Dioscorea cayenensis-rotundata* complex) of West Africa. *Euphytica* 47, 179–187.
Hanashiro, I., Tagawa, M., Shibahara, S., Iwata, K. and Takeda, Y. (2002) Examination of molar-based distribution of A, B and C chains of amylopectin by fluorescent labeling with 2-aminopyridine. *Carbohydrate Research* 337(13), 1211–1215. DOI: 10.1016/S0008-6215(02)00110-6.
Hollo, J. (1964) L'utilisation industrielle de l'igname, paper presented to the I[er] congres Internationale des Industries Agricoles et Alimentaires des Zones Tropicales et Sub-Tropicales. *Abid*, 537–557.
Hoover, R. (2000) Acid-treated starches. *Food Reviews International* 16(3), 369–392. DOI: 10.1081/FRI-100100292.
Hoover, R. (2010) The impact of heat-moisture treatment on molecular structures and properties of starches isolated from different botanical sources. *Critical Reviews in Food Science and Nutrition* 50(9), 835–847. DOI: 10.1080/10408390903001735.
Huang, C.-C. (2009) Physicochemical, pasting and thermal properties of tuber starches as modified by guar gum and locust bean gum. *International Journal of Food Science & Technology* 44(1), 50–57. DOI: 10.1111/j.1365-2621.2007.01634.x.
Huang, C.-C., Lin, M.-C. and Wang, C.-C.R. (2006) Changes in morphological, thermal and pasting properties of yam (*Dioscorea alata*) starch during growth. *Carbohydrate Polymers* 64(4), 524–531. DOI: 10.1016/j.carbpol.2005.11.009.
Jane, J.-lin., Ao, Z., Duvick, S.A., Wiklund, M., Yoo, S.-H. *et al.* (2003) Structures of amylopectin and starch granules: how are they synthesized? *Journal of Applied Glycoscience* 50(2), 167–172. DOI: 10.5458/jag.50.167.
Jayakody, L., Hoover, R., Liu, Q. and Donner, E. (2007) Studies on tuber starches. II. Molecular structure, composition and physicochemical properties of yam (*Dioscorea* sp.) starches grown in Sri Lanka. *Carbohydrate Polymers* 69(1), 148–163. DOI: 10.1016/j.carbpol.2006.09.024.
Jiang, Q., Gao, W., Li, X. and Zhang, J. (2011) Characteristics of native and enzymatically hydrolyzed *Zea mays* L., *Fritillaria ussuriensis* Maxim. and *Dioscorea opposita* Thunb. starches. *Food Hydrocolloids* 25(3), 521–528. DOI: 10.1016/j.foodhyd.2010.08.003.
Jiang, Q., Gao, W., Li, X., Wang, H., Xia, Y. *et al.* (2012a) Comparison of starches separated from different *Dioscorea bulbifera* Linn. cultivars. *Starch/Stärke* 64(7): 531–537. DOI: 10.1002/star.201100202
Jiang, Q., Gao, W., Li, X., Xia, Y., Wang, H. *et al.* (2012b) Characterizations of starches isolated from five different *Dioscorea* L. species. *Food Hydrocolloids* 29(1), 35–41. DOI: 10.1016/j.foodhyd.2012.01.011.
Jiang, Q., Gao, W., Shi, Y., Li, X., Wang, H. *et al.* (2013a) Physicochemical properties and *in vitro* digestion of starches from different *Dioscorea* plants. *Food Hydrocolloids* 32(2), 432–439. DOI: 10.1016/j.foodhyd.2013.02.001.
Jiang, Q., Gao, W., Shi, Y., Li, X., Xia, Y. *et al.* (2013b) Untraditional source of starches - rhizome of *Dioscorea nipponica* Makino. from different geographical origins. *Starch/Stärke* 65(5–6), 410–416. DOI: 10.1002/star.201200173.
Kasemsuwan, T. and Jane, J.L. (1996) Quantitative method for the survey of starch phosphate derivatives and starch phospholipids by 31P nuclear magnetic resonance spectroscopy. *Cereal Chemistry* 73, 702–707.
Ketiku, A.O. and Oyenuga, V.A. (1973) Changes in the carbohydrate constituents of yam tuber *(Dioscorea rotundata*, poir) during growth. *Journal of the Science of Food and Agriculture* 24(4), 367–373. DOI: 10.1002/jsfa.2740240403.
Kouadio, O.K., N'dri, D.Y., Nindjin, C., Marti, A., Casiraghi, M.C. *et al.* (2013) Effect of resistant starch on the cooking quality of yam (*Dioscorea* spp.) and cassava (*Manihot esculenta*) based paste products. *International Journal of Food Sciences and Nutrition* 64(4), 484–493. DOI: 10.3109/09637486.2012.749839.
Li, X., Gao, W., Jiang, Q., Wang, Y., Guo, X. *et al.* (2011a) Physicochemical, crystalline, and thermal properties of native, oxidized, acid, and enzyme hydrolyzed Chinese yam (*Dioscorea opposita* Thunb) starch. *Starch/Stärke* 63(10), 616–624.

Lindeboom, N., Chang, P.R. and Tyler, R.T. (2004) Analytical, biochemical and physicochemical aspects of starch granule size, with emphasis on small granule starches: a review. *Starch/Stärke* 56(34), 89–99. DOI: 10.1002/star.200300218.

Mali, S., Ferrero, C., Redigonda, V., Beleia, A.P., Grossmann, M.V.E. *et al.* (2003) Influence of pH and hydrocolloids addition on yam (*Dioscorea alata*) starch pastes stability. *LWT - Food Science and Technology* 36(5), 475–481. DOI: 10.1016/S0023-6438(03)00043-4.

Matheson, N.K. (1990) A comparison of the structures of the fractions of normal and high-amylose pea-seed starches prepared by precipitation with concanavalin A. *Carbohydrate Research* 199(2), 195–205. DOI: 10.1016/0008-6215(90)84261-R.

Miege, J. (1948) Le *Dioscorea esculenta* Burkill en Cote d'Ivoire. *Journal d'agriculture traditionnelle et de botanique appliquée* 28(313–314), 509–614.

Miège, J. (1957) Influence de quelques caractères des tubercules semences sur la levée et Le rendement des ignames cultivées. *Journal d'agriculture tropicale et de botanique appliquée* 4(7), 315–342. DOI: 10.3406/jatba.1957.2405.

Miles, M.J., Morris, V.J., Orford, P.D. and Ring, S.G. (1985) The roles of amylose and amylopectin in the gelation and retrogradation of starch. *Carbohydrate Research* 135(2), 271–281. DOI: 10.1016/S0008-6215(00)90778-X.

Moorthy, S.N. and Nair, S.G. (1989) Studies *on Dioscorea rotundata* starch properties. *Starch/Stärke* 41(3), 81–83. DOI: 10.1002/star.19890410302.

Moorthy, S.N. and Padmaja, G. (1991) Comparative study on digestibility of raw and cooked starch of different tuber crops. *Journal of Root Crops* 17, 255–258.

Moorthy, S.N. (1991) Extraction of starches from tuber crops using ammonia. *Carbohydrate Polymers* 16(4), 391–398. DOI: 10.1016/0144-8617(91)90057-J.

Moorthy, S.N. and Padmaja, G. (1996) A rapid titrimetric method for the determination of starch content of cassava tubers. *Journal of Root Crops* 28(1), 30–37.

Moorthy, S.N., Anderson, L., Eliasson, A.C., Santacruz, S. and Ruales, J. (2006) Determination of amylose content in starches using Modulated DSC. *Starch/Stärke* 58, 209–214.

Moorthy, S.N., Larsson, H. and Eliasson, A.-C. (2008) Rheological characteristics of different tropical root starches. *Starch/Stärke* 60(5), 233–247. DOI: 10.1002/star.200700708.

Moorthy, S.N., Sajeev, M.S. and Anish, R.J. (2018) Functionality of tuber starches. In: *Starch in Food: Structure, Function and Applications* (2nd edn). Woodhead Publishing Series in Food Science, Technology and Nutrition., pp. 421–508.

N'da Kouam'e, V., Handschin, S., Derungs, M., Amani, N.G.G., Conde-Petit, B. (2011) Thermal properties of new varieties of yam starches. *Starch/Stärke* 63, 747–753.

Nkala, D., Sibanda, S., Tomasik, P. and Palasiński, M. (1994) Isolation and properties of starch from wild yam from Zimbabwe. *Starch/Stärke* 46(3), 85–88. DOI: 10.1002/star.19940460303.

Otegbayo, B., Oguniyan, D. and Akinwumi, O. (2014) Physicochemical and functional characterization of yam starch for potential industrial applications. *Starch/Stärke* 66(3–4), 235–250. DOI: 10.1002/star.201300056.

Pérez, E., Gibert, O., Rolland-Sabaté, Agnès., Jiménez, Y., Sánchez, T. *et al.* (2011) Physicochemical, functional, and macromolecular properties of waxy yam starches discovered from "Mapuey" (*Dioscorea trifida*) genotypes in the Venezuelan Amazon. *Journal of Agricultural and Food Chemistry* 59(1), 263–273. DOI: 10.1021/jf100418r.

Pérez, E., Rolland-Sabaté, A., Dufour, D., Guzmán, R., Tapia, M. *et al.* (2013) Isolated starches from yams (*Dioscorea* SP) grown at the Venezuelan Amazons: structure and functional properties. *Carbohydrate Polymers* 98(1), 650–658. DOI: 10.1016/j.carbpol.2013.06.051.

Rao, P.S. and Beri, R.M. (1952a) Tubers of *Dioscorea hispida* Dennst. *Indian Forester* 78(3), 146–152.

Rao, P.S. and Beri, R.M. (1952b) Starches from the tubers of *Dioscorea hispida*. *Science and Culture* 17(4), 82–83.

Rao, P.S. and Beri, R.M. (1953a) Starches from the tubers of *Dioscorea bulbifera*. *Science and Culture* 18(1), 41–42.

Rao, P.S. and Beri, R.M. (1953b) Non-cereal foods: tubers of *Dioscorea* species. *Indian Forester* 79(10), 568–571.

Rao, P.S. and Beri, R.M. (1955) *Dioscorea* starches. *Science and Culture* 20(8), 397–399.

Rašper, V. and Coursey, D.G. (1967) Properties of starches of some West African yams. *Journal of the Science of Food and Agriculture* 18(6), 240–244. DOI: 10.1002/jsfa.2740180605.

Rašper, V. and Coursey, D.G. (1969) Investigations on starches from major starch crops grown in Ghana: II.—Swelling and solubility patterns: Amyloclastic susceptibility. *Journal of the Science of Food and Agriculture* 20(11), 642–646. DOI: 10.1002/jsfa.2740201103.

Riley, C.K., Wheatley, A.O., Hassan, I., Ahmad, M.H., Morrison, E.Y.S.A. *et al*. (2004) In vitro digestibility of raw starches extracted from five yam (*Dioscorea* spp.) species grown in Jamaica. *Starch/Stärke* 56(2), 69–73. DOI: 10.1002/star.200300195.

Riley, C.K., Wheatley, A.O. and Asemota, H.N. (2006) Isolation and characterization of starches from eight *Dioscorea alata* cultivars grown in Jamaica. *African Journal of Biotechnology* 5,1528–1536.

Rolland-Sabaté, A., Georges Amani, N'Guessan., Dufour, D., Guilois, S. and Colonna, P. (2003) Macromolecular characteristics of ten yam (*Dioscorea* spp) starches. *Journal of the Science of Food and Agriculture* 83(9), 927–936. DOI: 10.1002/jsfa.1410.

Seidemann, J. (1964) Mikroskopische Untersuchung verschiedenerDioscorea-Stärken. *Starch/Stärke* 16(8), 246–253. DOI: 10.1002/star.19640160803.

Sharlina, M.S., Yaacob, W.A., Lazim, A.M., Fazry, S., Lim, S.J. *et al*. (2017) Physicochemical properties of starch from *Dioscorea pyrifolia* tubers. *Food Chemistry* 220, 225–232. DOI: 10.1016/j.foodchem.2016.09.196.

Soni, P.L., Sharma, H.W., Dobhal, N.P., Bisen, S.S., Srivastava, H.C. *et al*. (1985) The starches of *Dioscorea ballophylla* and *Amorphophallus campanulatus*. *Comparison with tapioca. Starch/Stärke* 37, 6–9.

Srichuwong, S. and Jane, J.L. (2007) Physicochemical properties of starch affected by molecular composition and structures. *Food Science and Biotechnology* 16(5), 663–674.

Szylit, O., Borgida, L.P., Bewa, H., Charbonnière, R. and Delort-Laval, J. (1977) Valeur nutritionnelle pour le poulet en croissance, de cinq amylacés tropicaux en relation avec quelques caractéristiques physio-chimiques de leur amidon. *Annales de zootechnie* 26(4), 547–563.

Tamiru, M., Maass, B.L. and Pawelzik, E. (2008) Characterizing diversity in composition and pasting properties of tuber flour in yam germplasm (*Dioscorea* spp.) from Southern Ethiopia. *Journal of the Science of Food and Agriculture* 88(10), 1675–1685. DOI: 10.1002/jsfa.3263.

Tattiyakul, J., Naksriarporn, T., Pradipasena, P. and Miyawaki, O. (2006) Effect of moisture on hydrothermal modification of *yam Dioscorea hispida* Dennst starch. *Starch/Stärke* 58(3-4), 170–176. DOI: 10.1002/star.200500462.

Treche, S. (1989) Potentialites nutritionnelles designames (*Dioscorea* sp.) cultiveées au Cameroun. Vol. I: Texte. Vol. II: Annexes. These, Editions del'ORSTOM. Collection Etudes et Theses, Paris, p 595.

Valetudie, J.-C., Guadeloupe, L., Colonna, P., Bouchet, B. and Gallant, D.J. (1995) Gelatinization of sweet potato, tania and yam tuber starches. *Starch/Stärke* 47(8), 298–306. DOI: 10.1002/star.19950470804.

Vamadevan, V., Bertoft, E. and Seetharaman, K. (2013) On the importance of organization of glucan chains on thermal properties of starch. *Carbohydrate Polymers* 92(2), 1653–1659. DOI: 10.1016/j.carbpol.2012.11.003.

Vasanthan, T. (2001) *Overview of Laboratory Isolation of Starch from Plant Materials: Current Protocols in Food Analytical Chemistry*. John Wiley and Sons, New York, pp. E2.1.1–E2.1.6. DOI: 10.1002/0471142913.fae0201s00.

Wang, S., Gao, W., Liu, H., Chen, H., Yu, J. *et al*. (2006a) Studies on the physicochemical, morphological, thermal and crystalline properties of starches separated from different *Dioscorea opposita* cultivars. *Food Chemistry* 99(1), 38–44.

Wang, S., Liu, H., Gao, W., Chen, H., Yu, J. *et al*. (2006b) Characterization of new starches separated from different Chinese yam (*Dioscorea opposita* Thunb.) cultivars. *Food Chemistry* 99(1), 30–37.

Wang, S., Yu, J., Gao, W., Liu, H. and Xiao, P. (2006c) New starches from traditional Chinese medicine (TCM): Chinese yam (*Dioscorea opposita* Thunb.) cultivars. *Carbohydrate Research* 341(2), 289–293. DOI: 10.1016/j.carres.2005.10.022.

Wang, W.-M., Lai, V.M.-F., Chang, K.-F., Lu, S. and Ho, H.-H. (2006d) Effect of amylopectin structure on the gelatinization and pasting properties of selected yam (*Dioscorea* spp.) starches. *Starch/Stärke* 58(11), 572–579. DOI: 10.1002/star.200600531.

Wang, S., Yu, J., Yu, J., Chen, H., Pang, J. *et al*. (2008a) Partial characterization of starches from *Dioscorea opposita* Thunb. cultivars. *Journal of Food Engineering* 88(3), 287–293. DOI: 10.1016/j.jfoodeng.2006.11.016.

Wang, S., Yu, J., Yu, J. and Liu, H. (2008b) The partial characterization of C-type rhizoma *Dioscorea* starch granule during acid hydrolysis. *Food Hydrocolloids* 22, 531–537.

Wang, S., Yu, J., Zhu, Q., Yu, J. and Jin, F. (2009) Granular structure and allomorph position in C-type Chinese yam starch granule revealed by SEM, 13C CP/MAS NMR and XRD. *Food Hydrocolloids* 23(2), 426–433. DOI: 10.1016/j.foodhyd.2008.02.012.

Wickramasinghe, H.A.M., Takigawa, S., Matsuura-Endo, C., Yamauchi, H. and Noda, T. (2009) Comparative analysis of starch properties of different root and tuber crops of Sri Lanka. *Food Chemistry* 112(1), 98–103. DOI: 10.1016/j.foodchem.2008.05.046.

Woodward, E.G. (1982) Starch recovery from starchy roots without peeling them: using cassava, potato, yam ortaro roots. Patent. GB2103635-A; JP58023800-A; AU8286182-A; ZA8205179-A; BR8204318-A;GB2103635-B.

Yeh, A.-I., Chan, T.-Y. and Chuang, G.C.-C. (2009) Effect of water content and mucilage on physicochemical characteristics of yam (*Discorea alata* purpurea) starch. *Journal of Food Engineering* 95(1), 106–114. DOI: 10.1016/j.jfoodeng.2009.04.014.

Yuan, Y., Zhang, L., Dai, Y. and Yu, J. (2007) Physicochemical properties of starch obtained from *Dioscorea nipponica* Makino comparison with other tuber starches. *Journal of Food Engineering* 82(4), 436–442. DOI: 10.1016/j.jfoodeng.2007.02.055.

Zhang, L., Liu, P., Wang, Y. and Gao, W. (2011) Study on physico-chemical properties of dialdehyde yam starch with different aldehyde group contents. *Thermochimica Acta* 512(1–2), 196–201. DOI: 10.1016/j.tca.2010.10.006.

Zhang, W., Chen, H., Wang, J., Wang, Y., Xing, L. *et al.* (2014) Physicochemical properties of three starches derived from potato, chestnut, and yam as affected by freeze-thaw treatment. *Starch/Stärke* 66(3–4), 353–360. DOI: 10.1002/star.201200270.

Zhou, H., Wang, J., Fang, X., Sun, Y. and Dou, X. (2012) Physicochemical properties of new starches isolated from *Dioscorea opposita* Thunb. bulbils. *Starch/Stärke* 64(4), 290–296. DOI: 10.1002/star.201100117.

Zhou, Q., Shi, W., Meng, X. and Liu, Y. (2013) Studies on the morphological, crystalline, thermal properties of an under utilized starch from yam *Dioscoreae zingiberensis* C. H. Wright. *Starch/Stärke* 65(1–2), 123–133. DOI: 10.1002/star.201200058.

Zhu, F. (2016) Structure, properties, and applications of aroid starch. *Food Hydrocolloids* 52, 378–392. DOI: 10.1016/j.foodhyd.2015.06.023.

7

Other Starches

7.1 Introduction

A number of minor tuber crops are cultivated in different parts of the world and mainly used as food. Among them, arrowroot is used for starch extraction in view of the high demand for this starch especially in bakery products. Although the very widely used arrowroot biscuits contain very little arrowroot starch, overall demand for this starch is very high. Similarly, arrowroot starch is in high demand as a convalescent food with high digestibility. Starch extracted from tubers of *Curcuma* spp. on a cottage level, with simple equipment, is utilized as invalid and baby food similar to arrowroot starch.

7.2 Extraction

The manufacture of arrowroot starch is very similar to the process used for cassava starch, which was often referred to as Brazilian arrowroot. In India, arrowroot starch is often extracted by cassava starch production units when cassava tubers are not available for starch isolation. Because of their size and shape, roots of arrowroot require the use of special washers. After removing the tops, which are poor in starch, the roots are cut or sliced to yield smaller pieces in order to make them suitable for use in the automatic potato-washing equipment in which the roots are slowly rubbed together in a stream of water under conditions that remove as much of the root skin as possible.

The roots used to be skinned with knives of German silver and pulped in roller mills, but the former operation has been discontinued and the pulping is carried out in very efficient rasping machines. As the cells of the rhizome are very tough, very efficient rasping machines are required, and more than one run of the pulp through the rasping machine is carried out. The pulp passes through a series of mills equipped with saw-like blades to tear the cells apart, and through hammer mills to convert the pulp to slurry. The liberated starch is washed from the pulp by fine water sprays on reels, and the pulping and washing processes are repeated two or more times.

The starch suspension is sieved through a series of screens of decreasing mesh size (50, 80 and 100), and is finally allowed to settle in glazed vats. Resuspension and resettling are carried out and the deposited starch cake is freed from discoloured surface layers, drained on trays and, finally, dried.

In modern factories the refining and dewatering of the final starch pulp are performed by centrifugal separation. The old method of drying was to air-dry the broken starch cake on wire netting, through which it fell as it dried and crumbled, and then collect it on trays beneath the netting. More modern mills use low-temperature dryers of the

Thomas heater type. In these dryers the starch moves on an endless belt to regions of progressively higher temperature with decrease in water content, the final temperature being 55–60°C. Rotary dryers, flash dryers or vacuum dryers are also employed in some factories. In the conventional air-drying method, drying used to take three to four days during hot weather, but under adverse conditions it may need a fortnight during which souring and discolouring microrganisms could develop in the moist starch and spoil it. In the old method worked in St Vincent, the average recovery of 19% of starch, based on the weight of the roots used, has been claimed. But with modern methods the yield is in the range of 15–18%. In southern parts of India, *Curcuma* starch is often extracted using a portable starch extractor, which is carried to different fields where the tubers are cultivated and the starch is extracted and sun-dried.

Table 7.2. Chemical composition of chayote tuber starch (CHS). (From Jiménez-Hernández *et al.*, 2007)

Characteristics (%)	CHS
Moisture	7.02 ± 0.15
Protein (n x 6.25)	0.29 ± 0.00
Carbohydrates	98.56 ± 2.95
Lipid	0.16 ± 0.00
Ash	0.46 ± 0.04
Phosphorus	0.15 ± 0.00
Amylase	12.90 ± 0.64
Amylopectin	87.10 ± 0.23
Amylase:amylopectin ratio	0.13 ± 0.01
Ph	8.12 ± 0.37
Granule size (μm)	7–50

7.3 Physiochemical Properties

7.3.1 Biochemical properties

As with other starches, proximate composition studies of canna starch showed that the protein content in the canna starch samples varied between 0.069% and 0.078%, lipid between 0.014% and 0.019% and ash between 0.025–0.03%.

Canna starch had ash, protein, lipid, ash and P contents as 0.3%, 0.07%, 0.015% and 371 ppm, respectively (Thitipraphunkal *et al.*, 2003). Lipid, ash, fibre and protein contents of Chinese water chestnut were 0.9%, 0.4%, 0.3% and 0.5%, respectively (Lan *et al.*, 2008). For three varieties of the same crop, Lee *et al.* (2015) found 0.36–1.04%, 0.59–1.04% for fat and 0.04–0.46% for ash, respectively (Table 7.1), while Liu *et al.* (2013) reported values of 0.10–0.15%, 0.62–0.79% and 0.07–0.14% for three varieties. Lemos *et al.* (2015) has reported the chemical properties of two cultivars of chestnut starch as 0.48% and 0.60% for protein and 0.0011% and 0.0007% for phosphorus. In Indian chestnut starch, Singh *et al.* (2009) obtained 0.2% protein, 0.2% ash and 0.1% lipid. Correa *et al.* (2013) reported protein, lipid and ash content in Peruvian lily of 0.11%, 0.26% and 0.38%, respectively. *Pacchyrhizus ahipa* roots had total ash content of 3.3–3.7%, crude fat 0.47%, crude protein 7.9–9.1% and phosphorus content of 6.7–7.7% on dry-weight basis (Doporto *et al.*, 2014). Braga *et al.* (2006) reported 0.5% protein for turmeric and ginger starches. The ash content was 1.5% for turmeric starch and 0.16–0.91% for ginger starch. The values for minor constituents of arrowroot, yam, canna and ginger starches were reported by Peroni *et al.* (2006) see Table 7.3.

Table 7.1. Chemical composition and amylose content of isolated chestnut starches. (From Lee *et al.*, 2015)

Sample	Moisture (%)	Protein (%)	Fat (%)	Ash (%)	Amylose (%)
GJ	12.87 ± 0.24	1.04 ± 0.08	0.63 ± 0.09	0.45 ± 0.04	20.37 ± 2.77
GY	12.86 ± 0.10	0.50 ± 0.05	0.59 ± 0.02	0.06 ± 0.04	14.96 ± 1.85
HD	13.55 ± 0.15	0.36 ± 0.06	1.04 ± 0.27	0.04 ± 0.01	14.46 ± 0.31

The chemical composition of starch from chayote have been determined by Jiménez-Hernández *et al.*, 2007.
GJ, isolated chestnut starch from Gongju-si chestnut fruit; GY, isolated chestnut starch from Gwangyang-si chestnut fruit; HD, isolated chestnut starch from Hadong-gun chestnut fruit.

Table 7.3. Chemical constituents of starch from minor tubers. (From Peroni *et al.*, 2006)

Starch	Protein (%)	Lipid (%)	Ash (%)	Phosphorus (%)
Arrowroot	0.15	0.17	0.28	0.018
Canna	0.13	0.19	0.24	0.031
Ginger	0.32	0.24	0.35	0.007

Jimenez-Hernandez *et al.* (2007) showed the variation in chemical composition of chayote tuber (see Table 7.2).

Jyothi *et al.* (2003) have found high phosphorous content for two *Curcuma* sp. – *C zeodora* and *C malabarica* – of 450 ppm, which is similar to potato and canna starch. Hung and Duyen (2017) have studied the chemical composition of two *Curcuma* sp. and did not find much variation. The *C. longa* and *C. caesia* starches had different fine structures of amylose and amylopectin (see Table 7.4).

Guitérrez (2017) has analysed starch of guinea arrowroot (*Calthea allouia*) and found the following values: crude protein 0.28%, crude fat 0.09%, ash 0.15%. The coleus starch has reported values of 1–1.5% protein, 0.10% lipids, 0.4% fibre and 0.9% mineral matter (Moorthy, 1986). Eleazu *et al.* (2017) has compared the biochemical constituents of coleus starch.

7.3.2 Amylose and amylopectin content

The amylose content also shows considerable variation among the minor tuber crops. Canna starch showed higher amylose content than other tuber starches at 31%, out of which soluble amylose formed around 20% (Lan *et al.*, 2015). Wickramasinghe *et al.* (2009) found a value of 34% for amylose in canna starch. Santacruz *et al.* (2002) found very low amylose content for *Arracacha* (3–5%) and 23.4–24.2% and 17.4–19.4% for canna and *O tuberosa* starches, respectively. For Chinese water chestnut starch, values of 21.2–22.3% for amylose content of four varieties were found (Liu *et al.*, 2015). The absolute amylose content of canna from different cultivars ranged from 19% to 25% (Thitipraphunkul *et al.*, 2003). For Indian chestnut starch, the amylose content was 22.3% (Singh *et al.*, 2009). For lilium tuber starches, *L. lancifolium* had lower amylose content compared to *L. leucantham* and *L. rosthornii*. Fig.7.1. gives the GPC pattern of arrowroot and *Canna edulis* starches (Moorthy unpublished data).

Table 7.4. Chemical composition (%, w/w, db) of *C. longa* and *C. caesia* starches. (From Hung and Duyen, 2017)

Composition (%, db)	*C. longa*	*C. caesia*
Protein	0.65 ± 0.03	0.58 ± 0.05
Lipid	0.03 ± 0.01	0.05 ± 0.01
Ash	1.20 ± 0.05	1.05 ± 0.05
Total starch	98.12 ± 0.03	98.32 ± 0.05

For *Curuma* spp., there was only a small difference between *C. longa* and *C caesia* starches (Hung and Duyen, 2017) (see Table 7.5). For lilium tuber starches, *L. lancifolium* had lower amylose content compared to *L. leucantham* and *L. rosthornii*.

7.4 Structural Properties

7.4.1 Granule morphology

Pacchyrrhizus starch granules are round, cupoliform or polyhedral with size range of 6–35 μm. *Arracacha* starch granules are similar to cassava starch granules, and spherical or ovoid in shape, with size range of 5–27 μm. Canna starch granules are quite large (10–100 μm) (Moorthy, 2002). Thitipraphunkal *et al.* (2003) reported canna starch granules as spherical or oval and 20–40 μm in size (Puncha-arnon *et al.*, 2007). Chinese water chestnut starch grains, being round or crescent shape, have a size of up to 27 μm. Liu *et al.* (2015) found both large and small granules in four varieties of Chinese chestnut starches. The size determined by SEM varied from 1 μm to 16 μm. East Indian arrowroot starch grains are polyhedrons or hemispherical and 8–40 μm in size. Giant taro starch grains are

Table 7.5. Amylose content of *C. longa* and *C. caesia* starches. (From Hung and Duyen, 2017)

Sample	Blue value	Amylose content %	Apparent amylose content %
C. longa			
Starch	0.427 ± 0.007	26.3 ± 0.7	35.6 ± 0.6
Amylose	0.678 ± 0.010		
Amylopectin	0.337 ± 0.020		
C. caesia			
Starch	0.457 ± 0.004	24.9 ± 0.3	38.1 ± 0.3
Amylose	0.723 ± 0.010		
Amylopectin	0.369 ± 0.010		

The amylose content determined by gel permeation chromatography after debranching with isoamylase was 4% for *A. xanthorriza*, 18.4% for *O. tuberosa* and 23.8% for *C. edulis* (Santacruz *et al.*, 2002).

small, irregular and 1–5 µm in size. Coleus starch granules are round or oval with a size range of 4–46 µm. Jayakody *et al.* (2005) found the granules to be dome-shaped and hemispherical, 5–25 µm in size. Queensland arrowroot granules are oval or polyhedral in shape and have a size range of 5–44 µm. Shoti starch granules have an elliptical shape with size of 14–42 µm. Swamp taro starch possesses round or angular granules 4–18 µm in size. Tacca starch granules are tiny compared to many other starches, the average size being 3.5 µm, and they possess polyhedral edges. *C. longa* and *C. caesia* starches had similar morphology (Hung and Duyen, 2017).

Microstructure of starch particles showed a smooth, flat outer surface. The approximate length and breadth of isolated elliptical starches were 25 µm and 10 µm, while the thickness was about 5 µm (Kuttigounder *et al.*, 2011). Scanning electron microscopy investigations showed that granules of *A. xanthorriza* were small, with irregular shape and sizes between 7 µm and 23 µm, whereas granules of *O. tuberosa* and *C. edulis* were both oval-shaped, with granular sizes between 20 µm and 55 µm and between 35 µm and 101 µm, respectively (Santacruz *et al.*, 2002).

Umadevi and Wankhede (1981) have studied the properties of winged bean starch. The starch granules are of two types: oval (25–36 µm) and round (10–12 µm). Under polarized light, they show strong centric crosses. For Peruvian lily, Correa *et al.* (2013) obtained values of 14.16–79.62 µm. Yu *et al.* (2015) have studied the granular morphology of starch from bulbs and bulbils of *L. lancifolium* and found that it has smaller granules as shown in Table 7.6. Starches from the three *Lilium* species presented different granule size and shape.

Hung and Duyen (2017) have compared the granular properties of yellow and black turmeric rhizomes. The granules were triangular or oval in shape. Jyothi *et al.* (2003) have compared the starches from *Curcuma malabarica* and *Curcuma zedoaria*. Microscopic studies revealed that the starch granules do not exhibit variation in size among species. For *Curcuma malabarica* starch, about 40% of starch grains had a size range of 9–30 µm while 60% of the grains were between 30–45 µm with an average granule size of 34.3 µm. For *Curcuma zedoaria* starch, 50% of the granules were in the range of 8–30 µm and the rest were in the range 35–60 µm. The average granule size was 33.13 µm. The granules were slightly

Table 7.6. Size distribution and amylose content of lilum starches. (Source: Yu *et al.*, 2015)

Species	Minimum size (µm)	Maximum size (µm)	Average size (µm)	Amylose content (%)
L. leucanthum	5.19 ± 0.09	116.09 ± 4.21	35.70 ± 0.33	24.41 ± 0.20
L. rosthornii	5.26 ± 0.13	82.33 ± 3.86	35.25 ± 0.24	25.27 ± 0.16
L. lancifolium	2.39 ± 0.06	55.63 ± 3.03	23.13 ± 0.18	28.17 ± 0.21

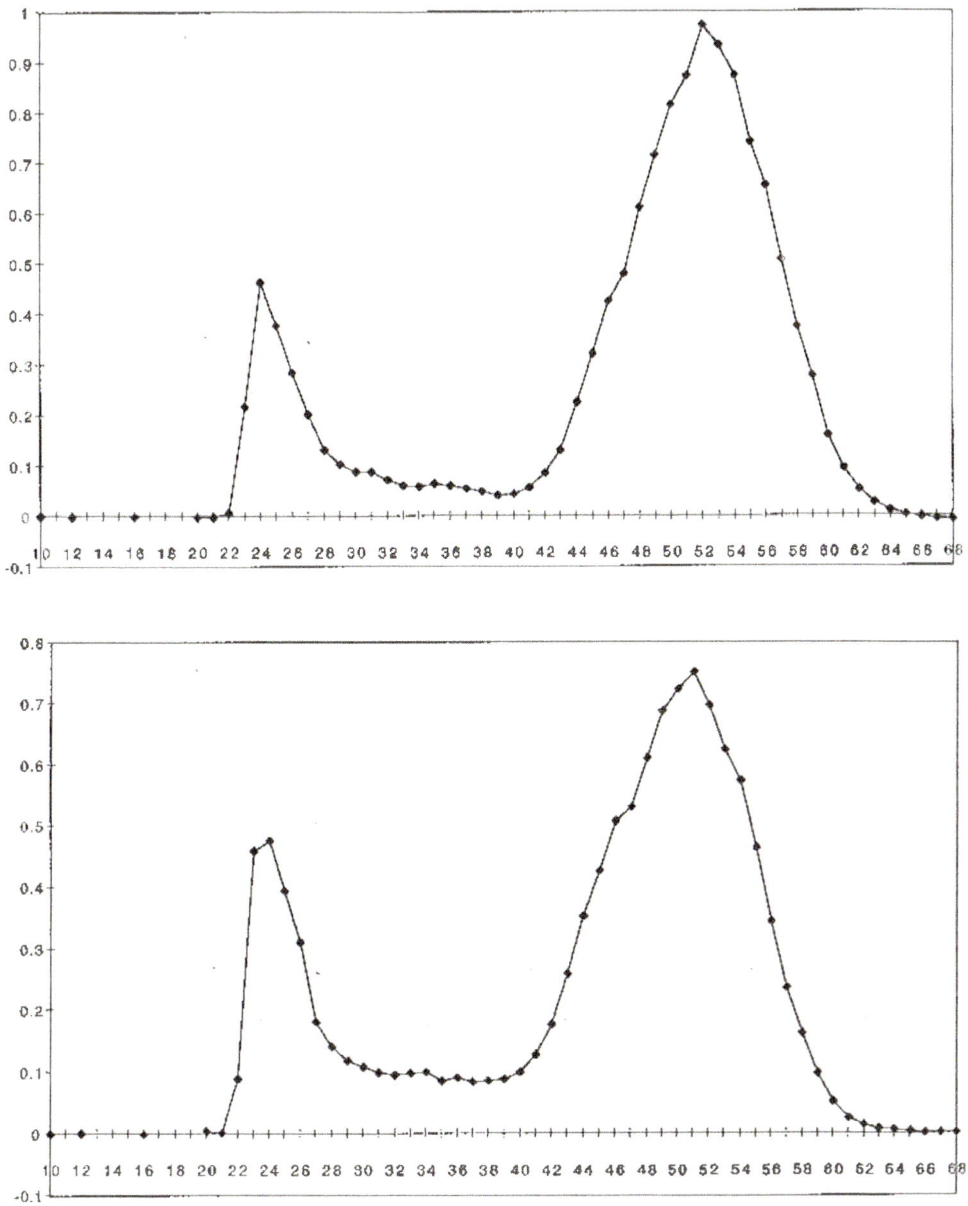

Fig. 7.1. GPC of arrowroot starch and *Canna edulis* starch. (Source: Moorthy, unpublished data)

larger than those of cassava, sweet potato and aroid starches, but similar to *Dioscorea rotundata*, *Dioscorea alata* and *Canna edulis*. *Curcuma* starch granules were elliptical in shape similar to *D. rotundata* and canna starch granules. The size and shape of three Andean species are determined by Santacruz *et al.* (2002) (see Table 7.7).

Scanning electron microscopy investigations showed that the starch granules from all cultivars of canna were oval-shaped granules

Table 7.7. Physico-chemical properties together with λ_{max} of the amylopectin fraction of *A. xanthorriza*, *C. edulis* and *O. tuberosa*. (Santacruz *et al*., 2002)

	Isolated amylopectin						
Source	Shape	Size (μm)	X-ray pattern	Amylase content (%)	Blue value	λ_{max} (nm)	
A. xanthorriza	Irregular	7–23	B	3–5	0.42 ± 0.02	559	551
C. edulis	Oval	35–101	B	23.4–24.2	0.92 ± 0.04	613	574
O. tuberosa	Oval	22–55	B	17.4–19.4	0.85 ± 0.04	598	560

with smooth surfaces and were around 10–100 μm in size (Thitipraphunkul *et al.*, 2003). The changes in size and morphology of starch granules of canna during rhizome maturation were studied. Granule size tended to increase with progressive development of segments from immature to mature stages and the full-size granules seemed to have a rounder shape. Starch from mother segments had typical characteristics in both granule size and surface features (Puncha-arnon *et al.*, 2007).

7.4.2 XRD and starch crystallinity

Canna edulis starch of three varities possessed B pattern (Thitipraphunkul *et al.*, 2003) and crystallinity of 22.5% (Lan *et al.*, 2008). Peruvian lily had B pattern and crystallinity of 27% (Correa *et al.*, 2013). Four varieties of Chinese chestnut starch had C pattern and it turned to B as water content increased (Liu *et al.*, 2015). The relative crystallinity ranged from 26.3% to 29% (Lemos *et al.*, 2015). Santacruz *et al.* (2002) reported B pattern for *Arracacha*, canna and *O tuberosa* starches.

The XRD data for traditional Chinese medicine starches obtained by Yu and Wang are given in Table 7.8.

X-ray diffractogram indicated turmeric starch to be of B type (Kuttigounder *et al.*, 2011). Yu *et al.* (2015) found similar XRD pattern for starch from bulbs and bulbils of three *Lilium* species using X-ray diffraction and FTIR patterns. The bulb starch had a higher degree of crystallinity. *L. lancifolium* had the lowest level of crystallinity. *C. longa* and *C. caesia* had B XRD patterns (Hung and Duyen, 2017). The *Curcuma* starch resembles yam and *C. edulis* starch in possessing B-type X-ray diffraction patterns (Jyothi *et al.*, 2003).

7.4.3 Amylose and amylopectin structure

Amylopectin showed different structures among the three Andean starches with a β-amylolysis limit of 67.6% for *C. edulis*, 64.5% for *O. tuberosa* and 56.6% for *A. xanthorriza*. The average chain length of the amylopectin was highest for *A. xanthorriza* (22.6), followed by *O. tuberosa* with 22.4, and lowest for *C. edulis* (21.9) (Santacruz *et al.*, 2002).

Supercritical fluid extraction of *Curcuma* and ginger starches did not alter the starch granules' morphology, as observed by SEM (Braga *et al.*, 2006).

Lemos *et al.* (2015) have studied the granular and structural properties of two cultivars of chestnut starch. Longal starch granules possess higher amylopectin molecular weight and degree of external organization. Longal amylopectin structural features, such as higher abundance of short chains, might explain the higher breakdown during pasting and lower gelatinization

Table 7.8. XRD pattern and crystallinity of some medicinal tubers. (Yu and Wang, 2008)

Sample	Crystallinity	Pattern
Purarea lobata	50.2	C
Polygonum multiforum	46	A
Alisma orientalis	38.1	A
F. uussurensis	44.7	B
F. cirrosa	42.6	B
Rhizoma dioscorea	45.7	C

Table 7.9. Fine structure of *C. longa* and *C. caesia* starches. (Hung and Duyen, 2017)

Sample	DP_n	CL	NC
C. longa			
Starch			
Amylose	451 ± 13		
Amylopectin	1093 ± 40	13.2 ± 0.1	82.0 ± 2.3
C. caesia			
Starch			
Amylose	263 ± 2		
Amylopectin	1156 ± 3	19.9 ± 0.4	57.1 ± 1.6

Table 7.10. Chain length distribution in Chinese chestnut starch. (Lemos *et al.*, 2015)

Property	Longal	Lada
DPw	17.8	18.6
Afp (6–8)	8.8	6.1
A + B1 a (BP 6–12)	31.7	24.7
A + B1 a (BP 13–24)	49.8	52.3
APC	0.39	0.32
DP 9–12	22.9	18.6
B1b (DP 25–36)	14.3	21
SC	95.8	98
B2 + B3 (DP >37)	4.2	2.0
LC	4.2	2.0

temperature. Lada's higher gelatinization temperature range is in line with a higher diversity of cluster with different double-helix stabilities that can be due to the heterogeneity of the side chains involved in the double helixes. They also observed the chain length distribution of two cultivars of chestnut starch and reported the values shown in Table 7.10 (Lemos *et al.*, 2015). The chain length of Peruvian lily starch is as follows: DP 12,19.6; DP 13, 24–30.3, DP 25, 36 17.1 DP>37, 32.8 (Correa *et al.*, 2013).

Hung and Duyen (2017) found that starch of *C. longa* had higher chain length of amylose and more short branching chains for amylopectin compared to *C. caesia* starch (Table 7.9). The starch had the largest proportion of ordered structure in granules' external region. The ordered structure in the external granular region of the bulb starch was similar to bulbil starch.

The HPSEC patterns of starch from arrowroot and *Canna edulis* are shown in Fig. 7.2.

7.5 Functional Properties

7.5.1 Swelling pattern and solubility

The values for the minor starches show a large range. For four varieties of Chinese chestnut starch, the swelling power increased with temperature (Liu *et al.*, 2015) (Table 7.11).

For Indian chestnut starch, the swelling power and solubility were 9.7 g/g and 12.5%, respectively (Singh *et al.*, 2009). Dried and cured-dried turmeric powder samples showed higher water-holding capacity (3.62 g/g and 4.78 g/g, respectively) compared to isolated starch (1.07 g/g) at 30°C (Kuttigounder *et al.*, 2011). The starch in the bulb of *Lilium* sp. had significantly higher solubility compared to bulbil starch (Yu *et al.*, 2015b in Table 7.12. The swelling property of Alstroemeria starch in different temperatures are given in Table 7.13 by Correa *et al.*, 2013.).

7.5.2 Viscosity

For chestnut starch, the pasting temperature ranged 69.4–70.6°C while peak viscosity was between 2658 and 3301 RVU for four varieties (Liu *et al.*, 2015). Leelavathi *et al.* (1987) have reported the viscosity properties of different tuber crops. Table 7.14 summarizes the pasting properties of different tuber starches (Peroni *et al.* 2006).

The pasting properties of starch of the three *Lilium* species were different (Yu *et al.*, 2015). The viscosity studies on *Curcuma* starch showed that variation exists between starches extracted from the two species. The *C. malabarica* starch had a higher peak viscosity of 3387 cP compared to *C. zedoaria* starch. Removal of curcumin led to increase in viscosity of *C. zedoaria* starch, to 3409 cP, which was close to that of *C. malabarica* starch. The breakdown in viscosity was quite low for both starches,

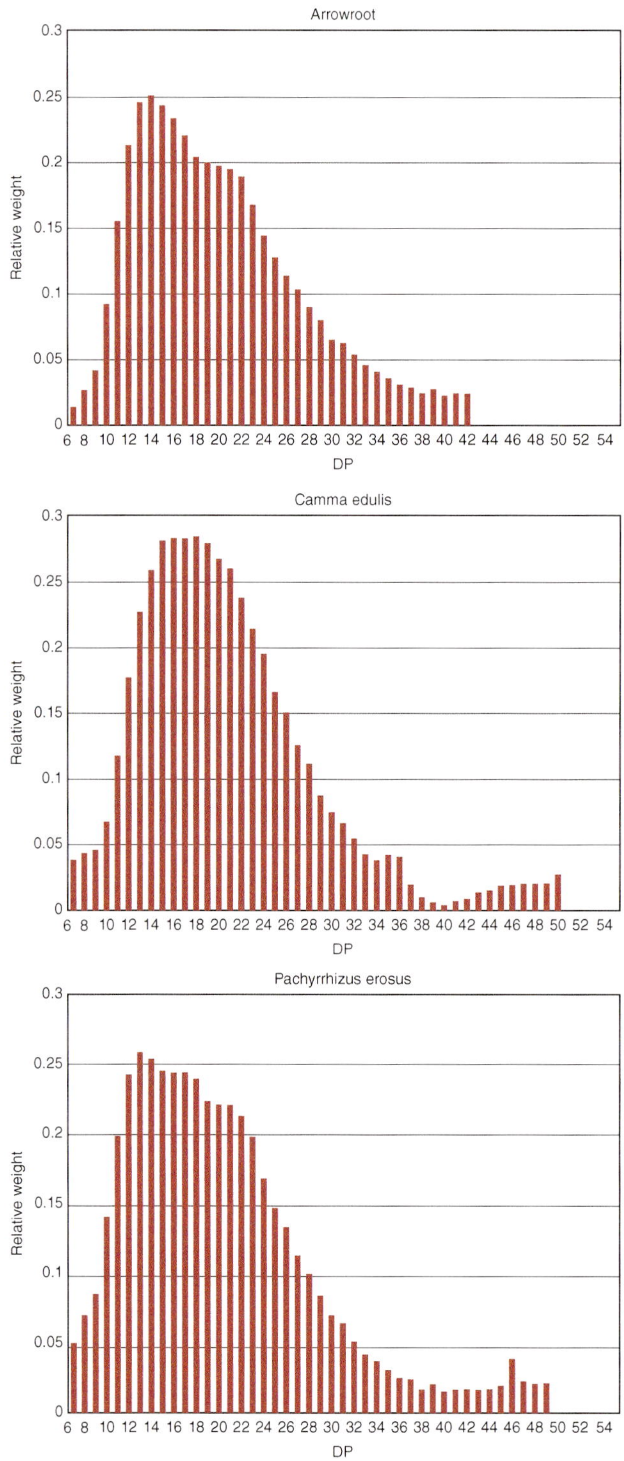

Fig. 7.2. HPSEC patterns of (a) arrowroot, (b) canna and (c) *Pacchyrhys* starches. (Source: Moorthy unpublished data)

Table 7.11. Swelling power of Chinese chestnut starch at different temperatures. (From Liu *et al.*, 2015)

Variety/temperature (°C)	50	60	70	80	92.5
Banhong	2.4	5.2	8.35	13.1	19.04
Yonlong	2.44	5.19	7.99	12.12	18.91
Yankui	2.25	4.78	7.59	12.34	18.45
Zaofeng	2.26	4.57	7.37	12.26	16.28

Table 7.12. Swelling power, solubility and gelatinization properties of bulb and bulbil starches from *L.lancifolium*. (Source: Yu *et al.*, 2015b)

Varieties	Swelling power (g/g)	Solubility (%)	T_o	T_p	T_c	ΔH	ΔT
Bulb	14.78 ± 0.12^a	17.46 ± 0.02^a	64.02 ± 0.12^a	68.30 ± 0.33^a	72.77 ± 0.30^a	12.09 ± 0.28^a	8.75 ± 0.26^a
Bulbil	14.37 ± 0.24^a	13.45 ± 0.03^b	62.85 ± 0.05^b	67.20 ± 0.11^b	72.43 ± 0.12^a	10.02 ± 0.12^b	9.58 ± 0.15^a

Table 7.13. Swelling properties of Alstroemeria. (Correa *et al.*, 2013)

	Alstroemeria		
Temperatures (°C)	WAI[b)]	WSI[c)]	SP[d)]
50	3.77 ± 0.26	1.28 ± 1.12	3.59 ± 0.53
60	8.11 ± 0.42	1.71 ± 0.15	8.17 ± 0.41
70	14.05 ± 0.25	6.18 ± 0.58	14.69 ± 0.34

Table 7.14. Viscosity characteristics of different tuber starches. (Source: Peroni *et al.* 2006)

Starch	Pasting temp	Peak viscosity RVU	Shear thinning RVU	Final viscosity RVU	Setback RVU
Arrowroot	71.7	339	204	197	62
Canna	71.2	413	196	335	120
Ginger	>95	78	0.2	119	41

showing that the granules are quite strong and resist breakdown under shear and heat (Jyothi *et al.*, 2003).

Pasting properties of starch were different among the three *Lilium* species (Yu *et al.*, 2015a). Lemos *et al.* (2015) compared the viscosity properties of starch from two chestnut varieties. Longal starch presented a lower gelatinization temperature range and higher peak height, suggesting more rigid swollen starch granules that contribute to the reinforcement of Longal gels (Lemos *et al.*, 2015). Supercritical fluid extraction of starch from *Curcuma* and ginger did not change the viscosity properties to any great extent compared to normal starch (Braga *et al.*, 2006). The viscograms of canna starches determined by Rapid viscoanalyser showed that Thai-green and Japanese-green starch paste were quite stable during cooking and had high setback (Thitipraphunkul *et al.*, 2003). The *C. longa* and *C. caesia* starches had different viscosity patterns (Hung and Duyen, 2017). Lee *et al.* (2015) have studied the viscosity properties of starches from chestnuts cultivated in Korea (see Table 7.15).

7.5.3 Rheological properties

Chestnut starch gels from two cultivars, Longal and Lada, showed different mechanical and rheological properties, being dependent on ageing temperature. Gels were more elastic at 5°C, evidenced by the higher intensity of relaxation spectra showing a slower relaxation process. The shape of the relaxation spectra was almost insensitive to ageing temperature suggesting a similar interaction at the two temperatures. Lada gels had the lowest equilibrium modulus, indicating a more fragile gel in the presence of shear disturbances. Under uniaxial compression, Longal gels were the most deformable before fracture and yield higher apparent Young modulus, showing also a higher degree of firmness (Lemos *et al.*, 2015).

Non-Newtonian shear-thinning characteristics were observed with turmeric powder dispersion containing 10% (w/w) solids. A power law model fitted well to correlate the shear-rate and shear-stress data (r = 0.993–0.999, $P \leq 0.01$) for both samples. Apparent viscosities of isolated turmeric starch and cured-dried turmeric powder dispersion containing 10% (w/w) solids were 1.29 ± 0.03 and 7.57 ± 0.39 mPa s, respectively (Kuttigounder *et al.*, 2011).

The rheological characteristics of starch extracted from canna and *Pacchyrhizus* starch have been examined by Moorthy *et al.* (2008) and show similar behaviour to other tropical tuber starches (Figs 7.3, 7.4 and 7.5); and Table 7.16 by Jiménez-Hernández 2007.

7.5.4 Retrogradation

The results obtained from freeze–thaw stability and light transmittance measurements indicated that starch from three canna varieties had a high tendency for retrogradation (Thitipraphunkul *et al.*, 2003). Lee *et al.* (2015) have compared the retrogradation properties of starch isolated from three cultivars of chestnut (see Table 7.17).

7.6 Thermal Properties

7.6.1 DSC

T_{onset} and T_{peak} for four chestnut starches were 59.5–69.2°C and 68.5–70.6°C and the ΔH was 9.7–10.8 J/g (Liu *et al.*, 2015). Thitipraphunkul *et al.* (2003) have studied three canna species and the values for T_{onset}, T_{end} and ΔH were 66°C, 71°C and 18 J/g, respectively. The DSC of Indian chestnut starch gave values of 69.6°C, 75.2°C and 7.3 J/g for T_{onset} T_{end} and ΔH, respectively (Singh *et al.*, 2009). The thermal properties of two *Curcuma* species were studied by Hung and Duyen (2017) (Table 7.18). Only slight

Table 7.15. RVA data of isolated chestnut starches and native corn starch. (Source: Lee *et al.*, 2015)

Sample	Pasting temp (°C)	Viscosity (RVU) Peak	Hold	Final	Breakdown	Setback
GJ	69.1 ± 0.2	587.0 ± 6.0	311.3 ± 13.3	477.7 ± 22.9	275.7 ± 10.3	166.3 ± 12.9
GY	70.3 ± 0.3	554.0 ± 8.2	357.0 ± 5.2	501.7 ± 15.9	197.0 ± 5.6	144.7 ± 12.7
HD	69.5 ± 0.1	528.7 ± 2.1	319.3 ± 3.1	451.3 ± 8.5	209.3 ± 2.5	132.0 ± 5.6

Breakdown = Peak Viscosity – Hold Viscosity; Setback = Final Viscosity - Hold Viscosity. GJ, isolated chestnut starch from Gongju-si chestnut fruit; GY, isolated chestnut starch from Gwangyang-si chestnut fruit; HD, isolated chestnut starch from Hadong-gun chestnut fruit.

Table 7.16. Rheological parameters of dispersions of chayote tuber starch (CHS). (Source: Jiménez-Hernández *et al.*, 2007)

Samples	(%)	η_{ap} (mPa s)	N*	K (Pa s)^n	R^2
CHS	1	75.09 ± 0.43^c	0.52	1.59	0.98
CHS	4	1715.20 ± 0.63^c	0.39	4.89	0.97

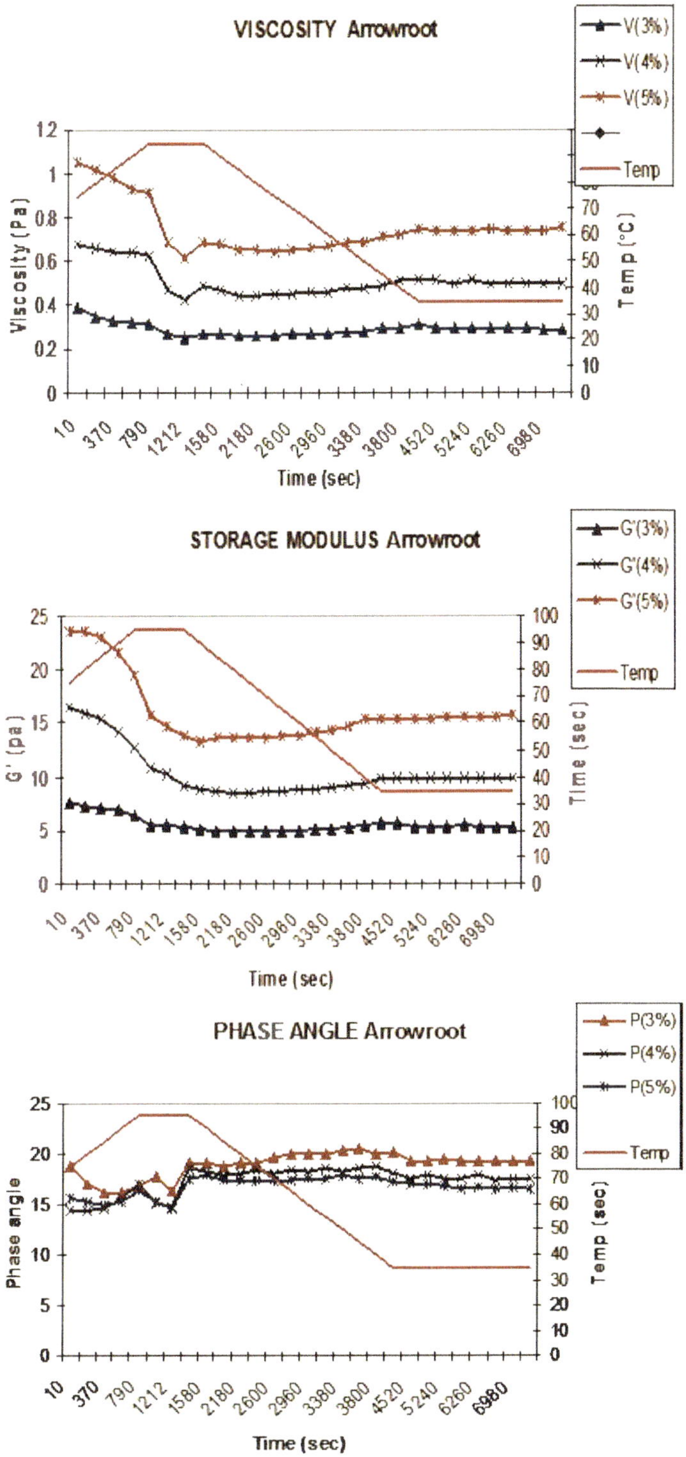

Fig. 7.3. Viscosity pattern, storage modulus and phase angle of arrowroot.

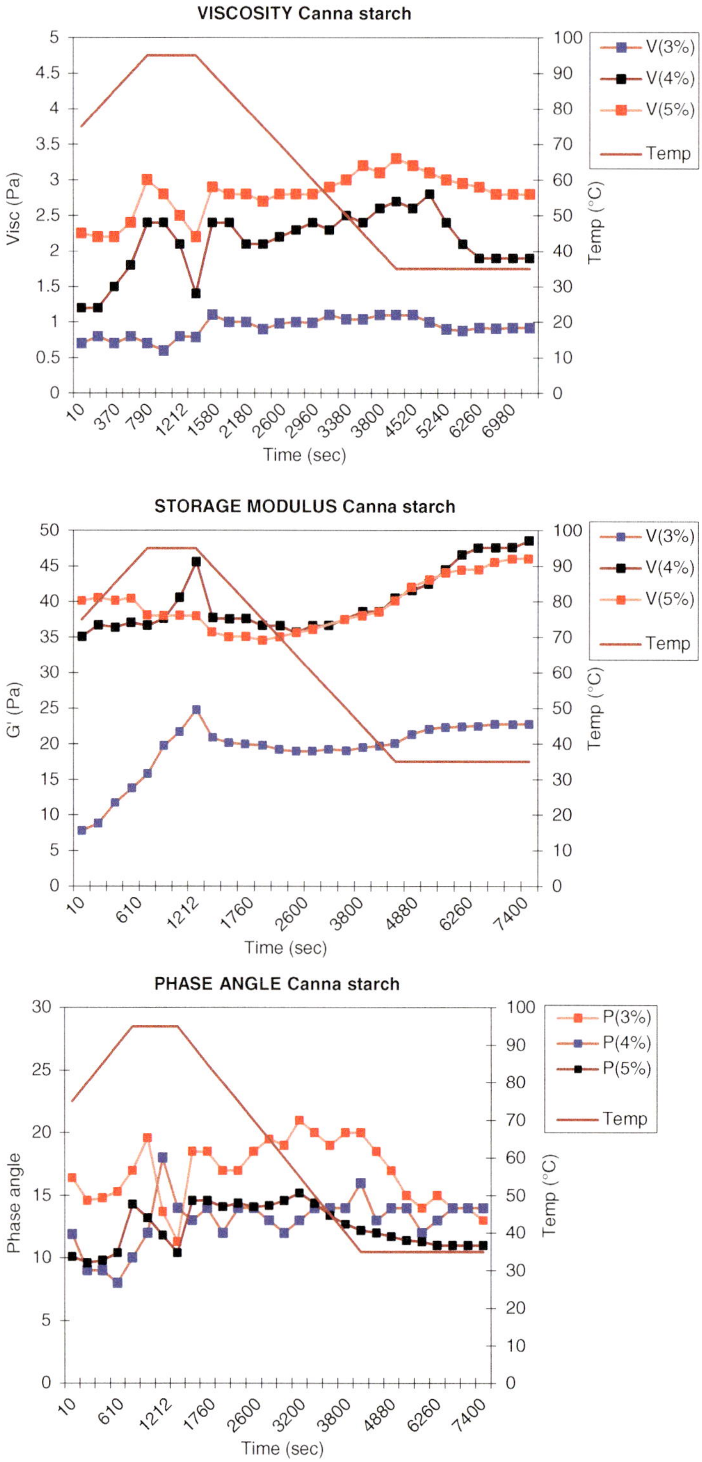

Fig. 7.4. Viscosity pattern, storage modulus and phase angle of canna starch (Moorthy, unpublished data).

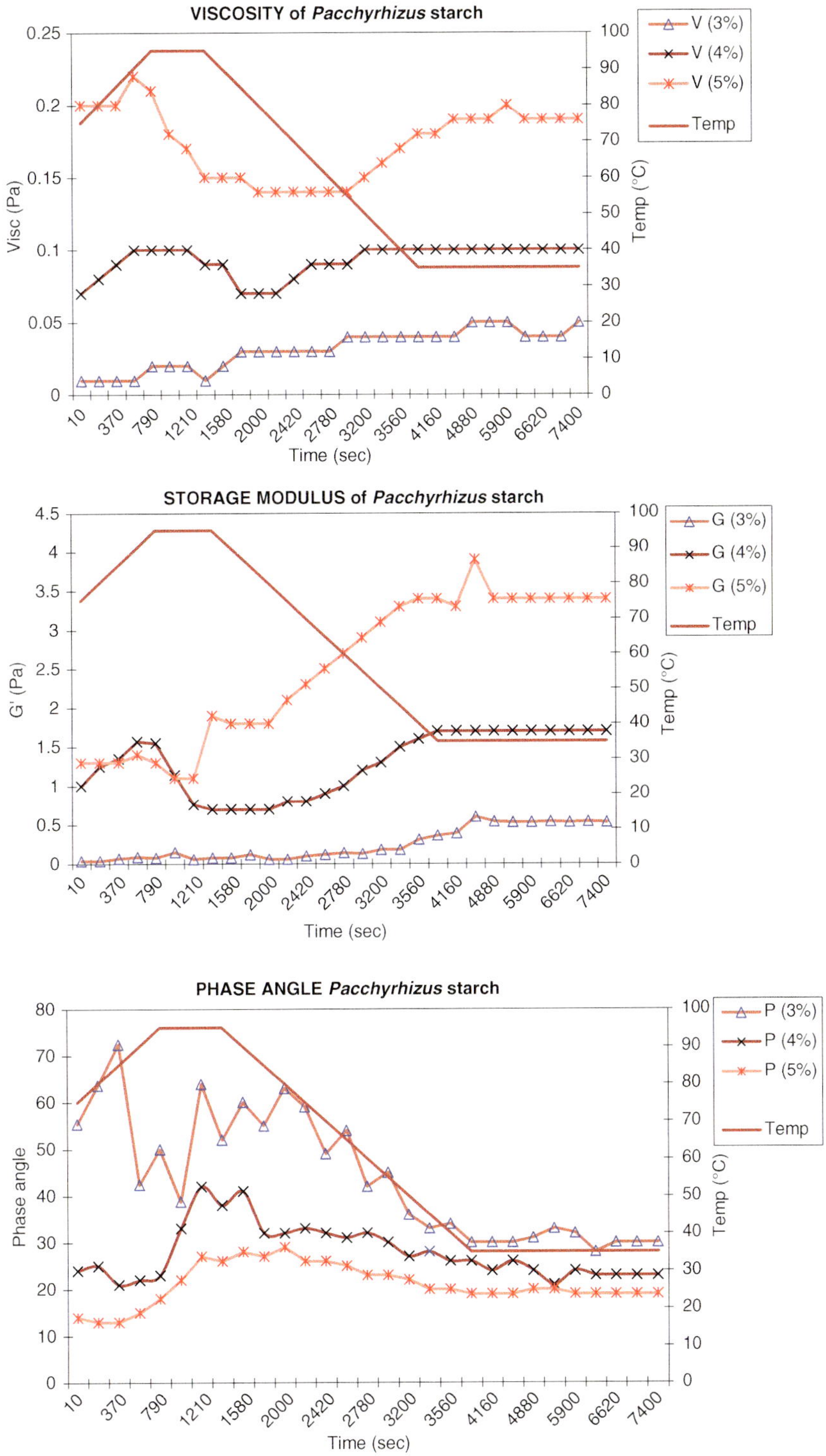

Fig. 7.5. Viscosity pattern, storage modulus and phase angle *Pacchyrhizus* starch (Moorthy, unpublished data).

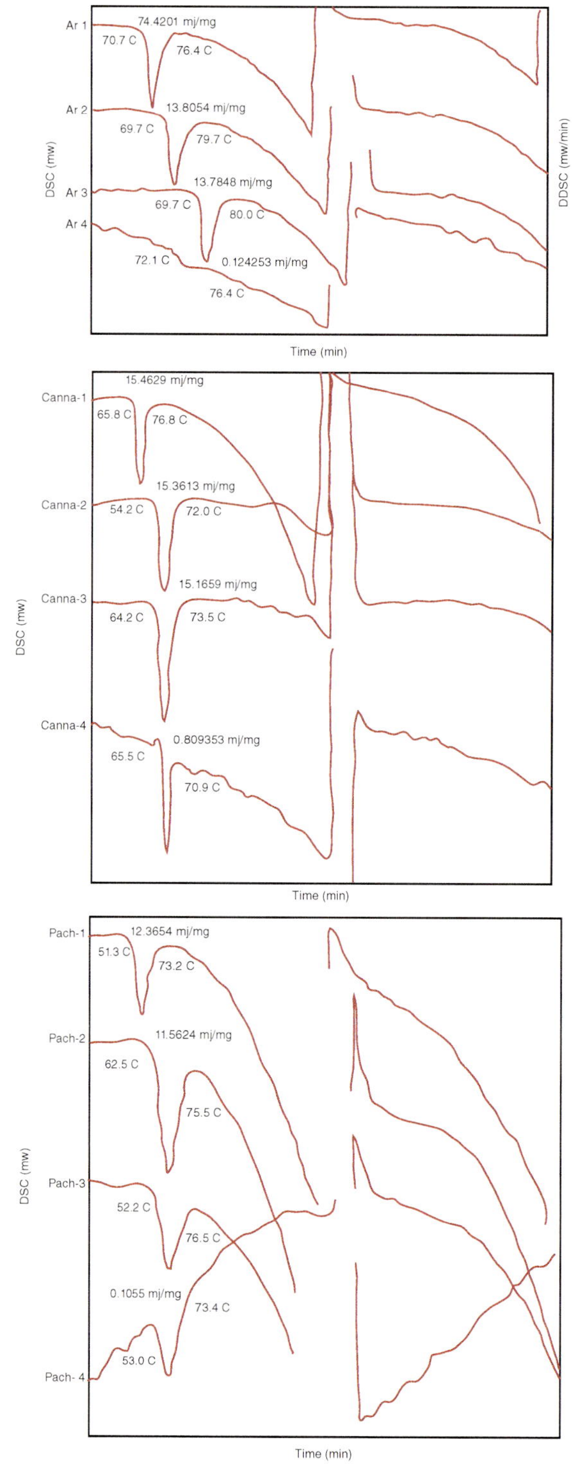

Fig. 7.6. MDSC of (a) arrowroot, (b) canna and (c) *Pacchyrhizus* starch (Source: Moorthy, unpublished data).

Table 7.17. Gelatinization and retrogradation properties of isolated chestnut starches analysed by DSC. (Source: Lee *et al*., 2015)

	Gelatinization				Retrogradation				Degree of
Sample	T_o (°C)	T_p (°C)	T_c (°C)	ΔH	T_o	T_p	T_c	ΔH	retrogradation
GJ	62.88 ± 0.02	67.37 ± 0.19	73.75 ± 0.17	11.11 ± 0.03	41.94 ± 1.74	56.50 ± 1.78	71.05 ± 0.92	6.02 ± 0.26	54.19 ± 2.51
GY	63.67 ± 0.17	68.36 ± 0.25	76.51 ± 0.65	9.97 ± 0.32	41.07 ± 0.17	55.51 ± 0.35	69.72 ± 1.01	5.26 ± 0.06	52.75 ± 2.30
HD	62.06 ± 0.07	67.25 ± 0.01	75.31 ± 0.22	9.40 ± 0.07	41.13 ± 0.20	54.84 ± 0.98	68.69 ± 1.08	4.57 ± 0.17	48.59 ± 2.15

The retrogradation percentage of *Lilium lancifolium* was lower compared to *L. leucanthum* and *L. rosthornii* (Yu *et al*., 2015).
GJ, isolated chestnut starch from Gongju-si chestnut fruit; GY, isolated chestnut starch from Gwangyang-si chestnut fruit; HD, isolated chestnut starch from Hadong-gun chestnut fruit.

Table 7.18. Thermal characteristics of *C. longa* and *C. caesia* starches. (Source: Hung and Duyen, 2017)

Sample	T_i (°C)	T_p (°C)	T_c (°C)	ΔH (J/g)
C. longa	74.12 ± 0.10	79.07 ± 0.18	84.57 ± 0.09	10.11 ± 0.05
C. caesia	76.61 ± 0.04	81.56 ± 0.00	86.37 ± 0.28	9.42 ± 0.83

Intial temperature (T_i), peak temperature (T_p), conclusion temperature (T_c), ΔH-enthalpy of the endothermic peak.
MDSC studies on arrowroot, canna and *Pacchyrhizus* have been carried out.

differences were noticed in gelatinization enthalpy of starch of *Lilium lancifolium*, *L. leucanthum* and *L. rosthornii* (Yu *et al*., 2015a). The same authors found that the T_{onset} and gelatinization enthalpy of starch of *Lilium* bulbs were higher than that from bulbils.

Differential scanning calorimetry thermogram data of the starches from the two *Curcuma* varieties have been compared (Jyothi *et al*., 2003). The gelatinization peak of *C. malabarica* starch was found to be a doublet. Multiplicity of peaks could arise due to deficiency of water during gelatinization. Since starch:water ratio of 1:3 was used, which provides enough water for gelatinization, the splitting of peak can be attributed to the structural differences in the starch. The T_{onset} was earlier for *C. malabarica* starch whereas T_{end} was nearly similar for both starches. The range of gelatinization of *Curcuma* starch was similar to yam and potato starches. The T_{onset} values of both the starches were similar to yam starches (75–80°C), but higher than cassava starch (65–69°C). The T_{peak} values also showed a similar trend. The ΔH values were similar.

Gelatinization temperatures were significantly lower, but gelatinization enthalpy was significantly higher, in *L. lancifolium* than in *L. leucanthum* and *L. rosthornii* (Yu *et al*., 2015). The enthalpy for gelatinization and dissociation of retrograded canna starches, investigated using differential scanning calorimeter, were 17.6–18.4 J/g and 12.3–15.0 J/g of starch, respectively (Thitipraphunkul *et al*., 2003).

The gelatinization enthalpy was 14.6 J/g of starch for *O. tuberosa*, 15.7 J/g for *C. edulis* and 17.6 J/g for *A. xanthorriza*, and the peak temperature of the endothermic DSC transition was 55.9°C, 61.2°C and 60.1°C, respectively (Santacruz *et al*., 2002). The *C. longa* and *C. caesia* starches had different gelatinization temperatures (Hung and Duyen, 2017).

7.7 Digestibility

Arrowroot starch is easily digestible and therefore finds wide use in convalescent foods, baby foods and bakery items, especially biscuits. *Arracacha* starch is easily digestible and is important in food applications. Chinese water chestnut starch finds use in canned foods because of its good digestibility. The digestibility of *Coleus* starch by porcine amylase was 11–16% for raw starch. The digestibility of the native winged bean starch by alpha amylase was very low (15%) but increased to 40% on gelatinization. However, glucoamylase completely hydrolysed and gelatinized water chestnut starch, which had digestibility of 15–40% depending on variety and enzyme used (Liu *et al.*, 2015).

The *In vitro* digestibility studies on *Lilium* starch showed that native and gelatinized bulb starches contained lower content of slowly digestible starch (SDS) and higher resistant starch than bulbil starch (Yu *et al.*, 2015b). In retrograded starch, the bulb and bulbils had equal values. The *C. longa* and *C. caesia* starches were also found to have high amounts of resistant starches (Hung and Duyen, 2017).

Starches from *L. leucanthum* and *L. rosthornii* showed higher resistance to porcine pancreatic α-amylase hydrolysis compared to that of *L. lancifolium* (Yu *et al.*, 2015b). *In vitro* digestion showed that native and gelatinized bulb starches contained lower slowly digestible starch (SDS) content and higher resistant starch (RS) content than bulbil starch. In retrograded starch, bulbs and bulbils had similar amounts of RDS, SDS and RS (Yu *et al.*, 2015b). Canna starch granules were predominantly superficially hydrolysed by α-amylase, even though some endocorrosion may have also occurred (Puncha-arnon *et al.*, 2007).

Thus the starches from minor tuber crops have diverse properties and need to be explored for food and industry applications.

References

Braga, M., Moreschi, S. and Meireles, M. (2006) Effects of supercritical fluid extraction on *Curcuma longa L.* and *Zingiber officinale* R. starches. *Carbohydrate Polymers* 63(3), 340–346. DOI: 10.1016/j.carbpol.2005.08.055.

Correa, Z., Zúñiga, A., Garfias, C. and Bello-Pérez, L.A. (2013) Isolation and characterization of *Alstroemeria hookeri* spp. hookeri starch in comparison with potato starch. *Starch/Stärke* 65(11–12), 991–998. DOI: 10.1002/star.201200265.

Doporto, M.C., Dini, C., Viña, S.Z. and García, M.A. (2014) *Pachyrhizus ahipa* roots and starches: composition and functional properties related to their food uses. *Starch/Stärke* 66(5-6), 539–548. DOI: 10.1002/star.201300196.

Eleazu, C.O., Eleazu, K.C., Iroaganachi, M.A. and Kalu, W. (2017) Starch digestibility and predicted glycemic indices of raw and processed forms of hausa potato (*S olenostemon rotundifolius* poir). *Journal of Food Biochemistry* 41(3), e12355–12356. DOI: 10.1111/jfbc.12355.

Guitiérrez, T.J. (2018) Characterization and in vitro digestibility of non-conventional starches from guinea arrowroot and La Armuña lentils as potential food sources for special diet regimens. *Starch/Stärke Bioengineering* 70(1–2), 1700124. DOI: 10.1002/star.201700124.

Hung, P.V. and Duyen, T.N.V. (2017) Structure, physicochemical characteristics, and functional properties of starches isolated from yellow (*Curcuma longa*) and black (*Curcuma caesia*) turmeric rhizomes. *Starch/Stärke* 5(6), 1–8.

Jayakody, L., Hoover, R., Liu, Q. and Weber, E. (2005) Studies on tuber and root starches. I. Structure and physicochemical properties of innala (*Solenostemon rotundifolius*) starches grown in Sri Lanka. *Carbohydrate Polymers* 38, 615–629.

Jiménez-Hernández, J., Salazar-Montoya, J.A. and Ramos-Ramírez, E.G. (2007) Physical, chemical and microscopic characterization of a new starch from chayote (*Sechium edule*) tuber and its

comparison with potato and maize starches. *Carbohydrate Polymers* 68(4), 679–686. DOI: 10.1016/j.carbpol.2006.07.035.
Jyothi, A.N., Moorthy, S.N. and Vimala, B. (2003) Physicochemical and functional properties of starch from two species of *Curcuma*. *International Journal of Food Properties* 6(1), 135–145. DOI: 10.1081/JFP-120016630.
Kuttigounder, D., Lingamallu, J.R. and Bhattacharya, S. (2011) Turmeric powder and starch: selected physical, physicochemical, and microstructural properties. *Journal of Food Science* 76(9), C1284–C1291. DOI: 10.1111/j.1750-3841.2011.02403.x.
Lan, W., Zhihua, Y., Yun, Z., Bijun, X. and Zhida, S. (2008) Morphological, physicochemical and textural properties of starch separated from Chinese water chestnut. *Starch/Stärke* 60(3-4), 181–191. DOI: 10.1002/star.200700645.
Lan, X., Li, Y., Xie, S. and Wang, Z. (2015) Ultrastructure of underutilized tuber starches and its relation to physicochemical properties. *Food Chemistry* 188, 632-640.
Lee, J.-W., Choi, H.-W., Seo, D.-H., Park, J.-D., Kum, J.-S. *et al.* (2015) Isolation and characterization of starches from chestnuts cultivated in three regions of Korea. *Starch/Stärke* 67(7-8), 585–594. DOI: 10.1002/star.201500034.
Leelavathi, K., Indrani, D. and Sidhu, J.S. (1987) Amylograph pasting behaviour of cereal and tuber starches. *Starch/Stärke* 39(11), 378–381. DOI: 10.1002/star.19870391103.
Lemos, A.M., Abraão, A.S., Cruz, B.R., Morgado, M.L., Rebelo, M. *et al.* (2015) Effect of granular characteristics on the viscoelastic and mechanical properties of native chestnut starch (*Castanea sativa* Mill). *Food Hydrocolloids* 51, 305–317. DOI: 10.1016/j.foodhyd.2015.05.021.
Liu, C., Wang, S., Chang, X. and Wang, S. (2015) Structural and functional properties of starches from Chinese chestnuts. *Food Hydrocolloids* 43, 568–576. DOI: 10.1016/j.foodhyd.2014.07.014.
Moorthy, S.N. (1986) Studies on Coleus starch. *Starch/Stärke* 38(3), 77–78. DOI: 10.1002/star.19860380303.
Moorthy, S.N., Vimala, B. and Mukherjee, A. (2002) Physicochemical and functional properties of *Canna edulis starch*. Tropical Science 42, 75–77.
Moorthy, S.N, Larsson, H. and Eliasson, A.C. (2008) Rheological characteristics of different tropical root starches, *Starch/Stärke* 60, 233–247.
Peroni, F.H.G., Rocha, T.S. and Franco, C.M.L. (2006) Some structural and physicochemical characteristics of tuber and root starches. *Food Science and Technology International* 12(6), 505–513. DOI: 10.1177/1082013206073045.
Puncha-arnon, S., Puttanlek, C., Rungsardthong, V., Pathipanawat, W. and Uttapap, D. (2007) Changes in physicochemical properties and morphology of Canna starches during rhizomal development. *Carbohydrate Polymers* 70(2), 206–217. DOI: 10.1016/j.carbpol.2007.03.020.
Santacruz, S., Koch, K., Svensson, E., Ruales, J. and Eliasson, A.C. (2002) Three underutilised sources of starch from the Andean region in Ecuador Part I. physico-chemical characterisation. *Carbohydrate Polymers* 49(1), 63–70. DOI: 10.1016/S0144-8617(01)00305-8.
Singh, G.D., Bawa, A.S., Riar, C.S. and Saxena, D.C. (2009) Influence of heat-moisture treatment and acid modifications on physicochemical, rheological, thermal and morphological characteristics of Indian water chestnut (*Trapa natans*) starch and its application in biodegradable films. *Starch/Stärke* 61(9), 503–513. DOI: 10.1002/star.200900129.
Thitipraphunkul, K., Uttapap, D., Piyachomkwan, K. and Takeda, Y. (2003) A comparative study of edible Canna (*Canna edulis*) starch from different cultivars. Part I. Chemical composition and physicochemical properties. *Carbohydrate Polymers* 53(3), 317–324. DOI: 10.1016/S0144-8617(03)00081-X.
Umadevi, S. and Wankhede, D.B. (1981) Isolation and physicochemical properties of starch from winged bean (*Psophocarpus* tetragonolobus). *Starch/Stärke* 33(1), 23–26. DOI: 10.1002/star.19810330108.
Wickramasinghe, H.A.M., Takigawa, S., Matsuura-Endo, C., Yamauchi, H., Noda, T. *et al.* (2009) Comparative analysis of starch properties of different root and tuber crops of Sri Lanka. *Food*112(1), 98–103. DOI: 10.1016/j.foodchem.2008.05.046.
Yu, J. and Wang, S. (2008) Morphological and crystalline properties of starches from new Sources-Traditional Chinese medicines (TCMs). *Starch/Stärke* 60(2), 110–114. DOI: 10.1002/star.200700661.

Yu, X., Zhang, J., Li, A., Wang, Z. and Xiong, F. (2015a) Morphology and physicochemical properties of 3 *Lilium* bulb starches. *Journal of Food Science* 80(8), C1661–C1669. DOI: 10.1111/1750-3841.12969.

Yu, X., Zhang, J., Shao, S., Yu, H., Xiong, F. *et al*. (2015b) Morphological and physicochemical properties of bulb and bulbil starches from *Lilium lancifolium*. *Starch/Stärke* 67(5–6), 448–458. DOI: 10.1002/star.201400209.

8

Starch Modifications

8.1 Introduction

Native starch has some limitations, especially for certain applications, and hence is often modified to overcome the undesirable properties while retaining the desirable ones. Nowadays, tailor-made starches are available to suit specific requirements; by modifying the structure and functionalities to suit diverse applications. The global modified starch market was valued at $7995 million in 2016 and is projected to reach $10,700 million by 2023, growing at a CAGR compound annual growth rate of 4.2% from 2017 to 2023. Modified starch is obtained by morphological or physicochemical changes in the structure of native starch via its treatment with heat, acid, alkali, chemical reagents or enzymes. These have improved water-holding capacity, heat resistance, binding characteristics and thickening properties. Modified starches are used widely in the food and beverage industry as thickening agents, stabilizing agents or emulsifying agents due to their partially digestible nature and their quality of being gluten-free, thus providing less fat than regular starches. The papermaking industry is another main application of modified starches, owing to their superior binding property for manufacturing high-strength and enhanced-quality paper with appropriate surface sizing, coating, and calendaring. Moreover, pharmaceutical, animal feed, textile, cosmetics, and other industries also make use of modified starch in various forms.

The global modified starch market is driven by increase in consumption of convenience foods and ease of modification in its functional properties. Furthermore, the rise in usage of modified starch in the pharmaceutical, cosmetics and paper industries fuels the growth. The newer and potential sources of modified starch and its untapped applications represent major opportunities for market expansion. All the major starches have been modified to suit to different applications. The modification of tuber starches, properties of modified starches and possible areas of application are outlined below.

8.2 Cassava

Physical, chemical and enzymatic modifications, or their dual/triple combinations, on cassava starch have been carried out to overcome the undesirable properties while retaining the useful ones (Zhu, 2015). The nature of modifications and their influence on starch's functional properties and structure have been summarized below.

8.2.1 Physical modifications

Physical modification of starch is very closely related to the emerging concept of 'green

Table 8.1. Pasting and rheological properties of native and octenyl succinylated cassava starch. (From Remya *et al*., 2017)

(Pa)

Sample	Peak viscosity (cP)	Pasting temperature(°C)	Storage modulus(G')	Loss modulus(GG'')	Phase angle (°)	Tan(δ)
Native cassava starch	1646 ± 36[g]	74.4 ± 0.05[c]	1270 ± 18.3[c]	1470 ± 11.8[e]	49.1 ± 1.12[e]	1.14[de]
S1	1791 ± 33[f]	74.9 ± 0.20[a]	1090 ± 8.8[e]	1450 ± 16.3[f]	53.0 ± 0.87[c]	1.33[b]
S2	3035 ± 46[b]	74.1 ± 0.25[a]	1360 ± 14.3[a]	1630 ± 8.4[b]	50.2 ± 0.96[d]	1.19[cd]
S3	2251 ± 83[c]	71.9 ± 0.15[f]	1230 ± 11.3[d]	1380 ± 6.4[g]	48.3 ± 0.78[f]	1.12[a]

technology' for environmentally friendly applications. It becomes all the more relevant in view of the hurdles in getting approval for genetic and chemically modified starches, especially for food applications (Hoover, 2010). Diverse physical modifications and also combined physical-cum-chemical modifications have been carried out to create the novel structure and properties of cassava starch.

Thermal modifications

Heat-moisture treatment of cassava starch has been carried out to modify the starch's properties. The treatment brought about considerable changes in physicochemical and functional properties. Moorthy (1985a) used steam pressure at different levels to modify starch. The final properties depended on the severity of treatment. Major changes occurred in the gelatinization temperatures, which were enhanced, while the swelling volume was lowered. The peak viscosity was reduced but the viscosity stability was improved, rendering it suitable for textile and paper applications. The results are explained as due to compression of the granules leading to closer and stronger bonding, and thus resulting in restricted swelling of the granules. Heating starch at 110°C for 16 hours in the presence of 20% moisture resulted in excellent freeze–thaw stability of starch, which was suitable for pie-filling (Abraham *et al*., 1993). Jyothi *et al*. (2005b) have studied the effect of annealing on cassava starch and identified the conditions required for achieving a product with desirable properties.

Electrical field modification

Electric current passed through starch at 33% or 45% moisture level to generate heat (termed ohmic heating) results in pre-gelatinized starch. The advantage of this method is that the solid loss is negligible, and it has been used to obtain modified starch with a range of degrees of gelatinization (Martínez-Bustos *et al*., 2005). In another study using a pulsed electrical field, the granules and crystallinity of the starch were disrupted, resulting in lowered viscosity, pasting temperatures and ΔH of gelatinization (Han *et al*., 2012).

High pressure-based modifications

High pressure-based treatments on starch represent a recent physical modification technique to impart novel functions and textures to food materials based on starch. High hydrostatic pressure, processing at 600 MPa, resulted in gelatinization and gel formation of starch (Vittadini *et al*., 2008). The physicochemical characteristics of the gel produced by high pressure indicated the presence of residual granules, higher ice-melting temperatures of frozen gel and less retrogradation. Oh *et al*. (2008) reported that pressure up to 600 MPa completely gelatinized the starch. Starch had lower gelatinization temperatures at 8 MPa but higher at 30 MPa in supercritical carbon dioxide medium (Francisco and Sivik, 2002). High pressure-based homogenization of starch suspension up to 100 MPa enhanced the temperature, leading to granular swelling and partial gelatinization, without affecting the wide-angle X-ray diffraction pattern (Che *et al.*, 2007a). High pressure-based microfluidization resulted in bigger starch granules with partial gelatinization, and induced a gel-like structure on the granule surface without affecting the X-ray diffraction pattern. Increasing pressure (up to 150 MPa) lowered the crystallinity percentage and the ΔH of gelatinization (Kasemwong *et al.*, 2011).

The effect of a high-speed jet on cassava starch water suspension was studied by Xia *et al.* (2015). The treatment affected the granule surface and crystallinity and increased the degree of gelatinization by over 30%. Thus it shows promise as a green technique to modify starch.

Extrusion

Extrusion using either single or twin screw extruders has been employed to prepare pre-gelatinized starch. Extrusion parameters such as moisture content, barrel temperature, and screw speed and number decided the properties of the resulting extruded starch (Leonel *et al.*, 2009; Nabeshima and Grossmann, 2001; Pérez-Sira and González-Parada, 1997). Pre-gelatinized starch with various degrees of gelatinization and degradation could be produced through extrusion.

Pre-gelatinization by drum drying

Pre-gelatinization of starch could be obtained by drum drying (Pérez-Sira and González-Parada, 1997; Moorthy, 2001). The resulting drum-dried starch possessed much higher water absorption, swelling and solubilization, but decreased apparent viscosity, and it can find uses in instant starches due to high miscibility in water and good tack.

UV-irradiation

UV-irradiation depolymerizes starch and reduces its intrinsic viscosity. ESR studies revealed that the reaction is brought about through free radicals (Bertolini *et al.*, 2001a). Studies on the influence of UV-irradiation coupled with lactic acid treatment of starch properties has assumed importance in accelerating the fermentation for production of the sour starch.

γ-Irradiation

γ-Irradiation decreased the swelling, viscosity and pasting temperature while increasing the solubility of starch, probably through free radicals, as revealed by electronic spin resonance (Bertolini *et al.*, 2001) and by the increased amount of reducing sugars (El Saadany *et al.*, 1976).

Ball milling

Ball milling involves a combination of interactions, namely constant friction of granules rubbing against one another, the ball and wall of the container mechanically breaking down granules into smaller pieces (Martínez-Bustos *et al.*, 2007a). The ball milling introduced breakdown of the molecules and the crystalline structure, partial gelatinization of the granules, reduced granule size, apparent viscosity, glass transition temperature and ΔH of gelatinization. It also increased the water absorption and cold water solubility without creating new chemical groups, as revealed by infrared spectroscopic analysis (Che *et al.*, 2007b; Huang *et al.*, 2007; Martínez-Bustos *et al.*, 2007a; Moraes *et al.*, 2013; Ren *et al.*, 2010; Sanguanpong *et al.*, 2003; Zhang *et al.*, 2013).

Cassava starch was ground in a ball mill for 4, 8, 16 and 32 hours and the structural and physicochemical characteristics were determined (Moraes *et al.*, 2013). Results obtained from HPAEC-PAD, GPC and amylose content determination indicated the breaking of hydrogen bounds and $\alpha - (1\rightarrow6)$ linkages of the starch molecules after treatment. X-ray diffractograms showed that the milling provided a reduction in the crystalline area of the starch granules. Most of the starch granules displayed agglomeration after four hours of milling, when observed under a scanning electron microscope. Solubility and water absorption capacity of the starch increased with an increase in the milling time, while RVA profiles showed a progressive reduction in the peak, breakdown and final viscosities, as well as the development of initial viscosity. Gelatinization temperatures and enthalpies were reduced. Prolonged ball milling accelerated the enthalpy of relaxation in the samples. These results confirmed a partial gelatinization of the starch (65.4%) after 32 hours of milling.

Blending with other starches

Simple blending of cassava starch with many other starches has been useful in modifying the functional properties of the starch. Maize starch and amorphophallus starch have been used in varying proportions to get desirable characteristics. Cassava starch was blended with konjac glucomannan and the mixture could be useful as tablet excipient in view of its superior binding and dispersing properties (Shanavas *et al.*, 2010).

A study was carried out to evaluate a mixed system based on native tapioca starch (NTS) with the addition of low methoxyl pectin (LMP) as a model for use in an acid environment (pH 3.0–3.2) (Agudelo *et al.*, 2014). This system was compared with a modified (highly cross-linked, hydroxypropylated and phosphated) waxy maize starch (MWCS) commonly used in industrial fruit fillings. The addition of LMP to replace 10% of the NTS significantly increased the peak viscosity (PV), hot paste viscosity (HPV), cold paste viscosity (CPV) and total relative setback of the NTS, and reduced relative breakdown. The mixed system also made the gel stronger and more elastic, achieving similar levels of viscoelastic properties to those of the control starch. At a frequency of 1 Hz in acid conditions and with 35% sugar, no significant differences in G', G" and tan G'/G" were found between the mixed system and the control (MWCS). The addition of pectin caused a significant increase in the firmness and consistency of the gels as analysed by extrusion tests.

8.2.2 Chemical modifications

Diverse types of chemical modifications have been reported. The modifications may be either by complexation of the chemical with starch, breakdown of starch structure or introducing various functional groups. Most of the chemical modifications have been well studied and applied in the food and non-food industry (Wurzburg, 1986). The increasing amount of research on chemical modifications of cassava starch for the last three decades underlines the importance of this methodology.

Complexation

Amylose fraction in starch can form strong complexes with lipids and surfactants. The hydrophobic long-chain portion of the surfactant or lipid can penetrate the interior portion of the starch helix while the hydrophilic end remains outside (Fig. 8.1). This brings about lowering of swelling by preventing easy entry of water into the starch molecules. Moorthy (1985b) has studied the effect of anionic, cationic and neutral surfactants of cassava starch. The viscosity properties varied widely depending on the surfactant used. GMS was very effective in complexing with the soluble amylose portion, thus offering a technique to reduce the cohesive texture of food products prepared using the starch. Radhika *et al.* (2008) have studied in detail the effect of various lipids on cassava starch properties.

Acid modification in aqueous solution

Hydrolysis of starch by mineral and organic acids (HCl, H_2SO_4, acetic or lactic acids) at various concentrations has been carried out to obtain acid-thinned starch, and the physicochemical, structural and morphological changes have been well documented. Lintnerization, a form of acid hydrolysis, occurs in two hydrolysis stages: first, the hydrolysis of the amorphous part of the granules and, second, of the crystalline region at a lower hydrolysis rate (Franco *et al.*, 2002). Acid treatment resulted in lower pasting viscosities and swelling power, higher solubility, alkali number, ΔH and broadened temperature of gelatinization, and decreased tendency to retrograde (Sanguanpong *et al.*, 2003; Beninca *et al.*, 2013b; Osunsami *et al.*, 1989; Atichokudom-chai *et al.*, 2001, 2002).

Studies on the surface morphological changes of granules eroded by acid solution (HCl, 6%) using scanning electron microscopy (SEM) and chemical force microscopy indicated that the erosion took place mostly on the surface without the formation of pores

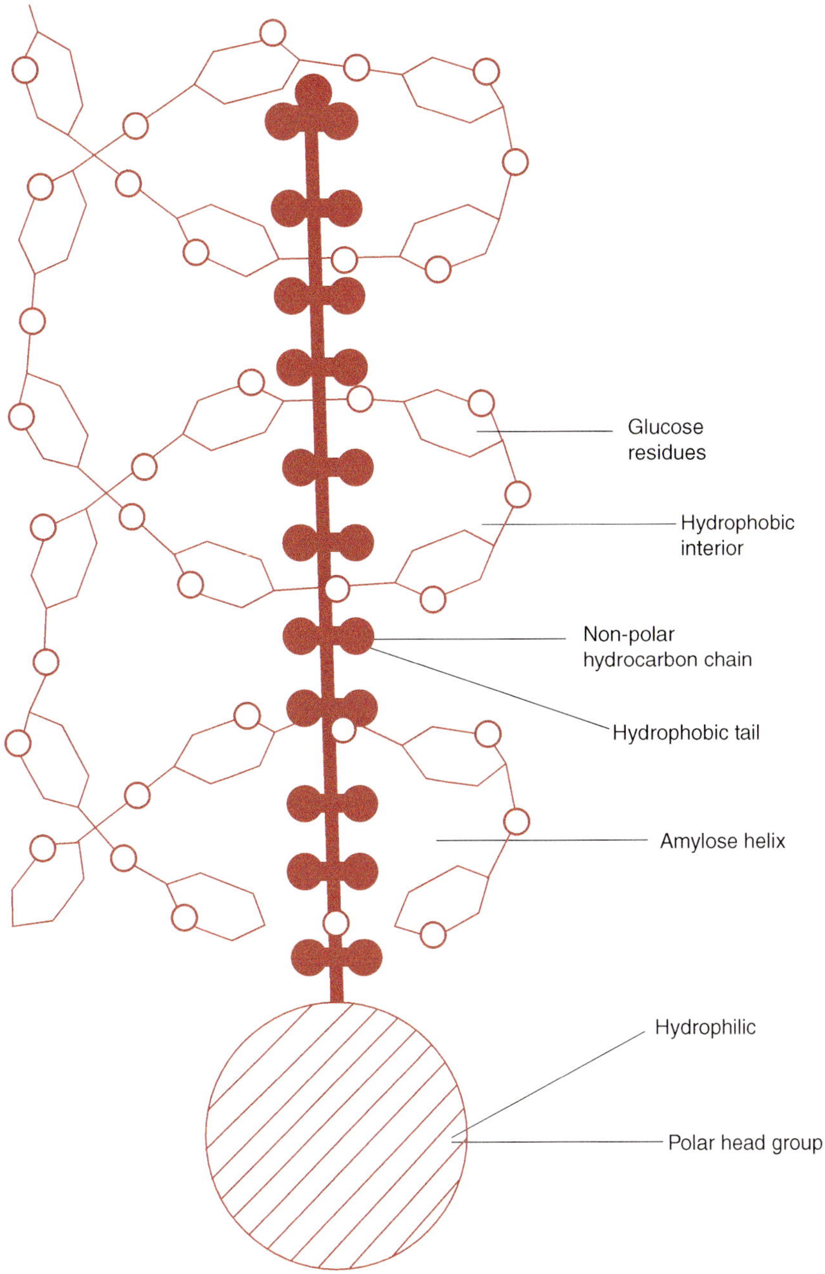

Fig. 8.1. Starch–surfactant complex.

(Atichokudomchai *et al.*, 2000; Wuttisela *et al.*, 2009). The carbon backbone of glucose units possessed a plate-like structure. The percentage of amylopectin and phosphate containing domains increased at the surface, with the increased acid concentration. The phosphate containing domains were packed in blocklets (Wuttisela *et al.*, 2009). The atomic force microscopy revealed that surface roughness of granules could be reduced by acid hydrolysis

(Beninca *et al.*, 2013b). Acid treatments (HCl or acetic acid) at low concentration altered the XRD pattern of starch from C- to A-type (Plata-Oviedo and Camargo, 1998) and increased the crystallinity percentage (Beninca *et al.*, 2013), while another study showed that acid hydrolysis (HCl) altered the A- or B-type crystallinity, depending on the specific genotypes with different structural properties (Rolland-Sabaté *et al.*, 2012, 2013). The ^{13}C cross-polarization magic angle spinning solid-state nuclear magnetic resonance (^{13}C CP/MAS solid-state NMR) spectroscopy and wide angle X-ray diffraction analyses have been used to study the physical properties (Atichokudomchai *et al.*, 2004). Relative crystallinity (%) increased initially before levelling off. The observed initial increase in relative double helix content and crystallinity was due to the destruction of the amorphous region, retrogradation of the partially hydrolysed amylose, and rearrangement of amylopectin double helices leading to crystallization (Atichokudomchai *et al.*, 2004). The amylose content decreased during hydrolysis, suggesting that on a molecular level most of the amylose chains were situated in the amorphous areas of starch granules (Atichokudomchai *et al.*, 2000; Franco *et al.*, 2002).

Acid modification in methanol/ethanol environment

Starch was treated with acids in methanol/ethanol medium to obtain a product with novel properties in high yield (Cavallini and Franco, 2010; Ferrini *et al.*, 2008). Treatment with HCl (0.36%) in anhydrous methanol resulted in greatly enhanced water solubility (up to 100% at 95°C), reduced granule size and lowered pasting viscosity. Gel-permeation chromatographic analysis showed that both amylose and amylopectin fractions were partially degraded during the hydrolysis (Ferrini *et al.*, 2008). Use of ethanol as a solvent during acid hydrolysis resulted in increased water solubility besides a high yield of ~98–99% (Cavallini and Franco, 2010).

Alkaline treatment

Treatment with NaOH solution resulted in lower amylose content, higher alkali number and increased enzyme susceptibility to amylase (*Bacillus* sp.) and this type of modification can be used to prepare maltodextrins of dextrose equivalent (DE) 20–23 (Raja, 1992).

Oxidation

Oxidation of cassava starch with hypochlorite, peroxide, acids, periodic acid or ozone resulted in oxidized starch with diverse properties depending on the type of oxidation as well as reaction conditions (Amorim *et al.*, 2011; Beninca *et al.*, 2013a; Klein *et al.*, 2014; Martínez-Bustos *et al.*, 2007b; Sangseethong *et al.*, 2010).

It was observed that when starch was oxidized with hypochlorite under pH of 8–11, use of pH 8 and 9 gave higher contents of carbonyl and carboxyl groups in the modified starch. Oxidation lowered gelatinization temperatures and ΔH of starch. Based on light-transmittance studies it was suggested that carboxyl groups retarded starch retrogradation (Sangseethong *et al.*, 2009). It was also found that starch oxidized by sodium hypochlorite solutions had increased ΔH and decreased pasting viscosity. Beninca *et al.* (2013) observed that the gelatinization temperatures and surface roughness were either increased or decreased depending on the specific concentration of hypochlorite solution. The glass transition temperature of oxidized starch was 3–6°C lower compared to control (Tran *et al.*, 2007).

A comparison of the structural properties of starch oxidized by peroxide and hypochlorite revealed that the former favoured the carboxyl group formation while the latter favoured the carbonyl group. Peroxide oxidized starch had a higher tendency for gelation with a firmer gel and increased the gelatinization temperatures, while hypochlorite oxidation had the opposite effect. Both types of oxidation lowered the ΔH of gelatinization, apparent viscosity and the molecular size of starch (Sangseethong *et al.*, 2010). Ozone introduced carbonyl and carboxyl groups onto starch, and rearranged amylopectin double helices leading to crystallization, thereby altering the sensory properties of starch (Amorim *et al.*, 2011; Klein *et al.*, 2014; Atichokudomchai *et al.*, 2004). The resulting product after lintnerization exhibited major peaks at DP of 13–14 with minor peaks at DP of 24–46 after debranching (Rolland-Sabaté *et al.*, 2012, 2013).

Cationization

Cationized starch has been produced by reaction of starch with 3-chloro-2-hydroxypropyl trimethyl ammonium chloride in an aqueous alcoholic-alkaline solvent (Han and Sosulski, 1998). Degree of substitution (DS) 0.03 was acheived at 50–60°C within two hours of reaction time. Cationization reduced the pasting temperature and viscosity of starch to a considerable extent.

Carboxymethylation

Carboxymethylated starch was obtained by treating starch with sodium monochloroacetate (SMCA) as an etherification agent with a reaction medium of a mixture of water and organic solvents (ethanol, isopropanol and tertiary butanol) (Qiu and He, 1999; Sangseethong *et al.*, 2005; Yao *et al.*, 2005). The product showed indentation and shrinkage on the granule surface and thus increased cold-water solubility, gel clarity, freeze–thaw stability, water-holding capacity, susceptibility to shear thinning, reduced gelatinization temperatures and retrogradation (Qiu and He, 1999; Sangseethong *et al.*, 2005). The preferred substitution position among anhydroglucose of starch was in the order C6 >C2>C3 (Yao *et al.*, 2005).

Succinylation

Starch succinate has been prepared by treatment with succinic anhydride in alkaline solution and the process could be rapidly achieved by microwave assistance (Jyothi *et al.*, 2005b). Starch succinate (DS up to 0.022) had increased swelling, viscosity, gel clarity and decreased enzyme susceptibility to amylase (Jyothi *et al.*, 2005a, 2005b). Starch octenyl succinate prepared by treatment with octenyl succinic anhydride through microwave radiation (Cova *et al.*, 2010) was more hydrophobic for the whole water activity range of 0–0.94.

Remya *et al.* (2017) have been successful in producing RS4-type resistant starch by chemical modification with octenyl succinic anhydride (OSA). The octenyl succinate ester exhibited significant changes in structural, functional and dietic properties in comparison to respective native counterparts. (Remya *et al.*, 2017) (see Table 8.1) The water-binding capacity of the starch increased upon modification. The cooked samples of octenyl succinylated cassava starch showed slowly digestible starch (SDS) and resistant starch (RS) in the range of 20.46–38.36% and 1.47–27.92%. RS and SDS of modified starches increased significantly and glycaemic index (GI) decreased in comparison to corresponding native starch (Remya *et al.*, 2017) (see Table 8.2).

Methylation

Using dimethyl carbonate as methyl group donor, disodium hydrogen phosphate/thiocarbamide as the catalyst and microwave irradiation as the energy source, methylated starch with DS from 0.033 to 1.087 was obtained (Hou *et al.*, 2011). The methylation was efficiently achieved using microwave assistance in a few minutes, and the whole process was safe, nontoxic and environmental friendly (Hou *et al.*, 2012). Reaction of cassava starch with benzaldehyde in the presence of zinc chloride provide

Table 8.2. Digestibility parameters of native and modified cassava starch. (From Remya *et al.*, 2017)

(%)

Sample	*In vitro* starch digestibility	Rapidly digestible starch (RDS)	Slowly digestible starch (SDS)	Resistant starch (RS)	Glycaemic index
Native cassava starch	423.0 ± 1.04[a]	78.1 ± 0.26[b]	20.5 ± 0.41[d]	1.5 ± 0.39[f]	91.2 ± 1.34[a]
S1	41.0 ± 0.12[a]	46.8 ± 0.32[c]	38.4 ± 0.57[a]	14.7 ± 0.64[d]	64.3 ± 0.94[c]
S2	36.1 ± 3.50[b]	40.9 ± 1.09[e]	38.4 ± 0.88[a]	20.8 ± 0.62[b]	62.1 ± 1.68[d]
S3	32.1 ± 1.57[c]	39.5 ± 0.09[e]	32.6 ± 1.12[c]	27.9 ± 1.52[a]	60.3 ± 0.44[e]

a modified starch with very high clarity and low retrogradation for its solutions (Johnston *et al.*, 2011).

Hydroxypropylation

Cassava starch was hydroxypropylated in water, 2-propanol and water in the presence of a phase transfer catalyst, tetrabutyl ammonium bromide. Phase transfer catalyst could increase the degree of substitution. The derivatives had high viscosity, swelling power, clarity, light transmittance, and water-binding capacity, but low retrogradation, PT, enzyme susceptibility and syneresis. The functionality of the derivatives like gel firmness, gumminess and springiness were superior to native starch and the derivative can have potential application in food industry.

Addition of long alkyl side chains

Esterification of starch with acyl imidazoles, acid chlorides and methyl ester derivatives of fatty acids having n-alkyl chains with 12–22 carbon atoms resulting in starch with much higher hydrophobicity at DS levels from 0.06 to 1.2 (Barrios *et al.*, 2013). Most of the derivatives from acyl imidazoles were stable up to 300°C. The substituted n-alkyl side chains of 16 or more carbon atoms were able to crystallize in separate paraffinic phases independent of the starch backbone (Barrios *et al.*, 2013).

Cassava starch was esterified using rosin acid in the presence of lipase to yield a product of DS 0.031–0.092%. The derivative possessed higher viscosity and emulsifying properties, lower swelling power, solubility and transparency in water, indicating applications in food and biomedical products (Lin *et al.*, 2015).

Citric acid modification

Pornsuksomboon *et al.* (2016) prepared starch-based foams using citric acid modified cassava starch and studied the properties of the foam in comparison to petroleum-based foams. They used a baking process for blending native and modified starches. The blended starch foam with a ratio 50:50 for native:modified starches revealed a uniform distribution of cell sizes. The baked foams could be useful to replace trays produced from petroleum-based foams.

Cross-linking (CL)

Cross-linked cassava starch was obtained by reacting starch granules with adipic and acetic mixed anhydrides, phosphorus oxychloride, sodium trimetaphosphate and epichlorohydrin in an alkaline medium. With increased degree of CL, the PV of pasting, swelling volume, solubility, light transmittance of starch paste and enzyme susceptibility to α-amylase and glucoamylase was reduced (Jyothi *et al.*, 2006; Yussof *et al.*, 2013), while ΔH of gelatinization was increased (Jyothi *et al.*, 2006). Cross-linking reduced the glass transition temperature of starch by 3–6°C (Tran *et al.*, 2007).

Cross-linked starch could also be produced using single-screw extruder as the reactor and sodium trimetaphosphate as the reagent. The phosphate cross-linked starch (DS up to 1.5×10^{-4}) increased gel strength, water absorption and resistance to high temperature and shear, but decreased gel cohesiveness, clarity and water solubility (Nabeshima and Grossmann, 2001). Enhancement of osmotic pressure by addition of sodium sulphate and sodium chloride increased the efficiency of cross-linking (Varavinit *et al.*, 2007). Wongsagonsup *et al.* (2014) cross-linked cassava starch with sodium trimetaphosphate and sodium tripolyphosphate and found that the high thermal stability and gel strength of the derivatives can be useful in soup-based products.

Starch foams from native cassava starch/cross-linked starch blends were prepared by baking in a hot mould (Pornsuksomboon *et al.*, 2014). The native cassava starch:cross-linked starch ratio was varied in the series of 100:0, 80:20, 60:40, 50:50, 40:60, 20:80 and 0:100. The effect of the native cassava starch:cross-linked starch ratio on the properties of foams, including density, morphology, water adsorption, impact strength and thermal stability, was investigated. The blending of cross-linked starch resulted in a higher density of the foams, which were between 232 J/m^2 and 268 J/m^2. The adsorption of water by the modified starch (7.3%) and the blend foams (11–12%) was lower than for the native cassava starch foam (13.9%). The size of the cells

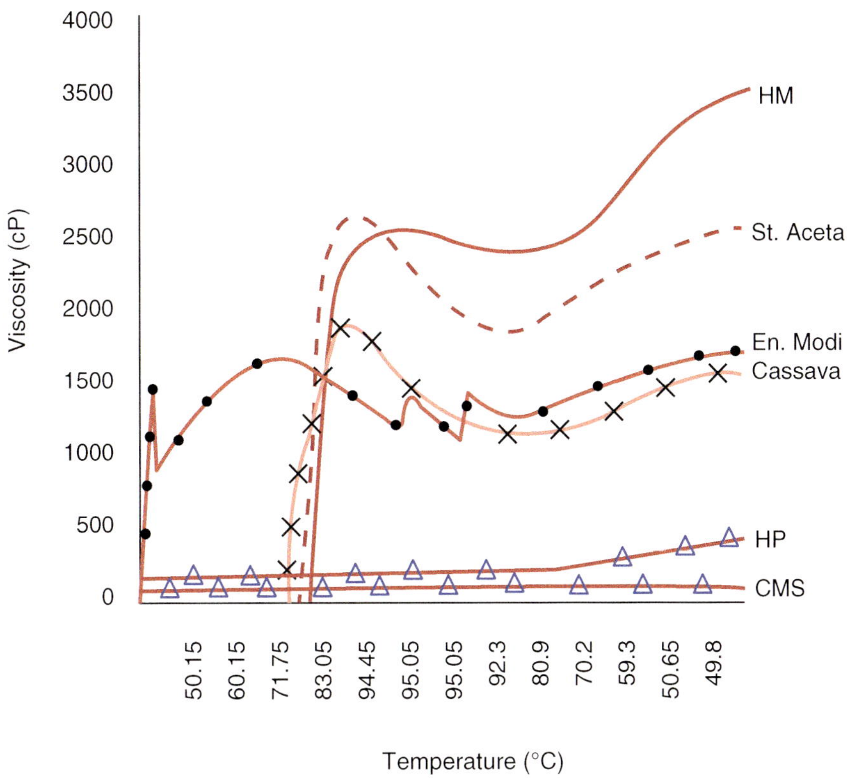

Fig. 8.2. Viscosity of various derivatives of cassava starch. (From Shanavas *et al*., 2010)

in the blend starch foam was between the cell size of the native cassava starch and the cross-linked starch foams.

Shanavas (2011) has studied the viscosity and storage modulus of physically, chemically and enzymatically modified cassava starch for possible application as tablet excipient (see Figs 8.2 and 8.3).

8.3 Dual Modifications

Some dual modifications have been used to produce products of desired properties.

8.3.1 Ball milling–aluminate coupling

Treatment of the ball-milled starch with aluminate coupling agent increases the hydrophobicity of starch. The coupling efficiency is facilitated by the ball milling (Che *et al*., 2007b).

8.3.2 Acid-treatment ball milling

When acid-treated starch was subjected to ball milling the total double helix content in the granules was not affected. The small starch particles produced during the treatment were structurally more heterogeneous as indicated by wider gelatinization range (Sanguanpong *et al*., 2003).

8.3.3 Carboxymethylation-thermal treatment

Modification of cassava starch by treatment with carboxymethyl cellulose followed by dry heating brought about vast changes in physicochemical and functional properties. Water and oil binding capacities and paste clarity were increased, while swelling and solubility decreased. Paste viscosity was also increased. The product

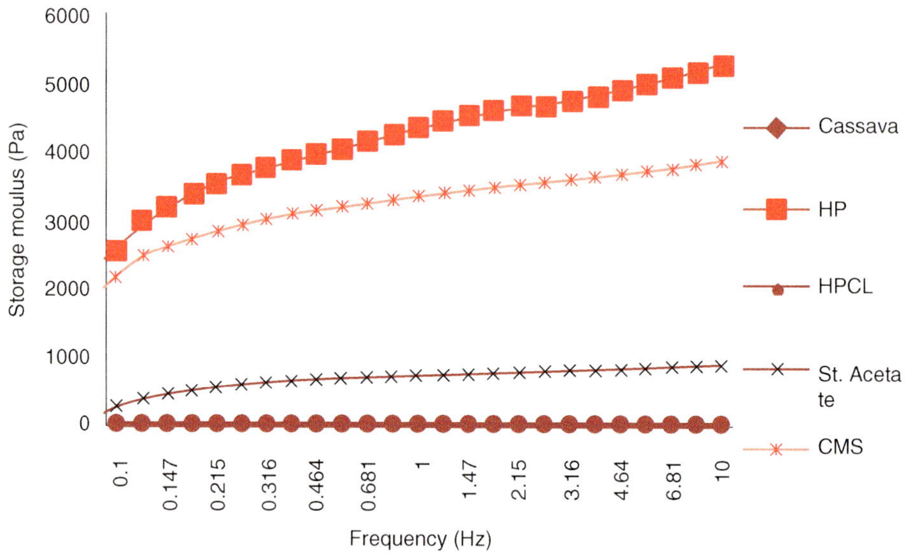

Fig. 8.3. Storage modulus of derivatives of cassava starch. (From Shanavas *et al.*, 2010)

can be used in food and non-food applications (Chandanasree *et al.*, 2016).

8.3.4 Carboxymethylation-long alkyl side chains addition

Ionic complexes produced by treatment of carboxymethylated starch with quaternary ammonium salts possessed more hydrophobicity at low water activity and more hydrophilicity at high water activity proved by moisture sorption isothermal analysis (Cova *et al.*, 2010).

8.3.5 Acetylation cross-linking

Dually modified starch produced by cross-linking acetylated starch had properties reflecting both reagents. The product showed increased PV and SB but decreased BD of pasting, glass transition temperature (T_g) up to 6°C and gel hardness (Tran *et al.*, 2007). Retrogradation of amylopectin was hardly affected, suggesting a rather low degree of modification.

8.4 Tertiary Modifications

8.4.1 Acid-ethanol treatment ball milling

Starch treated with acid in ethanol medium was further subjected to ball milling. The modification decreased the crystallinity, temperature, ΔH of gelatinization and granule size, while amylose content and pasting viscosity were not affected (Cavallini and Franco, 2010).

8.4.2 Oxidation cross-linking – phosphorylation

Oxidized cassava starch (OCS) was cross-linked with succinic acid (SA) and then phosphorylated with sodium tripolyphosphate to improve viscosity stability and usability of OCS for warp sizing (Zhu *et al.*, 2013). The level of cross-linking (LCL) and the degree of substitution (DS) of the phosphorylation were determined according to the stability, water dispersibility, adhesion to fibres, desizability, aerobic biodegradation and film properties of the cross-linked and phosphorylated OCS. The cross-linking exhibited significant effects on the stability, water dispersibility, adhesion, film properties and desizability, but was less sensitive to biodegradation. The phosphorylation after cross-linking not only showed no significant effect on the usability but also exhibited no adverse influence upon the stability. Suitable LCL and DS values resulted in stable viscosity, strong adhesion to fibre, tough starch film, good desizability and improved usability of OCS for warp sizing. Based on the stability and usability, the LCL and DS values were recommended to be one cross-link for 370–590 AGU units and 0.038, respectively.

8.5 Graft polymerization

Cassava starch has been graft polymerized with various monomers to give superabsorbent products. Superabsorbent polymers (SAPs) prepared from cassava starch by graft co-polymerization of acrylamide onto starch using ceric ammonium nitrate (CAN) as free radical initiator, followed by alkali saponification (Jyothi *et al.*, 2011a) were evaluated as a soil conditioner and their effect on physical, chemical and biological properties of the soil, as well as on plant growth parameters, was studied. Amendment of the soil with the starch-based SAP hydrogel significantly altered the physical and chemical properties of the soil. Bulk density of the soil was decreased, whereas water-holding capacity and porosity were increased on SAP treatment. There was an improvement in the soil status of major secondary micronutrients and organic carbon in the SAP-amended soil. An increase in bacteria (16%) and fungi counts (18%), as well as superior plant growth, were observed for the treatment with SAP and irrigated at intervals of three days, compared to those of control without SAP (Parvathy and Jyothi, 2014).

8.6 Enzymatic Modifications

Native starch has been subjected to hydrolysis by various enzymes to impart novel properties for diverse applications. Modification by bacterial and fungal lipases introduced fatty acid functional group (lauric acid and palmitic acid) onto the granules and the product possessed reduced amylase susceptibility, higher onset temperature of decomposition and enhanced hydrophobicity with potential as novel materials for thermoplastics (Rajan and Abraham, 2006; Rajan *et al.*, 2008).

Partial hydrolysis by thermo-stable amylase from *Bacillus licheniformis* broke down the granules into oligosaccharides with a broad molecular size distribution. Rheological analysis showed that the solid components in the slurry were considerably lowered by modification. The spray-dried hydrolysed starch could be used as fat substitute in view of the free-flowing properties (Khatoon *et al.*, 2009).

Starch was partially hydrolysed using branching enzyme (from *B. subtilis*) and maltogenic amylase (from *B. stearothermophilus*). The resulting modified starch had higher branching density and much lower enzyme susceptibility to amylase and glucoamylase as compared with native ones (Le *et al.*, 2009). Enzyme debranched starch was autoclaved and then stored at −18°C to obtain retrograded starch which had a typical B-type crystallinity and RS content of 35% (Hung *et al.*, 2012).

Starch was subjected to acid-hydrolysis-debranching-annealing/temperature-cycling/heat-moisture and their combined treatments produced recrystallized particles as resistant starch (Mutungi *et al.*, 2009a, 2009b; Onyango *et al.*, 2006; Onyango and Mutungi, 2008). The particles had mono- or bimodal size distribution (<5 m and >20 m) with the relative crystallinity of 31.1–56.1% and *in vitro* digestibility rate of 6.8–62.8% (Mutungi *et al.*, 2010). Upon digestion, average DP of the chains, percentage of crystallinity, ordered fractions and ΔH of resulted starch fractions increased. Raman spectroscopy could be feasibly employed to rapidly measure the starch crystallinity (Mutungi *et al.*, 2012). The moisture sorption behaviours of these modified starches were linked to their enzyme susceptibility.

8.6.1 Sour starch

Sour starch is a traditional food ingredient in Latin American countries produced by fermentation of native cassava starch followed by sun-drying cassava starch. During the process the granules are modified by enzymes and organic acids (mainly lactic acid) and the sunlight contributes to the oxidation of starch (Marcon *et al.*, 2009). In view of the superior expansion properties, sour starch is widely used in bakery products (Cardenas and Buckle, 1980). The very popular bread pan-de-yuca, made from sour starch, is a gluten-free product suitable for people with coeliac disease.

Fermentation (0 or 30 days) and drying (oven or sun) (dual modifications) decreased the pasting PV of sour starch. Both genotype and modification type contributed to the expansion and baking performance of the sour starch (Alvarado *et al.*, 2013; Onitilo *et al.*, 2007). Low values of intrinsic viscosity, hot-paste viscosities and hydrodynamic volumes of the components of sour starch could be related to higher baking expansion (Mestres and Rouau, 1997).

The microflora in the fermentation of starch were identified as species from *Streptococcus*, *Bacillus*, *Lactobacillus* and *Saccharomyces* with lactic acid bacteria being the most abundant (Parada *et al.*, 1996; Cardenas and Buckle, 1980; Figueroa *et al.*, 1995; George *et al.*, 1995). Introduction of glucose syrup during fermentation hastens the fermentation process while achieving the desired properties (Marcon *et al.*, 2006). Fermentation lowered pasting viscosity but increased the starch solubility (Nakamura and Park, 1975), whereas the chemical composition of the granules was little affected (Camargo *et al.*, 1988). Petruccelli *et al.* (1993) reported that fermentation lowered T_{peak} and T_{end} without affecting ΔH of gelatinization, and decreased paste viscosity. Fermentation increased apparent amylose content and ΔH of gelatinization, and decreased average granule diameter, solubility and swelling power, PV, BD and SB of pasting. The altered textural properties of starch paste were attributed to short amylose-like fragments formed by enzymatic hydrolysis of amylopectin (Numfor *et al.*, 1996).

Sour starch has been widely used in breadmaking. Expansion during baking of sour starch was attributed to water vaporization and paste fluidity, and was correlated with changes in thermomechanical properties of dough-crumb around the starch-melting temperature (Mestres *et al.*, 2001c). The carboxyl functional groups introduced during fermentation sun-drying contributed greatly to the expansion during baking. The good baking expansion activity has been associated with low paste viscosities and high swelling and solubilization (Mestres *et al.*, 2000).

Recent efforts have been initiated to find alternatives to natural fermentation and sun-drying with a view to improving the hygiene and increasing the fermentation rate. Controlled organic acid and enzyme hydrolysis has been suggested as an alternative for fermentation to produce sour starch (Camargo *et al.*, 1988). Sun-drying of the lactic acid treated starch gave good biscuit-making performance (Plata-Oviedo and Camargo, 1998). Hydrogen peroxide treatment imparted good expansion property to the sour starch and resulted in more homogeneous products with properties similar to the sour starch used in bakery (Dias *et al.*, 2007, 2011a). Similarly, dual-modified starch involving fermentation followed by sodium hypochlorite had increased expansion properties and reduced the biscuit hardness (Dias *et al.*, 2011b). Lactic acid and UV-irradiation treatments lowered the viscosities and setback of pasting with only minor effect on the DSC gelatinization parameters. Structural analysis indicated degradation of amylopectin and amylose molecules (Bertolini *et al.*, 1999; Franco *et al.*, 2010). The lactic acid treated starch irradiated with UV-B or UV-C possesses increased T_{peak} of gelatinization and the desired baking expansion ability. Chromatographic analysis showed that UV-B mostly degraded the amylose component while UV-C degraded both the amylose and amylopectin (Vatanasuchart *et al.*, 2005).

An inoculum-provided fermentation has been perfected for production of starchy flour from cassava tubers and the product was ideal for bakery items in view of superior expansion property (George *et al.*, 1995).

8.7 Sweet Potato

8.7.1 Physical treatments

Heat-moisture treatment (HMT) and annealing can facilitate starch chain interactions within the amorphous and crystalline domains, reflecting the structural defects in the granules (Hoover, 2010) and imparting characteristics such as reduced swelling, increased gelatinization temperatures and polymorphic transformation of B- to A-type crystallinity (Genkina *et al.*, 2004; Jyothi *et al.*, 2010). Electrolysis of starch in the presence of salt greatly enhanced the retrogradation, thereby increasing the amount of resistant starch (Lian *et al.*, 2012).

Blending sweet potato starch with other tubers has been practised as a simple technique of modification procedure for extending the range of functionality of the systems (Zhu and Corke, 2011). As physically modified starch appears 'greener' and presents healthier options to demanding modern consumers, more types of alterations, such as sonication and high-pressure treatments, need to be explored to obtain more diverse properties for specific application in food and industry.

8.7.2 Chemical modification

Acid modifications

Acid modification with HCl or H_2SO_4 solution has been carried out to alter the properties of starch, though it is reported that sweet potato starch is more resistant to acid hydrolysis compared with many other starches (Campanha and Franco, 2011). Genkina *et al.* (2009) developed novel genotypes that have an abundance of short unit chains of amylopectin, which would lead to more structural defects in the granules, facilitating faster degradation of starch.

Derivatization

Hydroxypropylation and acetylation studies have been carried out on sweet potato starch. Hydroxypropylation increased the swelling properties and susceptibility to α-amylolysis of the granules. Both the acetylated and hydroxypropylated starches lowered the temperatures and transitional enthalpy of gelatinization of native starches (Gunaratne and Corke, 2007; Lee and Yoo, 2009). Structural analysis revealed that the acetyl and hydroxypropyl groups were unevenly distributed over the amylopectin fractions owing to having different-sized granules and to the location of the molecules (Chen *et al.*, 2005; Zhao *et al.*, 2012). Acetyl groups were found to be located near the branching point, in the external and internal chain regions (Chen *et al.*, 2005). At higher levels of hydroxypropyl (HP), substituents were found to be present in the amorphous regions and periphery clusters of starch granules (Zhao *et al.*, 2012).

Sweet potato starches (SPS) were hydroxypropylated to evaluate the effect of molar substitution (MS, 0.042–0.153) on the rheological properties, freeze–thaw stability, paste clarity and gel strength (Lee and Yoo, 2009). The swelling power and solubility values of HPSPS were higher than those of native sweet potato starch (SPS) and increased with an increase in MS (see Table 8.3). The transition temperatures (T_{Onset} T_{Peak} and $T_{Conclusion}$), and enthalpy (H) of gelantinization of HPSPS, were lower than those of native SPS and significantly decreased with an increase in MS (see Table 8.4). Rheological properties of HPSPS pastes were measured under the conditions of steady and dynamic shear. Their consistency index (K), apparent viscosity (η_{a} ,100) Casson yield stress (σ_{oc}), complex (η^*) and dynamic moduli (G' and G") values decreased with an increase in MS; while their flow behaviour index (n) and tan δ (ratio of G' / G") values increased. The dependence of apparent viscosity on temperature followed the Arrehenius model for all samples. The paste clarity of HPSPS paste was enhanced with increasing MS of hydroxypropyl groups. The HPSPS gels exhibited lower gel strength and also better freeze–thaw stability with a significant decrease in syneresis (g/100 g) compared to native SPS (see Table 8.5).

Oxidation

By introducing carboxyl groups to replace the hydroxyl ones, starch was oxidized to produce low-viscosity pastes even at high concentrations, which were stable against retrogradation (Silva *et al.*, 2008).

Cross-linking

Cross-linking with phosphorus oxychloride ($POCl_3$) decreased the swelling power and amylose leaching, and increased shear stability and resistance to enzyme and acid hydrolysis. The rheological properties appeared much more affected by cross-linking compared to the thermal properties as measured by DSC (Gunaratne and Corke, 2007; Kim and Yoo, 2010). For example, cross-linking did not alter the melting enthalpy of sweet potato, but it decreased the peak viscosity of

Table 8.3. Hydroxypropyl content (g/100 g), molar substitution (MS), swelling power and solubility of hydroxypropyl sweet potato starches with different propylene oxide contents. (From Lee and Yoo, 2009)

Propylene oxide content (g/100 g)	Hydroxypropyl content (g/100 g)	MS (-)	Swelling power (g g^{-1})	Solubility (g g^{-1})
0 (native)	-	-	45.4 ± 0.2^{e}	15.2 ± 0.1^{e}
3	1.49 ± 0.01^{d}	0.042 ± 0.001^{d}	52.3 ± 0.3^{d}	19.6 ± 0.2^{d}
6	2.85 ± 0.01^{c}	0.082 ± 0.001^{c}	55.0 ± 0.2^{c}	21.6 ± 0.2^{c}
9	4.51 ± 0.01^{b}	0.132 ± 0.001^{b}	57.3 ± 0.4^{b}	23.7 ± 0.3^{b}
12	5.16 ± 0.01^{a}	0.153 ± 0.001^{a}	61.1 ± 0.4^{a}	25.2 ± 0.6^{a}

pasting four-fold (Gunaratne and Corke, 2007). The degree of cross-linking had a profound effect on the starch functional behaviours.

Dual modifications

Hydroxypropylation-acetylation and hydroxypropylation-cross-linking of sweet potato starch (Das *et al.*, 2010; Gunaratne and Corke, 2007) decreased the swelling factor and amylose leaching, melting enthalpy of DSC and peak viscosity of pasting, but increased the hot- and cold-paste viscosities of pasting (Gunaratne and Corke, 2007). Hydroxypropylation-cross-linking increased the starch gel strength, but decreased the rupture strength, gel elasticity and adhesiveness (Das *et al.*, 2010). Dual chemical modifications as well as single ones provide a wider range of functionalities for potential applications.

Enzymatic modification

Enzymatic modifications were less commonly employed compared with chemical and physical ones. One study reported the modification of sweet potato starch using 4-α-glucanotransferase, resulting in a more rigid and thermoreversible gel (Lee *et al.*, 2008). Using an inoculum-provided fermentation, sweet potato starch could be modified to improve its expansion properties and thereby its potential applications in bakery products (Jyothi *et al.*, 2001).

8.8 Yams

8.8.1 Physical modifications

Extrusion

Pregelatinized starch of *D. alata* could be produced by extrusion. The process disrupts the structure and decreased retrogradation compared to native starch. Since it possesses high cold-thickening capacity and gel strength, it can find applications in instant creams and pudding manufacture (Alves *et al.*, 1999).

Table 8.4. Thermal properties of hydroxypropylated sweet potato starches with different molar substitution (MS). (Lee and Yoo, 2009)

MS	T_o (°C)	T_p (°C)	T_c (°C)	(ΔH (J g^{-1})
0 (native)	66.6 ± 0.2^{a*}	76.9 ± 0.2^{a}	82.6 ± 0.2^{a}	16.5 ± 0.2^{a}
0.042	62.1 ± 0.0^{b}	74.0 ± 0.0^{b}	79.8 ± 0.1^{b}	15.6 ± 0.0^{b}
0.082	59.9 ± 0.1^{c}	70.9 ± 0.0^{d}	77.1 ± 0.1^{d}	15.0 ± 0.1^{c}
0.132	58.0 ± 0.1^{d}	67.0 ± 0.0^{d}	74.5 ± 0.1^{d}	13.9 ± 0.0^{d}
0.153	57.7 ± 0.1^{d}	67.3 ± 0.2^{d}	74.5 ± 0.3^{d}	12.3 ± 0.2^{c}

*Mean of three measurements. SD values in the same column with different letters are significantly different ($p<0.05$).

Table 8.5. Storage modulus (G') and loss modulus (G''), complex viscosity (η^* and tan δ at 6.28 rad s^{-1} for hydroxypropylated sweet potato starch pastes (5 g/ 100 g) with different molar substitution (MS) at 25°C . (Lee and Yoo, 2009)

MS	G' (Pa)	G' (Pa)	η^*(Pa s)	tanδ
0 (native)	29.5 ± 0.3[a*]	11.0 ± 0.0[a]	5.00 ± 0.04[a]	0.37 ± 0.01[c]
0.042	19.5 ± 0.4[b]	8.54 ± 0.28[b]	3.39 ± 0.08[b]	0.44 ± 0.01[b]
0.082	13.1 ± 0.3[c]	6.48 ± 0.24[c]	2.33 ± 0.02[c]	0.49 ± 0.02[a]
0.132	12.3 ± 0.2[d]	6.38 ± 0.10[c]	2.20 ± 0.03[d]	0.52 ± 0.02[a]
0.153	10.7 ± 0.1[e]	5.51 ± 0.02[d]	1.92 ± 0.01[e]	0.51 ± 0.01[a]

*Mean of three measurements. SD values in the same column with different letters are significantly different (p<0.05).

Heat-moisture treatment

Heat-moisture treatment (HMT) of yam starch at moisture levels less than 35% up to 16 hours at temperatures of 84–120°C has been examined. The treatment decreased the swelling, solubility and amylose leaching, and transformed B-type polymorph into A-type for the *D. hispida* and *D. alata* starches (Gunaratne and Hoover, 2002; Tattiyakul *et al.*, 2006, 2012; Trinh *et al.*, 2012). HMT altered the parameters of pasting and gelatinization (DSC), depending on the experimental conditions, namely: temperature, moisture content and duration of treatment. For example, HMT of *D. hispida* starch at a water-to-starch ratio of 13:100 (wet basis) changed the polymorph from B- to C-type, while a water-to-starch ratio of 18–30:100 hardly had any effect (Tattiyakul *et al.*, 2006). Slowly digestible starch fraction could be enhanced by selecting suitable experimental conditions (Trinh *et al.*, 2013). Favourable experimental conditions facilitate the mobility and interactions of chains to form new double helices as well as a more perfect crystal structure and compact granules, but too-harsh conditions disrupt the granular structure.

Annealing

Annealing decreased the gelatinization temperature range (DSC), amylose leaching, granular swelling, ΔH of retrogradation and susceptibility to acid hydrolysis, but increased temperatures and ΔH of gelatinization of starches of *D. alata* and *D. esculenta* (Jayakody *et al.*, 2009; Vamadevan *et al.*, 2013b). The process had a diverse effect on different genotypes, suggesting the influence of the composition and structure of starches (Jayakody *et al.*, 2009) and of cluster and building-block structure on the gelatinization behaviour of starch out of the uniqueness of yam starch (Vamadevan *et al.*, 2013a, b). Starch of *D. esculenta* exhibited one of the highest values for ΔH before and after ANN compared to other types of starches. This has been attributed to the longer IB-CL of *D. esculenta* amylopectin that facilitates the arrangement and packing of double helix and the crystals in the granules, resulting in a more perfect structure. The relatively minor effect on the ΔH of *D. esculenta* starch upon ANN suggested that the crystal packing in the native starch was highly ordered (Vamadevan *et al.*, 2013a, b).

Steam pressure treatment

Steam pressure treatment, using an autoclave, of starches from *D. alata* and *D. rotundata* reduced soluble amylose content, viscosity, swelling, paste stability and clarity without affecting the polymorph type. This type of modification appears to involve both altered pressure and heat moisture levels, with the moisture content of starch remaining as low as 7% (Moorthy, 1999), and proved to be a simple physical method for the modification of yam starches.

In another study, *Dioscorea alata* and *Dioscorea rotundata* starches were subjected to steam pressure treatment (0.35–0.7 kg cm^{-2} and 1.05 kg cm^{-2}) for different periods. The total amylose content in the treated starched was not significantly affected, but the soluble amylose content decreased three- to five-fold. Reducing values and A_{max} of the iodine complexes were unaffected. Viscosity of the starch paste was reduced by the treatment, and at higher pressures and longer time of treatment the peak

Table 8.6. DSC data on untreated and steam pressure-treated *D. rotundata* starch (From Moorthy, 1999)

Steam pressure (kg cm^{-2})	Treatment time (min)	T_{max} (°C)	ΔH (calg^{-1})
0	0	74.9	2.66
0.35	60	75.1	2.59
0.70	30	84.1	2.77
0.70	60	82.3	2.53
1.05	15	88.9	2.63
1.05	60	87.8	2.73

viscosity values were reduced to very low values. Pasting temperatures were enhanced considerably (see Table 8.6). Swelling volumes underwent reduction, but no change in solubility was noticed. Clarity and paste stability were markedly lowered. The results suggest that starch granules are compressed by the steam pressure treatment, leading to vast alterations in physicochemical properties.

Blending with other starches

Blending of different starches can lead to properties unique and different compared with those of individual starches. Starch of *D. alata* was mixed with maize and cassava starches and the textural, thermal and morphological properties indicated that blending starches at certain ratios reduced the exudate in yam and maize starch gels, as well as cohesiveness of cassava starch gel (Karam *et al.*, 2005, 2006).

To study the effect of the mixing, three types of starch from *Dioscorea* sp., *Dioscorea piperifoila* Humb. var Wild and *Dioscorea trifida* tubers were mixed in different proportions and their thermal, structural and rheological properties were characterized (Hornung *et al.*, 2017). Using thermogravimetric (TG) and differential thermal analysis (DTA), it was found that all the mixed samples present enhanced thermal stability. By differential scanning calorimetry (DSC) and viscoamylograph (RVA) techniques it was verified that the amylose content from the *Dioscorea* sp. starch influenced the gelatinization and paste properties of the mixed samples. Furthermore, using scanning electron micrograph (SEM) analysis it was possible to suggest a correlation between the size of the starch granules and their relative degree of crystallinity, which was calculated by the X-ray diffraction (XRD) patterns. The colour analysis revealed that, irrespective of the proportion, the strong tendency to retain the yellow colour of the native *Dioscorea piperfolia* predominated in the mixed samples.

8.8.2 Chemical modifications

Acid hydrolysis in aqueous environment

Acid hydrolysis of different genotypes of *D. opposita* using 2.2 mol/L hydrochloric acid up to 32 days at 35°C solubilized the starch, reduced particle size, increased the degree of crystallinity and changed polymorph from C- to A-type, confirming previous observations with other types of starch (Hoover, 2000; Wang *et al.*, 2007a, 2007b, 2008c, 2009; Yu *et al.*, 2009; Li *et al.*, 2011a). Structural studies indicated that during acid hydrolysis of *D. opposita* starch (C-type) the less crystalline B-type polymorph part was in the centre and the more organized A-type was in the peripheral part of the granules (Wang *et al.*, 2009a).

Acid hydrolysis in an alcohol medium

Acid hydrolysis of *D. opposita* starch in ethanol solvent (0.36% HCl in ethanol), up to 12 hours, lowered water-binding capacity, granule size, amylose content, temperatures and ΔH of gelatinization. The treatment also changed the polymorph from C- to A-type, indicating the preferential hydrolysis of the region having looser structure within the B-type polymorph. The starch yield was >92% and the use of ethanol solvent provided a higher starch yield compared to the aqueous medium (Jiang *et al.*, 2011).

Alkali treatment

The effect of alkali treatment for 0, 15 and 30 days at 300°C on amylose content, morphological properties, crystalline properties, swelling power, solubility absorbance spectra and *in vitro* digestibility of starch were investigated (Jiang *et al.*, 2014). The amylose content of all three starches decreased after 15 days of alkaline treatment and then increased after 30 days. There were similar changes in relative crystallinity for *D. zingiberensis* and *D. persimilis* starches.

It was observed that the three starches displayed a reducing trend in swelling power with a significant increase in solubility. Adhesion among some of the starch granules was observed after alkali treatment for 30 days in *D. zingiberensis* and *D. opposita* starches, while *D. persimilis* starch showed some hollows on the granule surface. The rise in the absorbance ratio of 1047:1035 cm and 1047:1022^{-1} cm from FTIR was observed during alkali treatment for *D. persimilis* and *D. opposita* starches. Alkali treatment elevated the *in vitro* digestibility, with resistant starch values climbing from 50.16% to 70.74% for *D. zingiberensis* and *D persimiliis* starches, respectively, but there was no significant change in resistant starch value for *D. opposita* starch.

Oxidation

Yam starch could be oxidized by various oxidants to create a range of structures and properties (Li *et al.*, 2011; Oladebeye *et al.*, 2013). Starch of *D. opposita* was oxidized by sodium periodate to produce dialdehyde starch, which had decreased molecular weight, degree of crystallinity, surface smoothness and thermal stability, but increased T_{onset} and T_{peak} of gelatinization as the content of aldehyde groups increased (up to 94%) (Zhang *et al.*, 2011). Oxidation of starch of *D. opposita* by sodium hypochlorite changed the polymorph from C- to A-type, increased solubility and decreased granule size, temperatures, ΔH of gelatinization and swelling (Li *et al.*, 2011a). Oxidation of starches of *D. cayenensis* and *D. rotundata* by ozone introduced the carbonyl and carboxyl groups (Oladebeye *et al.*, 2013), enhanced water solubility, weight-average molecular weight, amylose content and paste viscosity of pasting, and decreased number-average molecular weight, gel strength and setback of pasting. Ozone-based modification, being green and feasible, appears superior to oxidation by periodate or hypochlorite in solution, while at same time providing similar functional properties (Oladebeye *et al.*, 2013).

Native yam starch was modified by oxidation (with sodium hypochlorite), acid and enzyme (with glucoamylase), respectively (Xia *et al.*, 2011). Ash content, fat content, protein and AM contents were reduced following modifications. All the modifications reduced swelling capacity but increased the solubility. The SEM of the native starch revealed the presence of large oval or spherical, to small irregular-shaped granules; however, oxidation and hydrolysis caused collapse of the granules. From the XRD, the typical C-type XRD pattern of the native starch changed to the typical A-type XRD pattern of the three modified starches. The above results also demonstrated that the amorphous region of yam starch granules was preferentially degraded compared with the crystalline region of the granules. DSC showed that oxidation and acid/enzyme hydrolysis reduced the peak temperature of gelatinization (T_{peak}) of native starch significantly. Enthalpy of gelatinization (ΔH) of oxidation and enzyme starch was reduced but showed no obvious change following acid modification.

Substitution

Yam starches from several species have been subjected to substitution reactions, namely: carboxymethylation (Lawal *et al.*, 2008a; Nattapulwat *et al.*, 2009; Wang *et al.*, 2009), hydroxypropylation (Lawal *et al.*, 2008b; Wang *et al.*, 2009; Odeku and Picker-Freyer, 2009), acetylation (Wang *et al.*, 2008b) and succinylation (Lawal, 2012), and all the treatments increased the hydrophilic nature of the starch. Carboxymethylation of starch from *D. alata* was achieved using single- and multi-step approaches to obtain a range of degrees of substitution. The optimal DS of 0.98 was obtained using single-step modification. The DS could be increased to 2.24 by using nine step modifications, which decreased the crystallinity and improved the thermal stability (Lawal *et al.*, 2008a). Optimization of the reaction conditions to obtain maximum DS using single-step modification has been acheived (Nattapulwat *et al.*, 2009; Wang *et al.*, 2009b). Carboxymethylation improved the thermal stability and decreased the degree of crystallinity (Lawal *et al.*, 2008a; Nattapulwat *et al.*, 2009; Wang *et al.*, 2009b). Hydroxypropylation of starches from *D. alata*, *D. dumetorum*, *D. oppositifolia* and *D. rotundata* enhanced swelling, water solubility and susceptibility to α-amylase of *Bacillus licheniformis*, and decreased paste turbidity, syneresis, setback, retrogradation and DSC gelatinization parameters (Lawal *et al.*, 2008b; Odeku and Picker-Freyer, 2009b). The increased enzyme susceptibility has been explained as due to the disrupted granule and crystalline part (Lawal *et al.*, 2008b). Acetylation of *D. opposita* starch also disrupted the granules, increased starch thermal stability, and decreased glass transition temperature and degree of crystallinity (Wang *et al.*, 2008b). Succinylation of *D. cayenensis* starch increased swelling and water solubility, paste clarity, PV and BD of pasting, and lowered PT and SB of pasting and syneresis of paste (Lawal, 2012).

Cross-linking

Cross-linking of starches from *D. alata*, *D. dumetorum*, *D. oppositifolia* and *D. rotundata* in alkaline environment disrupt the granular structure, increased swelling power, solubility of starch in cold water (22°C) and decreased DSC gelatinization parameters (Odeku and Picker-Freyer, 2009b).

The effects of annealing, acid hydrolysis and citric acid modification on the physical and functional properties of starches of four Nigerian yam cultivars were studied (Falade and Ayetigbo, 2015) (see Table 8.7). Loose (0.3636–0.7658 g/cm^3) and packed bulk densities and CIE L* of native and modified starches decreased significantly (p<0.05) with annealing but increased while b* value induced modifications. The CIE a* and b* values of extracted starches ranged between 74.8–38.3 and 4.40–12.9, respectively. Among the native searches, L* ranked bitter yam>white yam>water yam >yellow yam. The starch granule shapes varied with cultivar but not with the modifications. Starch granule sizes reduced significantly with the modifications, with increased incidence of small and medium-sized granules and decrease in large-sized granules. Short and long length of native white, yellow, water and bitter yam starches ranged from 19.5 to 54.9, and from 15.6–42.9 to 27.30–5.0, respectively. Chemical modifications caused ething fragmentation and roughening of some granules' edges. The WAC (0.63 g/g) on annealing and OAC (0.437 g/g–1.053 g/g) was not significantly affected by the modifications. Generally, alkaline water retention (1.017–1.053) was not significantly affected by the modification of acid-hydrolysed yam starches. Emulsion index increased insignificantly with citric acid modification, while annealing and acid hydrolysis had variable effect.

8.8.3 Dual modifications

Hydroxypropylation cross-linking

Hydroxypropylation cross-linking of starches from *D. alata*, *D. dumetorum*, *D. oppositifolia* and *D. rotundata* increased swelling power and decreased gelatinization parameters and ΔH measured by DSC (Odeku and Picker-Freyer, 2009b).

Heat-moisture treatment debranching

Increasing the number of cycles of HMT (100°C, up to 24 hours) of debranched *D. alata* starch resulted in retrogradation starch having increased degree of crystallinity, temperatures and ΔH of melting, and RS content (up to 92% after 5 HMTs). The HMT transformed the polymorph from B- to C_A-type, and the boiling stable RS content reached 81% (Trinh *et al.*, 2013). This dual modification is a simple method that provides opportunities to increase the RS content in starch.

Ultrasonication-acetylation

Ultrasound treatment of *D. zingiberensis* starch prior to acetylation greatly enhanced the degree of acetylation and reaction efficiency. Through cavitation, ultrasound disrupts the granules and thereby allowing easier penetration of chemicals into the granules for reaction (Zhang *et al.*, 2012). It is expected that ultrasound treatment can be exploited to enhance the reactivity of starch for other types of modifications.

Enzymatic modifications

Physiochemical and biofunctional properties of three starches, including native starch (NS), the dual enzyme-treated starch (DES) and the cross-linked carboxymethyl starch (CCS), were investigated (Huang *et al.*, 2016). The resistant starch (RS) contents ranged from 33.91% to 43.71%, which were decreased by modification. For the first time, the anti-constipation and hypolipidemic effect of yam starches were demonstrated. The anti-constipation effect of the three starches was evaluated by intestine propulsion and defecation in constipated mice. There were more short-chain fatty acids (SCFAs) produced in the colon with DES and CCS than with native starches. Therefore, the

Table 8.7. Granule morphology and size distribution of native and modified yam starches. (From Falade and Ayetigbo, 2015)

Cultivar	Modification	Predominant shape(s)	Interval size distribution (μm)	Mean (μm)	Median (μm)	Mode (μm)	Small granules (%)	Medium granules (%)	Large granules (%)
White yam	Native	Triangle, oval, oval-triangular	15.60–54.90	33.89ab ± 8.77	35.10	39.00	0	21	79
	Annealed	Triangle, oval, oval-triangular	11.70–54.60	31.67bc ± 8.57	31.20	31.20	0	25	75
	Acid hydrolysed	Triangle, oval, oval-triangular	11.70–58.50	28.94d ± 9.76	27.30	23.40	0	43	57
	Citric modified	Triangle, oval, oval-triangular	11.70–58.50	27.65d ± 9.76	23.40	19.50	0	51	49
Water yam	Native	Oval, oval-triangular, triangular	15.60–58.50	32.21ab ± 9.42	31.20	23.40	0	30	70
	Annealed	Oval, oval-triangular, triangular	11.70–54.60	27.14d ± 9.63	25.35	19.50	0	50	50
	Acid hydrolysed	Oval, oval-triangular, triangular	7.80–97.50	28.59d ± 13.53	23.40	19.50	2	49	49
	Citric modified	Oval, oval-triangular, triangular	7.80–54.60	27.77d ± 10.59	23.40	23.40	1	50	49
Yellow yam	Native	Oval round	15.60–58.50	34.67a ± 8.66	35.10	35.10	0	13	87
	Annealed	Oval round	7.80–54.60	29.41cd ± 9.29	31.20	39.00	1	40	59
	Acid hydrolysed	Oval round	7.80–46.80	28.16d ± 8.56	29.25	31.20	2	36	62
	Citric modified	Oval round	7.80–46.80	28.16d ± 10.42	27.30	27.30	2	36	62
Bitter yam	Native	Random irregular polyhedral	2.50–7.50	4.35e ± 1.31	5.00	5.00	100	0	0
	Annealed	Random irregular polyhedral	2.50–7.50	4.45e ± 1.36	5.00	5.00	100	0	0

Continued

Table 8.7. Continued

Cultivar	Modification	Predominant shape(s)	Interval size distribution (µm)	Mean (µm)	Median (µm)	Mode (µm)	Small granules (%)	Medium granules (%)	Large granules (%)
	Acid hydrolysed	Random irregular polyhedral	2.50–5.00	4.25e ± 1.15	5.00	5.00	100	0	0
	Citric modified	Random irregular polyhedral	2.50–5.00	3.50e ± 1.23	2.50	2.50	100	0	0

Means in columns not followed by same letters are significantly different at 5% level ($p<0.05$). Small granules (1–10 µm), medium granules (11–25 µm) and large granules (>25 µm).

Table 8.8. Yam modifications. (From Zhu, 2015)

Type of modification	Species	Major features of result	References
Chemical			
Acid hydrolysis, A	*D. opposita*	Acid hydrolysis stabilized starch, reduced particle size, increased degree of crystallinity and changed polymorph from C- to A-type	Wang *et al.*, 2007a, 2007b, 2008a, 2008b; Li *et al.*, 2011a; Yu *et al.*, 2009
Acid hydrolysis in ethanol solvent, B	*D. opposita*	Acid treatment in ethanol environment gave high starch yield (>92%) while decreasing granule size, amylase content and temperatures and ΔH of gelatinization, and changing polymorph from C- to A-type	Jiang *et al.*, 2011b
Alkali treatment, C	*D. opposita, D. persimilis, D. zingiberensis*	Alkali treatment decreased amylase content (measured by an iodine spectrophotometer) first (15th day) before increasing again (30th day), increased water solubility, decreased swelling power and altered enzyme susceptibility	Jiang *et al.*, 2014
Oxidation by sodium periodate for dialdehyde	*D. opposita*	Dialdehyde starch had decreased molecular weight, degree of crystallinity, smoothness of granule surface, thermal stability of decomposition, ΔH, but increased To and Tp of gelatinization	Zhang *et al.*, 2011
Oxidation by sodium hypochlorite	*D. opposita*	Oxidation reduced swelling and increased solubility, damaged granules and decreased granule size, changed polymorph from C- to A-type, decreased temperatures and ΔH of gelatinization	Li *et al.*, 2011b
Oxidation by ozone	*D. cayenensis, D. rotundata*	Ozone oxidation increased contents of carbonyl and carboxyl groups, water solubility, weight-average molecular weight, amylase content, and cool-paste viscosity of pasting, and decreased number-average molecular weight, gel strength and setback of pasting	Oladebeye *et al.*, 2013
Carboxymethylation	*D. alata, D. esculenta, D. opposita*	Optimal degree of substitution (DS) of 0.98 was obtained using single-step modification. Nine-step modification increased DS to 2.24, while decreasing the crystallinity and improving thermal stability. Structural changes in starch through modification were followed by diverse methods such as infrared and nuclear magnetic resonance spectroscopy	Lawal *et al.*, 2008a; Nattapulwat *et al.*, 2009; Wang *et al.*, 2009b

Continued

Table 8.8. Continued

Type of modification	Species	Major features of result	References
Chemical			
Hydroxypropylation	*D. alata, D. dumetorum, D. oppositifolia, D. rotundata*	DS increased as volume of propylene oxide increased. Hydroxypropylation increased swelling, solubility and enzyme susceptibility to α-amylase of *Bacillus licheniformis*, and decreased paste turbidity, syneresis, setback and retrogradation, and gelatinization parameters measured by DSC	Lawal et al., 2008b; Odeku and Picker-Freyer, 2009
Acetylation, D	*D. opposita*	Acetylation increased thermal stability of starch, decreased glass-transition temperature and crystallinity and disrupted the granules	Wang *et al.*, 2008e
Succinylation	*D. cayenensis*	Succinylation increased the swelling and solubility, paste clarity, peak viscosity and breakdown of pasting, reduced pasting temperature and setback of pasting, and syneresis of paste	Lawal, 2012
Cross-linking	*D. alata, D. dumetorum, D. oppositifolia, D. rotundata*	Cross-linking increased swelling power and solubility of starch and decreased gelatinization parameters measured by DSC	Odeku and Picker-Freyer, 2009
Hydroxypropylation – cross-linking, E	*D. alata, D. dumetorum, D. oppositifolia, D. rotundata*	Dual modification of hydroxypropylation cross-linking of starch increased swelling power and decreased temperatures and ΔH of gelatinization measured by DSC	Odeku and Picker-Freyer, 2009
Physical modification			
Extrusion	*D. alata*	Pre-gelatinized starch with a range of properties (cold viscosity, gel-forming capacity and retrogradation) was prepared by altering the extrusion conditions	Alves *et al.*, 1999
Annealing	*D. alata, D.esculenta*	Annealing decreased the gelatinization temperature range (DSC), amylase leaching, granular swelling, ΔH of retrogradation, susceptibility toward acid hydrolysis, and increased temperatures and ΔH of gelatinization. Different starch genotypes showed different response to annealing	Jayakody *et al.*, 2009
Annealing	*D. esculenta*	Annealing increased To and ΔH of gelatinization while narrowing the range. Cluster structure of amylopectin was related to the annealing effect on starch.	Vamadevan *et al.*, 2013b

Continued

Table 8.8. Continued

Type of modification	Species	Major features of result	References
Chemical			
Blending yam starch with maize and cassava starches	*D. alata*	Textural, thermal and morphological properties of starch blends were studied. Blending starches at a certain ratio reduced the exudates in yam and maize starch gels, and cohesiveness of cassava starch gel.	Karam *et al.*, 2005, 2006
Heat-moisture treatment	*D. hispida*	HMT (water-to-starch ratio of 13:100 to 30:100) decreased swelling and amylase leaching while increasing gelatinization temperature. It either increased or decreased pasting viscosity, depending on moisture content of the system. At a starch-to-water ratio of 13:100, starch polymorph changed from B- to C-type.	Tattiyakul *et al.*, 2006
Heat-moisture treatment	*D. hispida*	HMT (moisture 13%, 90–130°C for 10 hours) changed polymorph of granules from B- to A- or C-type. HMT decreased swelling, starch solubility and amylase leaching. HMT at less than 100°C decreased degree of crystallinity, temperatures and ΔH of gelatinization, and increased the PV of pasting and the complex moduli over 0.001 to 10 Hz in dynamic shear test (7% starch content)	Tattiyakul *et al.*, 2012
Chemical heat-moisture treatment, F	*D. alata*	HMT changed B-type polymorph to A-type, decreased swelling power, amylase leaching, ΔH of gelatinization and susceptibility to acid retrogradation. HMT increased gelatinization temperatures and enzyme susceptibility.	Gunaratne and Hoover, 2002
Heat-moisture treatment of non-granular starch	*D. alata*	SDS content was enhanced by controlling the conditions (moisture content [25–35%] and reaction time [up to 24 h]) of HMT at100°C. The heat-moisture treatment led to formation of A-type polymorph, increased double helix content, and increased ΔH of retrograded starch.	Trinh *et al.*, 2012
Steam-pressure treatment	*D. alata, D. rotundata*	Steam-pressure treatment reduced soluble amylase content, viscosity, swelling, paste stability and clarity.	Moorthy, 1999

Continued

Table 8.8. Continued

Type of modification	Species	Major features of result	References
Chemical			
Dual modification			
Heat-moisture treatment after debranching	*D. alata*	HMT (100°C, up to 24 h) of debranched starch increased degree of crystallinity, melting temperature and ΔH of gelatinization, and RS content (up to 92% after 5HMT cycles) of retrograded starch, and changed polymorph from B-to CA-type; boiling-stable RS content reached 81%	Trinh *et al.*, 2013
Ultrasonication acetylation	*D. zingiberensis*	Ultrasound treatment before acetylation greatly enhanced degree of substitution and reaction efficiency	Zhang *et al.*, 2012

A = 2.2 mol/L hydrochloric acid, up to 32 days, at 35°C; B = 0.36% HCl in ethanol up to 12 h; C = 30 days at 35°C; D = granule, using glacial acetic acid/acetic anhydride with sulphuric acid as the catalyst; E = hydroxypropylation followed by cross-linking; F = 100°C for 10 h, moisture 30%.

modified starches were proved to be superior bioactive starches for improving the small intestine propulsion function by generating more SCFAs, which can be exploited in the pharmaceutical and food industries.

8.9 Aroids

8.9.1 Physical modifications

Aroid starches have been subjected to dry heat (Pramodrao and Riar, 2014) and heat-moisture treatment (Gunaratne and Hoover, 2002). Dry heating of *C. esculenta* starch at 130°C for up to four hours decreased the paste transmittance, increased the sediment volume and water-binding capacity, altered the colour and had minor effect on the swelling and solubility of starch (Pramodrao and Riar, 2014). Heat-moisture treatment of *X. sagittifolium* starch (30% moisture) at 100°C for 10 hours had no noticeable effect on the polymorph type (A-type) and retrogradation, but at the same time decreased swelling, amylose leaching, ΔH of gelatinization and susceptibility to acid hydrolysis, and increased temperatures of gelatinization and enzyme susceptibility (Gunaratne and Hoover, 2002).

Peruvian carrot (*A. xanthorrhiza*) starches were ground in a ball-mill for four, eight, 16 and 32 hours and the result indicated a breaking of hydrogen bounds and α-(1→6) linkages of the starch molecules after treatment. X-ray diffractograms showed that the milling provided a reduction in the crystalline region of the starch granules. After 16 hours a shapeless mass was obtained. Solubility and water-absorption capacity increased with the milling time, while RVA profiles showed a progressive reduction in peak, breakdown and final viscosities, as well as the development of initial viscosity. Gelatinization temperatures and enthalpies were lowered. Prolonged ball milling accelerated the enthalpy relaxation in both starches. These results confirmed a partial gelatinization of up to 82.6%. This starch was more affected by ball milling compared to cassava starch because of its lower amylose content and defects in its crystalline structure (Moraes *et al.*, 2013) (see Table 8.9).

Blending

The physicochemical properties of *Amorphophallus paeonifolius* (Dennst.) Nicolson starch (APS) and the properties of the starch konjac glucomannam (APS/KGM) were systematically studied (Meng *et al.*, 2015) (see Table 8.10). The pasting temperature decreased with KGM addition, while the peak viscosity increased. The storage modulus G′ and loss modulus G″ of APS/KGM were higher than those of APS alone.

Table 8.9. Branch-chain length distribution of amylopectins of native and milled Peruvian carrot starch. (From Moraes *et al.*, 2013)

Branch chain length distribution (%)						
Milling time (h)	DP 6–12	DP 13–24	DP 25–36	DP >37	DP	Highest detectable DP
Peruvian carrot						
0	33.53 ± 0.37 a	43.45 ± 0.94 a	11.57 ± 0.08 a	10.91 ± 0.10 a	19.22 ± 0.06 a	78
4	33.52 ± 0.11 a	43.53 ± 0.40 a	11.60 ± 0.01 a	10.91 ± 0.00 a	19.25 ± 0.09 a	78
8	33.59 ± 0.43 a	43.46 ± 0.10 a	11.54 ± 0.24 a	10.85 ± 0.20 a	19.19 ± 0.12 a	78
16	33.75 ± 025 a	43.44 ± 0.27 a	11.46 ± 0.09 a	10.81 ± 0.19 a	19.17 ± 0.11 a	78
32	33.74 ± 0.41 a	43.43 ± 0.29 a	11.56 ± 0.07 a	10.59 ± 0.56 a	19.09 ± 0.22 a	78

DP = degree of polymerization. Each value represents the mean of four replicates. Values followed by the same letter within the same column, for each starch, are significantly different (p<0.05).

Table 8.10. Brabender pasting characteristic of the starches and starch/KGM blends. (From Meng *et al.*, 2015)

Starch	KGM content (%)	PT/°C	PV/BU	CV/BU	BD/BU	SB/BU
APS	0	79.5 ± 02^{a}	108.3 ± 2.5^{d}	120.7 ± 1.5^{e}	34.7 ± 1.2^{d}	54.3 ± 0.56^{b}
	0.1	78.1 ± 0.1^{b}	151.7 ± 7.5^{c}	123.0 ± 3.0^{c}	72.3 ± 5.5^{c}	51.0 ± 1.7^{b}
	0.3	40.4 ± 0.4^{c}	231.3 ± 5.5^{b}	153.7 ± 5.0^{b}	134.7 ± 2.1^{b}	52.3 ± 3.5^{b}
	0.5	37.8 ± 0.2^{d}	292.0 ± 5.6^{a}	191.0 ± 4.0^{a}	168.0 ± 7.2^{a}	61.3 ± 5.5^{a}

8.9.2 Chemical modifications

Acid treatments

Starch of *X. sagittifolium* at 140°C for three hours in the presence of HCl (0.182%, w/w) pyrodextrinization of fully gelatinized starch decreased the molecule size and enzyme susceptibility to amyloglucosidase and α-amylase by up to 60% (Laurentin *et al.*, 2003; Laurentin and Edwards, 2004). An *in vitro* study employing human faecal bacteria showed that the fermented products of pyrodextrins had twice the amount of propionate and a lower amount of acetate, but did not alter the butyrate content, compared with the native starch (Laurentin and Edwards, 2004).

Starches of *X. sagittifolium* (Lawal, 2004) (one genotype) and *C. esculenta* (seven genotypes) (Mweta *et al.*, 2010) were acid-hydrolysed. Variation in the extents of acid hydrolysis (35°C, 2.2 M HCl, up to 16 days) (20.4–25.8%) was noted between seven genotypes of *C. esculenta*, revealing the structural diversity of the starches (Mweta *et al.*, 2010). Acid hydrolysis of *X. sagittifolium* starch decreased swelling and oil absorption, and increased solubility, PT and SB of pasting, T_{peak} and ΔH of gelatinization, and retrogradation tendency, without affecting the polymorphism (Lawal, 2004).

Substitutions

Starches from *X. sagittifolium* and *C. esculenta* were converted to acetyl (Lawal, 2004), carboxymethyl (Lawal *et al.*, 2007) and hydroxypropyl (Chattopadhyay *et al.*, 2015) derivatives for novel functionalities. Acetylation enhanced the swelling, and T_{peak} and ΔH of gelatinization, without affecting the morphology and polymorphism of *X. sagittifolium* starch (Lawal, 2004). Lawal *et al.* (2007) observed that the influence of chemical concentration, solvent type, reaction time and starch was explored, and carboxymethyl starch derivatives with a range of degree of substitution were

Table 8.11. Pasting properties of unfermented and fermented pink and white cocoyam starch. (From Obadina *et al.*, 2013)

	0 hours		24 hours		48 hours		72 hours		96 hours	
Properties	Pink	White	Pink	White	Pink	White	Pink	White	Pink	White
Peak 1	333.5 ± 40.3^{a}	4243 ± 56.6^{a}	3600.5 ± 24.7^{a}	5120.5 ± 19.1^{b}	5510 ± 53.7^{b}	5120.5 ± 48.1^{b}	5510.5 ± 8.9^{b}	5922.5 ± 27.6^{c}	4531.5 ± 214.3^{b}	6522.6 ± 45.9^{d}
Trough	2373.5 ± 14.8^{a}	2893.5 ± 26.2^{a}	2728.5 ± 16.3^{a}	3423.5 ± 0.7^{b}	3465±239.0^{h}	3423.5 ± 20.5^{b}	3465 ± 239.0^{h}	3382 ± 24.0^{b}	3148.5 ± 120.9^{h}	4139.5 ± 0.7^{d}
Breakdown	957 ± 24.^{b}	1349.5 ± 30.4^{a}	872 ± 8.5^{a}	1795 ± 19.0^{b}	2231 ± 65.1^{c}	2008.5 ± 68.6^{c}	1299 ± 545.9^{c}	2540.5 ± 3.5^{d}	1383 ± 93.3^{c}	2383 ± 45.3^{d}
Final	3934 ± 38.1^{a}	4489.5 ± 17.7^{a}	4562.5 ± 55.9^{b}	4805 ± 4.24 ^{b}	5183.5 ± 27.6^{c}	5054 ± 11.3^{c}	5319 ± 212.2^{c}	5054 ± 11.3^{c}	5052 ± 210.7^{c}	6301 ± 74.9^{d}
Viscosity										
Setback	1650.2 ± 53.0^{a}	1593 ± 8.5^{ab}	1834 ± 39.6^{b}	1479.5 ± 4.9^{a}	1904.5 ± 3.9^{h}	1630.5 ± 9.9^{b}	1857 ± 11.3_{b}	1821 ± 806^{c}	1903.5 ± 89.8^{h}	2161.5 ± 74.9^{d}
Peak time	4.86 ± 0.0^{b}	4.93 ± 0.0^{c}	5.26 ± 0.6^{d}	4.9 ± 0.0^{b}	4.73 ± 0.0^{c}	4.9 ± 0.0^{c}	5.1 ± 0.04^{d}	4.7 ± 0.00^{a}	4.9 ± 0.04^{c}	4.7 ± 0.1^{d}
Pasting	82.3 ± 0.0^{a}	84.1 ± 0.0^{b}	84.35 ± 0.63^{b}	83.4 ± 0.0^{a}	82.28 ± 0.03^{a}	83.3 ± 0.3^{a}	84.35 ± 0.63^{b}	83.3 0.2^{a}	84.38 ± 0.63^{h}	83.6 ± 0.6^{d}

obtained. Experimental conditions for the hydroxypropylation of *C. esculenta* starch were correlated to the swelling power of the derivatives using evolutionary operation (EVOP) factorial design, and the swelling of starch was maximized under the modification conditions (Chattopadhyay *et al.*, 2015).

Oxidation

Starch of *X. sagittifolium* has been oxidized (Lawal, 2004) to provide a product with increased solubility and oil absorption, and decreased swelling, PT and SB of pasting, T_{peak} and ΔH of gelatinization, and retrogradation tendency, without affecting the morphology and polymorphism of the granules.

8.9.3 Enzymatic modification

The effect of fermentation time on chemical and physiochemical properties of two varieties of cocoyam (*Colocasia esculenta* (white) and *Xanthosoma sagittifolium* (pink)) starch was evaluated (Obadina *et al.*, 2013; see Table 8.11). The cocoyams were fermented (for 24, 48, 72 and 96 hours) and dried in a cabinet dryer at 600°C for 24 hours, and subsequently milled into starch. Fermentation increased the peak viscosity of both varieties at 96 hours and the starch from white cocoyam had a higher peak viscosity (6301) compared to the pink (5052). The pasting temperatures of fermented starch were not significantly different from unfermented starch for both pink and white cocoyam starch (82.0–84.0°C). White cocoyam starch had higher swelling power than the pink starch (9.92–9.97) at 96 hours. There were no significant differences in the dispensability value, bulk density, fat, crude fibre and carbohydrate for both varieties of control and fermented samples. The water solubility index (WSI) for white starch at control level was higher than that of pink (7.5, 4.5) but at 96 hours the pink variety was higher (3, 1.5), and water-absorption capacity increased markedly due to the fermentation period. Protein content was observed to increase during fermentation.

8.9.4 Combined modifications for resistant starch production

A combination of modifications was used to enhance the type 3 (retrograded) resistant starch (RS) content in starches from *A. paeoniifolius*

(Mahmood *et al.*, 2006; Reddy *et al.*, 2014) and *C. esculenta* (Mahmood *et al.*, 2006; Simsek *et al.*, 2012). Gelatinized starch of *A. paeoniifolius* was subjected to pullulanase hydrolysis, autoclave, refrigeration and freeze-drying, resulting in an RS content of 36% (Reddy *et al.*, 2014). A similar strategy on *C. esculenta* starch produced an RS content of 35% (Simsek *et al.*, 2012). Commonly practised processing methods such as boiling and cooling enhanced the RS content of starches from *C. esculenta* and *A. paeoniifolius* by 14% to 19%, while cooking in an open pan for five minutes with an oil-to-sample ratio of 0.06:1 (v/w) decreased it by half (Mahmood *et al.*, 2006). Modifications of aroid starches have been reviewed by Zhu (2016)(see Table 8.12).

8.10 Other Starches

John *et al.* (2002) have examined the effect of acid treatment on arrowroot starch. Increasing the time of treatment reduced the amylose (blue value), intrinsic viscosity and swelling power. XRD data indicated a change in d values due to changes in the lattice parameters of crystalline phase. Acid treatment lowered the DSC gelatinization parameters, including the enthalpy of gelatinization.

Canna edulis L starch was modified with calcium chloride in order to form controlled released matrix. Four different formulations of ondansetron granules were used to provide a dissolution profile of controlled release. Two formulae consisted of 15% and 30% modified starch; one formula utilized matrix reference standards; and in the third the granules had negative control. Methocel-hydroxypsropyl methyl cellulose was used as a controlled released matrix reference standard in the third formula. Calcium starch was synthesized in the presence of sodium hydroxide to form gelatinized mass and calcium chloride was the cross-linking agent. Modified starch has a higher swelling index, water solubility and compressibility index. The profiles indicated that granules that employed calcium *Canna edulis* L starch as a matrix are able to resemble a controlled drug-released profile of matrix reference (Putri *et al.*, 2017).

Thus the results show that a wide variation in different starches by different modification techniques is obtained.

Aroids modifications

Table 8.12. Modification of aroid starches. (From Zhu, 2016)

Modification type	Species	Major findings	References
Physical modification			
Dry heating	*C. esculenta*	Dry heating (130°C up to 4 hours) of starch decreased the paste transmittance, increased the sediment volume and water-binding capacity, and altered the colour, without affecting the swelling and solubility.	Pramodrao and Riar, 2014
Heat-moisture treatment	*X. sagittifolium*	(HMT conditions, 100°C, 10 hours, 30% moisture). Polymorph type (A) was not affected; HMT decreased swelling, amylose leaching, DH of gelatinization, and susceptibility to acid hydrolysis, increased the temperatures of gelatinization and enzyme susceptibility, and had little effect on retrogradation.	Gunaratne and Hoover, 2002

Continued

Table 8.12. Continued

Modification type	Species	Major findings	References
Chemical modification			
Pyrodextrinization	*X. sagittifolium*	The pyrodextrinization was done at 140°C for 3 hours in the presence of HCl (0.182%, w/w). The modification drastically reduced the size of molecules, completely gelatinized the starch, and altered the colour, and decreased the enzyme susceptibility of starch to amyloglucosidase and α-amylase up to 60%. An *in vitro* study employing human faecal bacteria showed that the fermented products of pyrodextrins had twice the amount of propionate and a lower amount of acetate without altering the butyrate content, as compared with the native starch.	Laurentin *et al*., 2003; Laurentin and Edwards, 2004
Acid hydrolysis	*X. sagittifolium*	Acid hydrolysis decreased the swelling and increased the solubility; decreased oil absorption; increased the retrogradation, pasting temperature and setback, and T_{peak} and DH of gelatinization; morphology and polymorphism of starch were not affected.	Lawal, 2004
	C. esculenta	Diversity in the extent of acid hydrolysis (35°C, 2.2 M HCl, up to 16 days) (20.4–25.8%) was observed among seven genotypes.	Mweta *et al*., 2010
Oxidation	*X. sagittifolium*	Oxidation decreased the swelling and increased the solubility and oil absorption; reduced the pasting temperature and setback, T_{peak} and ΔH of gelatinization, and retrogradation; morphology and polymorphism of starch were not affected.	Lawal, 2004
Acetylation	*X. sagittifolium*	Acetylation increased the swelling and hydrophilic tendency, oil absorption; reduced the pasting temperature and setback, retrogradation, and T_{peak} and ΔH of gelatinization; morphology and polymorphism of starch were not affected.	Lawal, 2004
Carboxymethylation	*X. sagittifolium*	NaOH and sodium monochloroacetate concentrations, solvent type, reaction time and temperature were related to degree of substitution and reaction efficiency; 20 carboxymethyl starch derivatives with a range of structure were synthesized.	Lawal *et al*., 2007
Hydroxypropylation	*C. esculenta*	Evolutionary operation factorial design technique was used to optimize the experimental conditions for maximum swelling power of the resulting starch.	Chattopadhyay *et al*., 2015
Resistant starch production	*A. paeoniifolius*	Type-3 resistant starch (up to 36%) was produced by enzymatic hydrolysis (pullulanase), autoclave, refrigeration, and freeze-drying.	Reddy *et al*., 2014

Continued

Table 8.12. Continued

Modification type	Species	Major findings	References
	C. esculenta	Type-3 resistant starch (up to 35%) was produced by heating, autoclaving, enzymatic debranching, retrogradation and drying. The resistant starch decreased the expected glycaemic index and had the binding capacity of bile acids.	Simsek *et al.*, 2012
	C. esculenta, A. paeoniifolius	Boiling and cooling increased the resistant starch content by 14–19%; Sautéing plus cooking decreased the resistant starch content by 46–56%.	Mahmood *et al.*, 2006

References

Abraham, T.E., Alves, R.M.L., Grossmann, M.V.E. and Silva, R.S.S.F. (1993) Stabilization of paste viscosity of cassava starch by heat moisture treatment. *Starch/Stärke* 45(4), 131–135. DOI: 10.1002/star.19930450404.

Agudelo, A., Varela, P., Sanz, T. and Fiszman, S.M. (2014) Tapioca starch as a potential thickener for fruit fillings. Evaluation of mixed models containing low-methoxyl pectin. *Food Hydrocolloids* 35, 297–304.

Alves, R.M.L., Grossmann, M.V.E. and Silva, R.S.S.F. (1999) Gelling properties of extruded yam (*Dioscorea alata*) starch. *Food Chemistry* 67(2), 123–127. DOI: 10.1016/S0308-8146(99)00064-3.

Amorim, E.O.C., Doval, V.C. and Cristianini, M. (2011) Effect of ozonation on the sensory characteristics and pasting properties of cassava starch. *Procedia Food Science* 1, 914–919. DOI: 10.1016/j.profoo.2011.09.138.

Atichokudomchai, N., Shobsngob, S. and Varavinit, S. (2000) Morphological properties of acid-modified Tapioca starch. *Starch/Stärke* 52(8-9), 283–289. DOI: 10.1002/1521-379X(20009)52:8/9<283::AID-STAR283>3.0.CO;2-Q.

Atichokudomchai, N., Shobsngob, S., Chinachoti, P. and Varavinit, S. (2001) A study of some physicochemical properties of high-crystalline tapioca starch. *Starch/Stärke* 53(11), 577–581. DOI: 10.1002/1521-379X(200111)53:11<577::AID-STAR577>3.0.CO;2-0.

Atichokudomchai, N., Varavinit, S. and Chinachoti, P. (2002) Gelatinization transitions of acid-modified Tapioca starches by differential scanning calorimetry (DSC). *Starch/Stärke* 54(7), 296–302. DOI: 10.1002/1521-379X(200207)54:7<296::AID-STAR296>3.0.CO;2-W.

Atichokudomchai, N., Varavinit, S. and Chinachoti, P. (2004) A study of ordered structure in acid-modified tapioca starch by 13C CP/MAS solid-state NMR. *Carbohydrate Polymers* 58(4), 383–389. DOI: 10.1016/j.carbpol.2004.07.017.

Barrios, S.E., Giammanco, G., Contreras, J.M., Laredo, E. and López-Carrasquero, F. (2013) Characterization of esterified cassava starch with long alkyl side chains and different substitution degrees. *International Journal of Biological Macromolecules* 59, 384–390. DOI: 10.1016/j.ijbiomac.2013.04.079.

Beninca, C., Colman, T.A.D., Lacerda, L.G., da Silva Carvalho Filho, M.A., Demiate, I.M. *et al.* (2013a) Thermal, rheological, and structural behaviors of natural and modified cassava starch granules, with sodium hypochlorite solutions. *Journal of Thermal Analysis and Calorimetry* 111(3), 2217–2222. DOI: 10.1007/s10973-012-2592-z.

Beninca, C., Colman, T.A.D., Lacerda, L.G., Filho, M.A.S.C., Bannach, G. *et al.* (2013b) The thermal, rheological and structural properties of cassava starch granules modified with hydrochloric acid at different temperatures. *Thermochimica Acta* 552, 65–69. DOI: 10.1016/j.tca.2012.10.020.

Bertolini, A.C., Mestres, C., Colonna, P., Lerner, D. and Della Valle, G. (1999) Comprehensive studies of molecular changes occurring in sour cassava starch. In: Colonna, P. and Guilbert, S. (eds) *Biopolymer Science: Food and Nonfood Applications*. FRA: INRA Editions, Paris, pp. 87–91.

Bertolini, A.C., Mestres, C., Lourdin, D., Valle, G.D., Colonna, P. et al. (2001a) Relationship between thermomechanical properties and baking expansion of sour cassava starch (Polvilho azedo). *Journal of the Science of Food and Agriculture* 81(4), 429–435.

Bertolini, A.C., Mestres, C., Colonna, P. and Raffi, J. (2001b) Free radical formation in UV- and gamma-irradiated cassava starch. *Carbohydrate Polymers* 44(3), 269–271. DOI: 10.1016/S0144-8617(00)00268-X.

Bertolini, A.C., Mestres, C., Lourdin, D., Valle, G.D., Colonna, P. *et al.* (2001c) Relationship between thermomechanical properties and baking expansion of sour cassava starch (*Polvilho azedo*). *Journal of the Science of Food and Agriculture* 81(4), 429–435. DOI: 10.1002/1097-0010(200103)81:4<429::AID-JSFA833>3.0.CO;2-2.

Camargo, C., Colonna, P., Buléon, A. and Richard-Molard, D. (1988) Functional properties of sour cassava (*Manihot utilissima*) starch: Polvilho azedo. *Journal of the Science of Food and Agriculture* 45(3), 273–289. DOI: 10.1002/jsfa.2740450311.

Campanha, R.B. and Franco, C.M.L. (2011) Gelatinization properties of native starches and their Näegeli dextrins. *Journal of Thermal Analysis and Calorimetry* 106(3), 799–804. DOI: 10.1007/s10973-011-1682-7.

Cardenas, O.S. and Buckle, T.S. (1980) Sour cassava starch production: a preliminary study. *Journal of Food Science* 45(6), 1509–1512. DOI: 10.1111/j.1365-2621.1980.tb07551.x.

Cavallini, C.M. and Franco, C.M.L. (2010) Effect of acid-ethanol treatment followed by ball milling on structural and physicochemical characteristics of cassava starch. *Starch/Stärke* 62(5), 236–245. DOI: 10.1002/star.200900231.

Chandanasree, D., Gul, K. and Riar, C.S. (2016) Effect of hydrocolloids and dry heat modification on physicochemical, thermal, pasting and morphological characteristics of cassava (*Manihot esculenta*) starch. *Food Hydrocolloids* 52, 175–182. DOI: 10.1016/j.foodhyd.2015.06.024.

Chattopadhyay, J., Banerjee, S. and Bhattacharyya, A.K. (2015) Comparative optimization study for chemical synthesis of hydroxypropyl starch from native corn and taro starch through evolutionary operation (EVOP) factorial design technique. *Starch/Stärke* 67(1-2), 154–162. DOI: 10.1002/star.201400106.

Che, L., Li, D., Wang, L., Özkan, N., Chen, X.D. *et al.* (2007a) Effect of high-pressure homogenization on the structure of cassava starch. *International Journal of Food Properties* 10(4), 911–922. DOI: 10.1080/10942910701223315.

Che, L.-M., Li, D., Wang, L.-J., Dong Chen, X. and Mao, Z.-H. (2007b) Micronization and hydrophobic modification of cassava starch. *International Journal of Food Properties* 10(3), 527–536. DOI: 10.1080/10942910600932982.

Chen, Z., Huang, J., Suurs, P., Schols, H. and Voragen, A. (2005) Granule size affects the acetyl substitution on amylopectin populations in potato and sweet potato starches. *Carbohydrate Polymers* 62(4), 333–337. DOI: 10.1016/j.carbpol.2005.07.035.

Cova, A., Sandoval, A.J., Balsamo, V. and Müller, A.J. (2010) The effect of hydrophobic modifications on the adsorption isotherms of cassava starch. *Carbohydrate Polymers* 81(3), 660–667. DOI: 10.1016/j.carbpol.2010.03.028.

Das, A.B., Singh, G., Singh, S. and Riar, C.S. (2010) Effect of acetylation and dual modification on physicochemical, rheological and morphological characteristics of sweet potato (*Ipomoea batatas*) starch. *Carbohydrate Polymers* 80(3), 725–732. DOI: 10.1016/j.carbpol.2009.12.018.

Dias, A.R.G., Elias, M.C., Oliveira, M. and Helbig, E. (2007) Oxidation of fermented cassava and corn starches: development of the expansion property. *Ciência e Tecnologia de Alimentos* 27, 794–799.

Dias, A.R.G., Zavareze, EdaR., Elias, M.C., Helbig, E., da Silva, D.O. *et al.* (2011a) Pasting, expansion and textural properties of fermented cassava starch oxidised with sodium hypochlorite. *Carbohydrate Polymers* 84(1), 268–275. DOI: 10.1016/j.carbpol.2010.11.033.

Dias, A.R.G., Zavareze, EdaR., Helbig, E., Moura, F.Ade., Vargas, C.G. *et al.* (2011b) Oxidation of fermented cassava starch using hydrogen peroxide. *Carbohydrate Polymers* 86(1), 185–191. DOI: 10.1016/j.carbpol.2011.04.026.

El Saadany, R.M.A., El Saadany, F.M. and Foda, Y.H. (1976) Transformation of tapioca starch by gamma irradiation. *Starch/Stärke* 28(5), 169–172. DOI: 10.1002/star.19760280506.

Falade, K.O. and Ayetigbo, O.E. (2015) Effects of annealing, acid hydrolysis and citric acid modifications on physical and functional properties of starches from four yam (*Dioscorea* spp.) cultivars. *Food Hydrocolloids* 43, 529–539. DOI: 10.1016/j.foodhyd.2014.07.008.

Ferrini, L.M.K., Rocha, T.S., Demiate, I.M. and Franco, C.M.L. (2008) Effect of acid-methanol treatment on the physicochemical and structural characteristics of cassava and maize starches. *Starch/Stärke* 60(8), 417–425. DOI: 10.1002/star.200700712.

Figueroa, C., Davila, A.M. and Pourquié, J. (1995) Lactic acid bacteria of the sour cassava starch fermentation. *Letters in Applied Microbiology* 21(2), 126–130. DOI: 10.1111/j.1472-765X.1995.tb01023.x.

Francisco, JdaC. and Sivik, B. (2002) Gelatinization of cassava, potato and wheat starches in supercritical carbon dioxide. *The Journal of Supercritical Fluids* 22(3), 247–254. DOI: 10.1016/S0896-8446(01)00135-8.

Franco, C.M.L., Cabral, R.A.F. and Tavares, D.Q. (2002) Structural and physicochemical characteristics of lintnerized native and sour cassava starches. *Starch/Stärke* 54(10), 469–475. DOI: 10.1002/1521-379X(200210)54:10<469::AID-STAR469>3.0.CO;2-Z.

Franco, C.M.L., Ogawa, C., Rabachini, T., Rocha, TdeS., Cereda, M.P. *et al.* (2010) Effect of lactic acid and UV irradiation on the cassava and corn starches. *Brazilian Archives of Biology and Technology* 53(2), 443–454. DOI: 10.1590/S1516-89132010000200025.

Genkina, N.K., Wasserman, L.A., Noda, T., Tester, R.F. and Yuryev, V.P. (2004) Effects of annealing on the polymorphic structure of starches from sweet potatoes (Ayamurasaki and Sunnyred cultivars) grown at various soil temperatures. *Carbohydrate Research* 339(6), 1093–1098. DOI: 10.1016/j.carres.2004.01.009.

Genkina, N.K., Kiseleva, V.I. and Noda, T. (2009) Comparative investigation on acid hydrolysis of sweet potato starches with different amylopectin chain-length. *Starch/Stärke* 61(6), 321–325. DOI: 10.1002/star.200800097.

George, M., Moorthy, S.N. and Padmaja, G. (1995) Biochemical changes in cassava tuber during fermentation and its effect on extracted starch and residue. *Journal of the Science of Food and Agriculture* 69(3), 367–371. DOI: 10.1002/jsfa.2740690314.

Gunaratne, A. and Hoover, R. (2002) Effect of heat–moisture treatment on the structure and physicochemical properties of tuber and root starches. *Carbohydrate Polymers* 49(4), 425–437. DOI: 10.1016/S0144-8617(01)00354-X.

Gunaratne, A. and Corke, H. (2007) Functional properties of hydroxypropylated, cross-linked, and hydroxypropylated cross-linked tuber and root starches. *Cereal Chemistry Journal* 84(1), 30–37. DOI: 10.1094/CCHEM-84-1-0030.

Han, H.L. and Sosulski, F.W. (1998) Cationization of potato and tapioca starches using an aqueous alcoholic-alkaline process. *Starch/Stärke* 50(11-12), 487–492. DOI: 10.1002/(SICI)1521-379X(199812)50:11/12<487::AID-STAR487>3.0.CO;2-P.

Han, Z., Zeng, X.A., Fu, N., Yu, S.J., Chen, X.D. *et al.* (2012) Effects of pulsed electric field treatments on some properties of tapioca starch. *Carbohydrate Polymers* 89(4), 1012–1017. DOI: 10.1016/j.carbpol.2012.02.053.

Hoover, R. (2000) Acid-treated starches. *Food Reviews International* 16(3), 369–392. DOI: 10.1081/FRI-100100292.

Hoover, R. (2010) The impact of heat-moisture treatment on molecular structures and properties of starches isolated from different botanical sources. *Critical Reviews in Food Science and Nutrition* 50(9), 835–847. DOI: 10.1080/10408390903001735.

Hou, C., Chen, Y., Chen, W. and Li, W. (2011) Microwave-assisted methylation of cassava starch with dimethyl carbonate. *Carbohydrate Research* 346(9), 1178–1181. DOI: 10.1016/j.carres.2011.04.001.

Hou, C., Chen, Y. and Li, W. (2012) Thiocarbamide and microwave-accelerated green methylation of cassava starch with dimethyl carbonate. *Carbohydrate Research* 355, 87–91. DOI: 10.1016/j.carres.2012.04.017.

Huang, Z., Lu, J., Li, X. and Tong, Z. (2007) Effect of mechanical activation on physico-chemical properties and structure of cassava starch. *Carbohydrate Polymers* 68(1), 128–135. DOI: 10.1016/j.carbpol.2006.07.017.

Huang, H., Jiang, Q., Chen, Y., Li, X., Mao, X. et al. (2016) Preparation, physico-chemical characterization and biological activities of two modified starches from yam (*Dioscorea Opposita* Thunb). *Food Hydrocolloids* 55, 244–253.

Hung, P.V., Lan Phi, N.T. and Vy Vy, T.T. (2012) Effect of debranching and storage condition on crystallinity and functional properties of cassava and potato starches. *Starch/Stärke* 64(12), 964–971.

Jayakody, L., Hoover, R., Liu, Q. and Donner, E. (2009) Studies on tuber starches. III. Impact of annealing on the molecular structure, composition and physicochemical properties of yam (*Dioscorea* sp.) starches grown in Sri Lanka. *Carbohydrate Polymers* 76(1), 145–153. DOI: 10.1016/j.carbpol.2008.10.003.

Jiang, Q., Gao, W., Li, X., Zhang, J. and Huang, L. (2011) Effect of acid-ethanol on the physicochemical properties of *Dioscorea opposita* Thunb. and *Pueraria thomsonii* Benth. starches. *Starch/Stärke* 63(5), 302–310. DOI: 10.1002/star.201000159.

Jiang, Q., Gao, W., Li, X., Man, S., Shi, Y. et al. (2014) Comparative susceptibilities to alkali-treatment of A-, B- and C-type starches of *Dioscorea zingiberensis, Dioscorea persimilis* and *Dioscorea opposita*. *Food Hydrocolloids* 39, 286–294.

Jinglin, Y., Shujun, W., Fengmin, J., Sun, L. and Yu, J. (2009) The structure of C-type Rhizoma Dioscorea starch granule revealed by acid hydrolysis method. *Food Chemistry* 113(2), 585–591. DOI: 10.1016/j.foodchem.2008.08.040.

John, J.K., Raja, K.C.M., Rani, S., Moorthy, S.N. and Eliasson, A. (2002) Properties of arrowroot starch treated with aqueous HCl at ambient temperature. *Journal of Food Science* 67(1), 10–14. DOI: 10.1111/j.1365-2621.2002.tb11350.x.

Johnston, D.A., Mukerjea, R. and Robyt, J.F. (2011) Preparation and characterization of new and improved soluble-starches, -amylose, and -amylopectin by reaction with benzaldehyde/zinc chloride. *Carbohydrate Research* 346(17), 2777–2784. DOI: 10.1016/j.carres.2011.10.003.

Jyothi, A.N., Rajasekharan, K.N., Moorthy, S.N. and Sreekumar, J. (2005a) Synthesis and characterization of low DS succinate derivatives of cassava (*Manihot esculenta* Crantz) starch. *Starch/Stärke* 57(7), 319–324. DOI: 10.1002/star.200400374.

Jyothi, A.N., Rajasekharan, K.N., Moorthy, S.N. and Sreekumar, J. (2005b) Microwave-assisted synthesis and characterization of succinate derivatives of cassava (*Manihot esculenta* Crantz) starch. *Starch/Stärke* 57(11), 556–563. DOI: 10.1002/star.200500429.

Jyothi, A.N., Moorthy, S.N. and Rajasekharan, K.N. (2006) Effect of cross-linking with epichlorohydrin on the properties of cassava (*Manihot esculenta* Crantz) starch. *Starch/Stärke* 58(6), 292–299. DOI: 10.1002/star.200500468.

Jyothi, A.N., Sajeev, M.S. and Sreekumar, J.N. (2010) Hydrothermal modifications of tropical tuber starches. 1. Effect of heat-moisture treatment on the physicochemical, rheological and gelatinization characteristics. *Starch/Stärke* 62(1), 28–40. DOI: 10.1002/star.200900191.

Karam, L.B., Grossmann, M.V.E., Silva, R.S.S.F., Ferrero, C. and Zaritzky, N.E. (2005) Gel textural characteristics of corn, cassava and yam starch blends: a mixture surface response methodology approach. *Starch/Stärke* 57(2), 62–70. DOI: 10.1002/star.200400328.

Karam, L.B., Ferrero, C., Martino, M.N., Zaritzky, N.E. and Grossmann, M.V.E. (2006) Thermal, microstructural and textural characterisation of gelatinised corn, cassava and yam starch blends. *International Journal of Food Science & Technology* 41(7), 805–812. DOI: 10.1111/j.1365-2621.2005.01110.x.

Kasemwong, K., Ruktanonchai, U.R., Srinuanchai, W., Itthisoponkul, T. and Sriroth, K. (2011) Effect of high-pressure microfluidization on the structure of cassava starch granule. *Starch/Stärke* 63(3), 160–170. DOI: 10.1002/star.201000123.

Khatoon, S., Sreerama, Y.N., Raghavendra, D., Bhattacharya, S. and Bhat, K.K. (2009) Properties of enzyme modified corn, rice and tapioca starches. *Food Research International* 42(10), 1426–1433. DOI: 10.1016/j.foodres.2009.07.025.

Kim, B.-Y. and Yoo, B. (2010) Effects of cross-linking on the rheological and thermal properties of sweet potato starch. *Starch/Stärke* 62(11), 577–583. DOI: 10.1002/star.201000035.

Klein, B., Vanier, N.L., Moomand, K., Pinto, V.Z., Colussi, R. *et al.* (2014) Ozone oxidation of cassava starch in aqueous solution at different pH. *Food Chemistry* 155, 167–173. DOI: 10.1016/j.foodchem.2014.01.058.

Laurentin, A., Cárdenas, M., Ruales, J., Pérez, E. and Tovar, J. (2003) Preparation of indigestible pyrodextrins from different starch sources. *Journal of Agricultural and Food Chemistry* 51(18), 5510–5515. DOI: 10.1021/jf0341518.

Laurentin, A. and Edwards, C.A. (2004) Differential fermentation of glucose-based carbohydrates in vitro by human faecal bacteria: a study of pyrodextrinised starches from different sources. *European Journal of Nutrition* 43, 183–189.

Lawal, O.S. (2004) Composition, physicochemical properties and retrogradation characteristics of native, oxidised, acetylated and acid-thinned new cocoyam (*Xanthosoma sagittifolium*) starch. *Food Chemistry* 87(2), 205–218. DOI: 10.1016/j.foodchem.2003.11.013.

Lawal, O.S. (2012) Succinylated *Dioscorea cayenensis* starch: effect of reaction parameters and characterisation. *Starch/Stärke* 64(2), 145–156. DOI: 10.1002/star.201100110.

Lawal, O.S., Lechner, M.D., Hartmann, B. and Kulicke, W.-M. (2007) Carboxymethyl cocoyam starch: synthesis, characterisation and influence of reaction parameters. *Starch/Stärke* 59(5), 224–233. DOI: 10.1002/star.200600594.

Lawal, O.S., Lechner, M.D. and Kulicke, W.M. (2008a) Single and multi-step carboxymethylation of water yam (*Dioscorea alata*) starch: synthesis and characterization. *International journal of Biological Macromolecules* 42(5), 429–435. DOI: 10.1016/j.ijbiomac.2008.02.006.

Lawal, O.S., Ogundiran, O.O., Adesogan, E.K., Ogunsanwo, B.M. and Sosanwo, O.A. (2008b) Effect of hydroxypropylation on the properties of white yam (*Dioscorea rotundata*) starch. *Starch/Stärke* 60(7), 340–348. DOI: 10.1002/star.200700681.

Le, Q.-T., Lee, C.-K., Kim, Y.-W., Lee, S.-J., Zhang, R. *et al.* (2009) Amylolytically-resistant tapioca starch modified by combined treatment of branching enzyme and maltogenic amylase. *Carbohydrate Polymers* 75(1), 9–14. DOI: 10.1016/j.carbpol.2008.06.001.

Lee, H.-L. and Yoo, B. (2009) Dynamic rheological and thermal properties of acetylated sweet potato starch. *Starch/Stärke* 61(7), 407–413. DOI: 10.1002/star.200800109.

Lee, S.H., Choi, S.J., Shin, S.I., Park, K.H. and Moon, T.W. (2008) Structural and rheological properties of sweet potato starch modified with 4-a-glucanotransferase from *Thermus aquaticus*. *Food Science and Biotechnology* 17, 705–712.

Leonel, M., Freitas, T.Sde. and Mischan, M.M. (2009) Physical characteristics of extruded cassava starch. *Scientia Agricola* 66(4), 486–493. DOI: 10.1590/S0103-90162009000400009.

Li, X., Gao, W., Jiang, Q., Wang, Y., Guo, X. *et al.* (2011) Physicochemical, crystalline, and thermal properties of native, oxidized, acid, and enzyme hydrolyzed Chinese yam (*Dioscorea opposita* Thunb) starch. *Starch/Stärke* 63, 616–624.

Li, X., Gao, W., Wang, Y., Jiang, Q. and Huang, L. (2011b) Granule structural, crystalline, and thermal changes in native Chinese yam starch after hydrolysis with two different enzymes – α-amylase and gluco-amylase. *Starch/Stärke* 63, 75–82.

Lian, X.J., Zhang, K.S., Luo, Q.F., Zhang, X. and Zhao, S.Y. (2012) The effects of electrolysis at room temperature on retrogradation of sweet potato starch. *International Journal of Biological Macromolecules* 50, 38–42.

Lin, R., Li, H., Long, H., Su, J. and Huang, W. (2015) Structure and characteristics of lipase-catalyzed rosin acid starch. *Food Hydrocolloids* 43, 352–359. DOI: 10.1016/j.foodhyd.2014.06.008.

Mahmood, I., Ghugre, P.S. and Udipi, S.A. (2006) Resistant starch in raw and processed roots and tubers. *Journal of Food Science and Technology* 43, 282–285.

Marcon, M.J.A., Vieira, M.A., Santos, K., De Simas, K.N., Amboni, R.D.D.C. *et al.* (2006) The effect of fermentation on cassava starch microstructure. *Journal of Food Process Engineering* 29(4), 362–372. DOI: 10.1111/j.1745-4530.2006.00073.x.

Marcon, M.J.A., Kurtz, D.J., Raguzzoni, J.C., Delgadillo, I., Maraschin, M. *et al.* (2009) Expansion properties of sour cassava starch (*Polvilho* Azedo): variables related to its practical application in bakery. *Starch/Stärke* 61(12), 716–726. DOI: 10.1002/star.200900132.

Martínez-Bustos, F., López-Soto, M., Zazueta-Morales, J.J. and Morales-Sánchez, E. (2005) Preparation and properties of pregelatinized cassava (*Manihot esculenta* Crantz) and Jícama (*Pachyrhizus erosus*) starches using Ohmic heating. *Agrociencia* 39, 275–283.

Martínez-Bustos, F., Amaya-Llano, S.L., Carbajal-Arteaga, J.A., Chang, Y.K. and Zazueta-Morales, JdeJ. (2007a) Physicochemical properties of cassava, potato and jicama starches oxidised with organic acids. *Journal of the Science of Food and Agriculture* 87(7), 1207–1214. DOI: 10.1002/jsfa.2805.

Martínez-Bustos, F., López-Soto, M., San Martín-Martínez, E., Zazueta-Morales, J.J. and Velez-Medina, J.J. (2007b) Effects of high energy milling on some functional properties of jicama starch (*Pachyrrhizus erosus* L. Urban) and cassava starch (*Manihot esculenta* Crantz). *Journal of Food Engineering* 78(4), 1212–1220. DOI: 10.1016/j.jfoodeng.2005.10.043.

Meng, F., Liang, Y., Shuai, T., Xiong, J. and Zhong, G. (2015) Physicochemical properties of *Amorphophallus paeoniifolius* (Dennst.) Nicolson starch and its blend with konjac glucomannan. *Starch/Stärke* 67(9-10), 820–828. DOI: 10.1002/star.201500019.

Mestres, C. and Rouau, X. (1997) Influence of natural fermentation and drying conditions on the physicochemical characteristics of cassava starch. *Journal of the Science of Food and Agriculture* 74(2), 147–155. DOI: 10.1002/(SICI)1097-0010(199706)74:2<147::AID-JSFA781>3.0.CO;2-J.

Mestres, C., Boungou, O., Akisso, N. and Zakhia, N. (2000) Comparison of the expansion ability of fermented maize flour and cassava starch during baking. *Journal of the Science of Food and Agriculture* 80(6), 665–672. DOI: 10.1002/(SICI)1097-0010(20000501)80:6<665::AID-JSFA585>3.0.CO;2-K.

Moorthy, S.N. (1985a) Acetylation of cassava starch using perchloric acid catalysis. *Starch/Stärke* 37(9), 307–308. DOI: 10.1002/star.19850370906.

Moorthy, S.N. (1985b) *Casssava starch and its modifications*. Technical bulletin No. 4. CTCRI, Trivandrum, India.
Moorthy, S.N. (1999) Effect of steam pressure treatment on the physicochemical properties of *Dioscorea* starches. *Journal of Agricultural and Food Chemistry* 47(4), 1695–1699. DOI: 10.1021/jf971082m.
Moorthy, S.N. (2001) Technical Bulletin. 18. CTCRI, Trivandrum, India.
Moraes, J., Álves, F.S. and Franco, C.M.L. (2013) Effect of ball milling on structural and physicochemical characteristics of cassava and Peruvian carrot starches. *Starch/Stärke* 65(3-4), 200–209. DOI: 10.1002/star.201200059.
Mutungi, C., Onyango, C., Jaros, D., Henle, T. and Rohm, H. (2009a) Determination of optimum conditions for enzymatic debranching of cassava starch and synthesis of resistant starch type III using central composite rotatable design. *Starch/Stärke* 61(7), 367–376. DOI: 10.1002/star.200800119.
Mutungi, C., Rost, F., Onyango, C., Jaros, D. and Rohm, H. (2009b) Crystallinity, thermal and morphological characteristics of resistant starch type III produced by hydrothermal treatment of debranched cassava starch. *Starch/Stärke* 61(11), 634–645. DOI: 10.1002/star.200900167.
Mutungi, C., Onyango, C., Rost, F., Doert, T., Jaros, D. *et al.* (2010) Structural and physicochemical properties and in vitro digestibility of recrystallized linear α-d-(1→4) glucans derived from mild-acid-modified cassava starch. *Food Research International* 43(4), 1144–1154. DOI: 10.1016/j.foodres.2010.02.008.
Mutungi, C., Passauer, L., Onyango, C., Jaros, D. and Rohm, H. (2012) Debranched cassava starch crystallinity determination by Raman spectroscopy: correlation of features in Raman spectra with X-ray diffraction and ^{13}C CP/MAS NMR spectroscopy. *Carbohydrate Polymers* 87(1), 598–606. DOI: 10.1016/j.carbpol.2011.08.032.
Mweta, D.E., Labuschagne, M.T., Bonnet, S., Swarts, J. and Saka, J.D.K. (2010) Isolation and physicochemical characterisation of starch from cocoyam (*Colocasia esculenta*) grown in Malawi. *Journal of the Science of Food and Agriculture* 112, 1886–1896. DOI: 10.1002/jsfa.4029.
Nabeshima, E.H. and Grossmann, M.V.E. (2001) Functional properties of pregelatinized and cross-linked cassava starch obtained by extrusion with sodium trimetaphosphate. *Carbohydrate Polymers* 45(4), 347–353. DOI: 10.1016/S0144-8617(00)00273-3.
Nakamura, I.M. and Park, Y.K. (1975) Some physico-chemical properties of fermented cassava starch (*Polvilho* Azedo). *Starch/Stärke* 27(9), 295–297. DOI: 10.1002/star.19750270904.
Nattapulwat, N., Purkkao, N. and Suwithayapan, O. (2009) Preparation and application of carboxymethyl yam (*Dioscorea esculenta*) starch. *AAPS Pharm Sci Tech* 10(1), 193–198. DOI: 10.1208/s12249-009-9194-5.
Numfor, F.A., Walter,, W.M. and Schwartz, S.J. (1996) Effect of emulsifiers on the physical properties of native and fermented cassava starches. *Journal of Agricultural and Food Chemistry* 44(9), 2595–2599. DOI: 10.1021/jf950610w.
Obadina, A.O., Fegha, G.D. and Olugbile, A.O. (2013) Effect of fermentation periods on chemical and physicochemical properties of cocoyam starch. *Starch/Stärke* 65(9–10), 747–752. DOI: 10.1002/star.201200246.
Odeku, O.A. and Picker-Freyer, K.M. (2009a) Characterization of acid modified Dioscorea starches as direct compression excipient. *Pharmaceutical Development Technology* 14, 259–270.
Odeku, O.A. and Picker-Freyer, K.M. (2009b) Evaluation of the material and tablet formation properties of modified forms of *Dioscorea* starches. *Drug Development and Industrial Pharmacy* 35(11), 1389–1406. DOI: 10.3109/03639040902960185.
Oh, H.E., Pinder, D.N., Hemar, Y., Anema, S.G. and Wong, M. (2008) Effect of high-pressure treatment on various starch-in-water suspensions. *Food Hydrocolloids* 22(1), 150–155. DOI: 10.1016/j.foodhyd.2007.01.028.
Oladebeye, A.O., Oshodi, A.A., Amoo, I.A. and Karim, A.A. (2013) Functional, thermal and molecular behaviours of ozone-oxidised cocoyam and yam starches. *Food Chemistry* 141(2), 1416–1423. DOI: 10.1016/j.foodchem.2013.04.080.
Onitilo, M.O., Sanni, L.O., Oyewole, O.B. and Maziya-Dixon, B. (2007) Physicochemical and functional properties of sour starches from different cassava varieties. *International Journal of Food Properties* 10(3), 607–620. DOI: 10.1080/10942910601048994.
Onyango, C., Bley, T., Jacob, A., Henle, T. and Rohm, H. (2006) Influence of incubation temperature and time on resistant starch type III formation from autoclaved and acid-hydrolysed cassava starch. *Carbohydrate Polymers* 66(4), 494–499. DOI: 10.1016/j.carbpol.2006.04.002.

Onyango, C. and Mutungi, C. (2008) Synthesis and *in vitro* digestion of resistant starch type III from enzymatically hydrolysed cassava starch. *International Journal of Food Science & Technology* 43(10), 1860–1865. DOI: 10.1111/j.1365-2621.2008.01764.x.

Osunsami, A.T., Akingbala, J.O. and Oguntimein, G.B. (1989) Effect of storage on starch content and modification of cassava starch. *Starch/Stärke* 41(2), 54–57. DOI: 10.1002/star.19890410205.

Parada, J.L., Zapata, E., de Fabrizio, S.V. and Martínez, A. (1996) Microbiological and technological aspects of cassava-starch fermentation. *World Journal of Microbiology and Biotechnology* 12(1), 53–56. DOI: 10.1007/BF00327801.

Parvathy, P.C. and Jyothi, A.N. (2014) Rheological and thermal properties of saponified cassava starch- *g* -poly(acrylamide) superabsorbent polymers varying in grafting parameters and absorbency. *Journal of Applied Polymer Science* 131(11), n/a. DOI: 10.1002/app.40368.

Pérez-Sira, E. and González-Parada, Z. (1997) Functional properties of cassava (Manihot esculenta Crantz) starch modified by physical methods. *Starch/Stärke* 49(2), 49–53. DOI: 10.1002/star.19970490203.

Petruccelli, S., Zamponi, R.A., Jovanovich, G. and Añón, M.C. (1993) Characterization of fermented cassava starches. *Journal of Food Biochemistry* 17(3), 161–172. DOI: 10.1111/j.1745-4514.1993.tb00465.x.

Plata-Oviedo, M. and Camargo, C. (1998) Effect of acid treatments and drying processes on physicochemical and functional properties of cassava starch. *Journal of the Science of Food and Agriculture* 77(1), 103–108. DOI: 10.1002/(SICI)1097-0010(199805)77:1<103::AID-JSFA10>3.0.CO;2-2.

Pornsuksomboon, K., Mészáros Szécsényi, K., Holló, B. and Kaewtatip, K. (2014) Preparation of native cassava starch and cross-linked starch blended foams. *Starch/Stärke* 66, 9–10, 818–823.

Pornsuksomboon, K., Holló, B.B., Szécsényi, K.M. and Kaewtatip, K. (2016) Properties of baked foams from citric acid modified cassava starch and native cassava starch blends. *Carbohydrate Polymers* 136, 107–112. DOI: 10.1016/j.carbpol.2015.09.019.

Pramodrao, K.S. and Riar, C.S. (2014) Comparative study of effect of modification with ionic gums and dry heating on the physicochemical characteristic of potato, sweet potato and taro starches. *Food Hydrocolloids* 35, 613–619. DOI: 10.1016/j.foodhyd.2013.08.006.

Putri, A.P., Ridwan, M., Darmawan, T.A., Darusman, F. and Gadri, A. (2017) Calcium modified edible Canna (*Canna edulis* L) starch for controlled released matrix. *IOP Conference Series: Materials Science and Engineering* 223, 012041. DOI: 10.1088/1757-899X/223/1/012041.

Qiu, H. and He, L. (1999) Synthesis and properties study of carboxymethyl cassava starch. *Polymers for Advanced Technologies* 10(7), 468–472. DOI: 10.1002/(SICI)1099-1581(199907)10:7<468::AID-PAT898>3.0.CO;2-W.

Radhika, G.S., Shanavas, S. and Moorthy, S.N. (2008) Influence of lipids isolated from soybean seed on different properties of cassava starch. *Starch/Stärke* 60(9), 485–492. DOI: 10.1002/star.200800207.

Raja, K.C.M. (1992) Studies on alkali-modified cassava starch - changes of structural and enzyme (α-Amylase) susceptibility properties. *Starch/Stärke* 44(4), 133–136. DOI: 10.1002/star.19920440404.

Rajan, A. and Abraham, T.E. (2006) Enzymatic modification of cassava starch by bacterial lipase. *Bioprocess and Biosystems Engineering* 29(1), 65–71. DOI: 10.1007/s00449-006-0060-5.

Rajan, A., Sudha, J.D. and Abraham, T.E. (2008) Enzymatic modification of cassava starch by fungal lipase. *Industrial Crops and Products* 27(1), 50–59. DOI: 10.1016/j.indcrop.2007.07.003.

Reddy, C.K., Haripriya, S., Noor Mohamed, A. and Suriya, M. (2014) Preparation and characterization of resistant starch III from elephant foot yam (*Amorphophallus paeonifolius*) starch. *Food Chemistry* 155, 38–44. DOI: 10.1016/j.foodchem.2014.01.023.

Remya, R., Jyothi, A.N. and Sreekumar, J. (2017) Comparative study of RS4 type resistant starches derived from cassava and potato starches via octenyl succinylation. *Starch/Stärke* 69(7–8), 1600264. DOI: 10.1002/star.201600264.

Ren, G.-yue., Li, D., Wang, L.-jun., Özkan, N. and Mao, Z.-huai. (2010) Morphological properties and thermoanalysis of micronized cassava starch. *Carbohydrate Polymers* 79(1), 101–105. DOI: 10.1016/j.carbpol.2009.07.031.

Rolland-Sabaté, A., Sanchez, T., Buléon, A., Colonna, P., Ceballos, H. *et al.* (2013) Molecular and supra-molecular structure of waxy starches developed from cassava (*Manihot esculenta* Crantz). *Carbohydrate Polymers* 92(2), 1451–1462. DOI: 10.1016/j.carbpol.2012.10.048.

Rolland-Sabaté, A., Sánchez, T., Buléon, A., Colonna, P., Jaillais, B. *et al.* (2012) Structural characterization of novel cassava starches with low and high-amylose contents in comparison with other commercial sources. *Food Hydrocolloids* 27(1), 161–174. DOI: 10.1016/j.foodhyd.2011.07.008.
Sangseethong, K., Ketsilp, S. and Sriroth, K. (2005) The role of reaction parameters on the preparation and properties of carboxymethyl cassava starch. *Starch/Stärke* 57(2), 84–93. DOI: 10.1002/star.200400302.
Sangseethong, K., Lertphanich, S. and Sriroth, K. (2009) Physicochemical properties of oxidized cassava starch prepared under various alkalinity levels. *Starch/Stärke* 61(2), 92–100. DOI: 10.1002/star.200800048.
Sangseethong, K., Termvejsayanon, N. and Sriroth, K. (2010) Characterization of physicochemical properties of hypochlorite- and peroxide-oxidized cassava starches. *Carbohydrate Polymers* 82(2), 446–453. DOI: 10.1016/j.carbpol.2010.05.003.
Sanguanpong, V., Chotineeranat, S., Piyachomkwan, K., Oates, C.G., Chinachoti, P. *et al.* (2003) Hydration and physicochemical properties of small-particle cassava starch. *Journal of the Science of Food and Agriculture* 83(2), 123–132. DOI: 10.1002/jsfa.1285.
Shanavas, S. (2011) Studies on modified starches useful for tablets and capsules. Kannur University, Kerala.
Shanavas, S., Moorthy, S.N., Sajeev, M.S., Misral, R.S. and Sundazeem, A.S. (2010) Viscoelastic properties and excipient functionality of cassava starch and konjac glumannan mixtures. *Trends in Carbohydrtae Research* 2, 11–22.
Silva, R.M., Ferreira, G.F., Shirai, M.A., Haas, A., Scherer, M.L. *et al.* (2008) Physicochemical characteristics of starches modified with potassium permanganate/lactic acid and sodium hypochlorite/lactic acid. *Ciencia e Tecnologia de Alimentos* 28, 66–77.
Simsek, S. El, S.N. and Nehir, E.S. (2012) Production of resistant starch from taro (*Colocasia esculenta* L. Schott) corm and determination of its effects on health by in vitro methods. *Carbohydrate Polymers* 90(3), 1204–1209. DOI: 10.1016/j.carbpol.2012.06.039.
Tattiyakul, J., Naksriarporn, T., Pradipasena, P. and Miyawaki, O. (2006) Effect of moisture on hydrothermal modification of yam *Dioscorea hispida* Dennst starch. *Starch/Stärke* 58(3–4), 170–176. DOI: 10.1002/star.200500462.
Tattiyakul, J., Naksriarporn, T. and Pradipasena, P. (2012) X-ray diffraction pattern and functional properties of *Dioscorea hispida* Dennst starch hydrothermally modified at different temperatures. *Food and Bioprocess Technology* 5(3), 964–971. DOI: 10.1007/s11947-010-0424-3.
Tran, T., Piyachomkwan, K. and Sriroth, K. (2007) Gelatinization and thermal properties of modified cassava starches. *Starch/Stärke* 59(1), 46–55. DOI: 10.1002/star.200600514.
Trinh, K.S., Lee, C.J., Choi, S.J. and Moon, T.W. (2012) Hydrothermal treatment of water yam starch in a non-granular state: slowly digestible starch content and structural characteristics. *Journal of Food Science* 77(6), C574–C582. DOI: 10.1111/j.1750-3841.2012.02703.x.
Trinh, K.S., Choi, S.J. and Moon, T.W. (2013) Structure and digestibility of debranched and hydrothermally treated water yam starch. *Starch/Stärke* 65(7-8), 679–685. DOI: 10.1002/star.201200149.
Vamadevan, V., Bertoft, E. and Seetharaman, K. (2013a) On the importance of organization of glucan chains on thermal properties of starch. *Carbohydrate Polymers* 92(2), 1653–1659. DOI: 10.1016/j.carbpol.2012.11.003.
Vamadevan, V., Bertoft, E., Soldatov, D.V. and Seetharaman, K. (2013b) Impact on molecular organization of amylopectin in starch granules upon annealing. *Carbohydrate Polymers* 98(1), 1045–1055. DOI: 10.1016/j.carbpol.2013.07.006.
Varavinit, S., Paisanjit, W., Tukomane, T. and Pukkahuta, C. (2007) Effects of osmotic pressure on the crosslinking reaction of tapioca starch. *Starch/Stärke* 59(6), 290–296. DOI: 10.1002/star.200700610.
Vatanasuchart, N., Naivikul, O., Charoenrein, S. and Sriroth, K. (2005) Molecular properties of cassava starch modified with different UV irradiations to enhance baking expansion. *Carbohydrate Polymers* 61(1), 80–87. DOI: 10.1016/j.carbpol.2005.02.012.
Vittadini, E., Carini, E., Chiavaro, E., Rovere, P. and Barbanti, D. (2008) High pressure-induced tapioca starch gels: physico-chemical characterization and stability. *European Food Research and Technology* 226(4), 889–896. DOI: 10.1007/s00217-007-0611-2.
Wang, S., Yu, J., Gao, W., Pang, J., Liu, H. and Yu, J. (2007a) Granule structural changes in native Chinese Yam (*Dioscorea opposita* Thunb var. Anguo) starch during acid hydrolysis. *Carbohydrate Polymers* 69, 286–92.

Wang, S., Yu, J., Yu, J., Chen, H., Pang, J. (2007b) The effect of acid hydrolysis on morphological and crystalline properties of *Rhizoma Dioscorea* starch. *Food Hydrocolloids* 21, 1217–1222.

Wang, S., Yu, J., Yu, J. and Liu, H. (2008a) Granule structure of C-type Chinese yam (*Dioscorea opposita* Thunb var. Zhongbowen) starch by acid hydrolysis. *Food Hydrocolloids* 22(4), 538–542.

Wang, S., Yu, J., Yu, J., Liu, H. (2008b) The partial characterization of C-type rhizoma Dioscorea starch granule during acid hydrolysis. *Food Hydrocolloids* 22, 531–537.

Wang, S., Yu, J., Yu, J. and Liu, H. (2008c) Granule structure of C-type Chinese yam (*Dioscorea opposita* Thunb var. Zhongbowen) starch by acid hydrolysis. *Food Hydrocolloids* 22(4), 538–542. DOI: 10.1016/j.foodhyd.2007.01.006.

Wang, X., Gao, W., Zhang, L., Xiao, P., Yao, L. *et al.* (2008e) Study on the morphology, crystalline structure and thermal properties of yam starch acetates with different degrees of substitution. *Science in China Series B: Chemistry* 51(9), 859–865. DOI: 10.1007/s11426-008-0089-1.

Wang, S., Yu, J., Zhu, Q., Yu, J. and Jin, F. (2009) Granular structure and allomorph position in C-type Chinese yam starch granule revealed by SEM, 13C CP/MAS NMR and XRD. *Food Hydrocolloids* 23(2), 426–433. DOI: 10.1016/j.foodhyd.2008.02.012.

Wongsagonsup, R., Pujchakarn, T., Jitrakbumrung, S., Chaiwat, W., Fuongfuchat, A. *et al.* (2014) Effect of cross-linking on physicochemical properties of tapioca starch and its application in soup product. *Carbohydrate Polymers* 101, 656–665. DOI: 10.1016/j.carbpol.2013.09.100.

Wurzburg, O.B. (ed.) (1986) *Modified Starches: Properties and Uses*. Boca Raton, CRC Press, Florida.

Wuttisela, K., Triampo, W. and Triampo, D. (2009) Chemical force mapping of phosphate and carbon on acid-modified tapioca starch surface. *International Journal of Biological Macromolecules* 44(1), 86–91. DOI: 10.1016/j.ijbiomac.2008.10.005.

Xia, L., Wenyuan, G., Qianqian, J., Yanli, W., Xinhua, G., Luqi, H. (2011) Physicochemical, crystalline, and thermal properties of native, oxidized, acid, and enzyme hydrolyzed Chinese yam (*Dioscorea opposite* Thunb) starch. *Starch/Stärke*. 63(10), 616–624.

Xia, W., Wang, F., Li, J., Wei, X., Fu, T. *et al.* (2015) Effect of high speed jet on the physical properties of tapioca starch. *Food Hydrocolloids* 49, 35–41. DOI: 10.1016/j.foodhyd.2015.03.010.

Yao, J., You, H., Bao, Z. and Shen, J. (2005) Investigation of substituted pattern of carboxymethylation cassava starch. *Chinese Journal of Analytical Chemistry* 33, 201–206.

Yu, J., Wang, S., Jin, F., Sun, L. and Yu, J. (2009) The structure of C-type rhizoma Dioscorea starch granule revealed by acid hydrolysis method. *Food Chemistry* 113, 585–591.

Yussof, N.S., Utra, U. and Alias, A.K. (2013) Hydrolysis of native and cross-linked corn, tapioca, and sweet potato starches at sub-gelatinization temperature using a mixture of amylolytic enzymes. *Starch/Stärke* 65(3–4), 285–295. DOI: 10.1002/star.201200002.

Zhang, L., Liu, P., Wang, Y. and Gao, W. (2011) Study on physico-chemical properties of dialdehyde yam starch with different aldehyde group contents. *Thermochimica Acta* 512(1–2), 196–201. DOI: 10.1016/j.tca.2010.10.006.

Zhang, L., Zuo, B., Wu, P., Wang, Y. and Gao, W. (2012) Ultrasound effects on the acetylation of *Dioscorea* starch isolated from *Dioscorea zingiberensis* C.H. Wright. *Chemical Engineering and Processing: Process Intensification* 54, 29–36. DOI: 10.1016/j.cep.2012.01.005.

Zhang, Y., Huang, Z., Yang, C., Huang, A., Hu, H. *et al.* (2013) Material properties of partially pregelatinized cassava starch prepared by mechanical activation. *Starch/Stärke* 65(5–6), 461–468. DOI: 10.1002/star.201200174.

Zhao, J., Schols, H.A., Chen, Z., Jin, Z., Buwalda, P. *et al.* (2012) Substituent distribution within cross-linked and hydroxypropylated sweet potato starch and potato starch. *Food Chemistry* 133(4), 1333–1340. DOI: 10.1016/j.foodchem.2012.02.021.

Zhu, F. (2015) Isolation, composition, structure, properties, modifications, and uses of yam starch. *Comprehensive Reviews in Food Science and Food Safety* 14(4), 357–386. DOI: 10.1111/1541-4337.12134.

Zhu, F. (2016) Structure, properties, and applications of aroid starch. *Food Hydrocolloids* 52, 378–392. DOI: 10.1016/j.foodhyd.2015.06.023.

Zhu, F. and Corke, H. (2011) Gelatinization, pasting, and gelling properties of sweetpotato and wheat starch blends. *Cereal Chemistry Journal* 88(3), 302–309. DOI: 10.1094/CCHEM-10-10-0145.

Zhu, Z., Zheng, H. and Li, X. (2013) Effects of succinic acid cross-linking and mono-phosphorylation of oxidized cassava starch on its paste viscosity stability and sizability. *Starch/Stärke* 65(9–10), 854–863. DOI: 10.1002/star.201200256.

9

Applications of Tuber Starches

9.1 Cassava Starch

Cassava starch is the most easily extractable starch, and since the tubers are low in proteins, lipids, fibre etc., the extracted starch is pure white in colour. Because the starch possesses excellent colour and brightness, bland taste, high viscosity, good clarity and fair sol stability, it is preferred in many food and non-food applications. Unlike cereal starches, it has no 'lipid' taste and is used in food items where taste should be neutral. Similarly, its paste is very clear and suits application in pie fillings which require the fruit components to be visible. Its high viscosity makes it the most sought-after starch in puddings and gravies. The relatively low retrogradation makes it very suitable for canned foods (Fig. 9.1).

Sago is an important food item in India, usually consumed during fasting on religious occasions and as a convalescent food. It has been found that cassava starch gives the best quality sago. Sago produced from other starches does not have the same brightness or colour.

The sago industry has developed in India in the last 50 years in and around Salem in Tamil Nadu (south India), where it was produced on a large scale. Presently, sago production units are located in Tamil Nadu and Andhra Pradesh in south India. A major portion of extracted cassava starch in India is utilized for sago production.

The tubers received from the farms are hygienically cleaned in water, the skin is peeled, tubers are crushed and the milk is allowed to pass through a sieve. The starch milk is then settled in tanks for three to eight hours, thus all residual impurities float to the top of the tank and are drained out of the settled milk. From this settled milk cake, globules are made by using vibrating screens. After sizing the globules by sieving, they are roasted on hot plates or heated in steam, depending on the desired final globular shape, and then dried under direct sunlight in big platforms. Roasted sago is known as common sago, while boiled sago is known as nylon sago.

Sago, commonly known as 'sabudana', is used as a food during fasting among Hindus in western and central parts of India (Gujarat and Maharashtra). In northern India, during the festival season, it is usually consumed during fasting, either prepared as *khichadi* (savory) or *kheer* (sweet). Sago is also used to prepare many other food items such as laddu, uppuma, pappad, vadai etc. The tapioca sago is very high in calories but very low in fat and protein (Table 9.1).

Cassava starch finds widespread uses in the textile and paper industries. The high viscosity and good adhesive properties are quite suitable in these applications. In the textile industry, the starch finds a use in sizing and finishing operations since the high viscosity is advantageous. However, the unstable viscosity of the paste

Fig. 9.1. Sago starch (source: M.S. Sajeev).

during heating and stirring is not desirable in these applications, but modification by cross-linking has been found to overcome this problem. In the paper industry also, cassava starch finds a use as a sizing and binding agent. Here also, modification by oxidation is carried out to reduce viscosity and improve film-forming capacity. In the adhesive industry, dextrins from cassava starch are preferred due to high tack and excellent colour. In the sweetener industry, the starch has advantages like lower gelatinization temperatures and easy processing to glucose syrup.

Table 9.1. Nutritional information for tapioca sago (approx. per 100 g of product). (Source: S.N. Moorthy.)

Energy	347.80 kcal
Protein	0.24 g
Carbohydrate	86.54 g
Fat	0.05 g

Cassava starch is the most preferred starch for bioethanol production and maltodextrins. It has wide application as fillers in soaps, detergents and explosives. Studies at ICAR-CTCRI and Elsevier have shown that the undesirable properties of cassava starch, like unstable viscosity, could be modified through several physical and chemical methods. Physically modified starches, i.e. heat-moisture-treated starches, annealed starches, steam-pressure-treated starches and microwave-irradiated starches, exhibit improved viscosity stability, higher pasting temperature, lower swelling power and lower *in vitro* digestibility in comparison with their native counterparts. Since no chemicals are involved in these modifications, these starches are especially useful in food applications. Cold water miscible starch (pre-gelatinized starch) could be prepared by a drum-drying technique and used as an instant binder and disintegration agent in tablet manufacturing, as a thickener in the food industries, as an additive to control water in oil well-drilling, for textile finishing and sizing, and as a wet-end additive in the paper industry.

Cassava starch factory residue (CSFR) is a solid fibrous by-product of starch factories, discharged at a rate of 60,000 tonnes annually. Presently, the major mode of utilization is as animal feed. The dry residue has about 56–60% starch, 15–18% cellulose, 4–5% hemicellulose and 2–3% lignin. Due to the low content of cellulose and hemicellulose, and high starch content, CSFR could be considered as a cellulo-starch by-product, which may be easily processed through less costly conversion techniques to several value-added products such as ethanol. A very high ethanol yield of approximately 390 L/tonne of dry CSFR has been obtained in the new technology, making CSFR an alternative substrate for ethanol production.

Research has shown that cassava starch incorporated into polypropylene granules can yield products having good strength and enhanced biodegradability. Starch-incorporated plastic films (up to 25–40%) possess adequate mechanical strength and flexibility and can be processed just like normal plastics, i.e. heat-sealed, printed, coloured etc. The granules and finished products can be stored almost like synthetic plastics and are biodegradable under burial in soil. From the outdoor weathering and soil-burial tests, the biodegradation time has been reported to vary from six months to five years depending on its composition and soil conditions.

ICAR-CTCRI has developed a technology for the production of a semi-synthetic superabsorbent polymer (SAP) based on tapioca starch. The water-absorption capacity is about 350–400 g/g of the dry sample. The process was perfected under laboratory conditions and the product tested for its water conservation potential as well as its effect on soil quality. The polymer was found to be effective in soil-moisture retention and the watering interval could be prolonged. It also improved the soil properties such as porosity, water-holding capacity and nutrient status. The pilot study indicated that application of this hydrogel can result in better plant growth through improvement in nutrient availability, including moisture retention. The polymer can be recommended for use as a soil additive especially under the prevailing thread of reduced moisture availability. It can be used effectively under controlled conditions such as in greenhouses for nurseries, and in ornamental and medicinal plants.

Eco-friendly biodegradable composite films based on cassava starch-chitosan-konjac glucomannan composites were developed for food-wrapping applications. Nanosilver has been incorporated in the films to achieve antimicrobial properties. The films exhibited desirable properties such as high-tensile strength and moisture resistance, low water solubility, swelling capacity, oxygen transmission rate, water vapour transmission rate and contact angle. The microbial growth was significantly lower in the food samples wrapped using nanosilver-incorporated films. Food contact studies showed that the amount of silver migration to the packaged food sample (4.3–4.5 μg/L) was significantly lower than the permitted levels (10 mg/L).

In place of using simple loose packaging, edible films can be applied directly on fruits and vegetables. Starch-based edible films are odourless, tasteless, colourless, non-toxic and biodegradable. They display very low permeability to oxygen and are suitable for food protection to improve quality and shelf life without impairing consumer acceptability. Cassava starch-konjac glucomannan blend was prepared and used successfully for liquid coating of carrot slices to increase shelf life. Water loss from the coated carrot slices was slower compared to that from uncoated samples. Coating significantly maintained the visual quality of carrot compared to uncoated samples throughout the storage period at low temperature. At room temperature, there was no significant change in the colour of coated samples until 21 days of storage, and thereafter a gradual change was observed.

A novolac-like phenolic resin was prepared by co-polymerization with phenol and hydrolysed starch as the substitute for formaldehyde (toxic to environment) under catalysis of sulphuric acid. Liquefied phenolated lignosulphonate was introduced to the phenol-starch resin as a partial phenol substitute. The phenol-formaldehyde resin, phenol-starch resin and lignin-phenol-starch resin were characterized with laboratory standard analysis for their physicochemical specifications such as solid content, gel time, softening point, viscosity and free phenol content. Mechanical properties of modified and unmodified resin core sand were investigated. Structures of modified and unmodified resins

Table 9.2. Mechanical properties of the resin core sand (Zhang *et al.*, 2013).

Sample	Cold tensile strength ± SD (MPa)	Hot tensile strength ± SD (MPa)	Cold bending strength ± SD (MPa)	Hot bending strength ± SD (MPa)
Phenol-formaldehyde	2.20 ± 0.02	1.45 ± 0.07	4.20 ± 0.09	2.71 ± 0.22
Phenol-starch (2.4:1))	1.70 ± 0.12	1.52 ± 0.16	4.60 ± 0.17	3.48 ± 0.24
Phenol-starch (2.0:1)	1.76 ± 0.06	1.55 ± 0.18	4.68 ± 0.07	3.48 ± 0.15
Phenol-starch (1.6:1)	1.73 ± 0.18	1.50 ± 0.06	4.70 ± 0.15	3.44 ± 0.02
Phenol-starch (1.2:1)	1.69 ± 0.07	1.49 ± 0.11	4.51 ± 0.18	3.43 ± 0.04
0 wt% lignin)	1.76 ± 0.06	1.55 ± 0.18	4.68 ± 0.07	3.48 ± 0.15
10 wt% lignin)	1.88 ± 0.06	1.61 ± 0.08	5.64 ± 0.11	3.60 ± 0.02
20 wt % lignin	1.90 ± 0.10	1.64 ± 0.05	5.69 ± 0.02	3.88 ± 0.17
30 wt % lignin	2.23 ± 0.05	1.85 ± 0.07	5.60 ± 0.17	3.79 ± 0.02
40 wt % lignin	2.06 ± 0.02	1.80 ± 0.26	5.01 ± 0.06	3.77 ± 0.09
50 wt % lignin	1.81 ± 0.26	1.75 ± 0.17	4.99 ± 0.05	3.53 ± 0.17
60 wt% lignin	1.75 ± 0.14	1.26 ± 0.17	4.11 ± 0.17	3.10 ± 0.06
70 wt % lignin	1.50 ± 0.22	1.05 ± 0.08	3.81 ± 0.25	2.87 ± 0.05
GB/T2684-2009	2.0–4.0	1.4–2.12	≥5.0	1.5–4.0

were characterized with infrared spectrum analysis (FTIR) and carbon-13 NMR spectrometry (13CNMR), while their thermal properties were studied with thermogravimetric analysis (TGA) and DSC. The modified resin and its core sand have acceptable physicochemical specifications, thermal resistance and mechanical properties as compared with unmodified resin and phenol-formaldehyde resin (Table 9.2).

Another potential application of the starch is in construction materials like cement, concrete and gypsum boards. Based on the studies on the use of cassava starch in cement, the following conclusions have been drawn:

1. Starch delays the setting time of cement, which may be advantageous when longer time is required for casting.
2. Cassava starch adheres more strongly to cement grains and hence leads to higher viscosity and less slump compared to maize starch.
3. Incorporation of starch improves durability properties.
4. Maximum compressive strength recorded for cassava starch at 28 days occurs at 0.05% concentration with lower initial setting time.
5. Cassava starch has a beneficial effect of significantly affecting concrete slump increase at a concentration of 0.06%, advantageous in lightweight aggregate in roof tiles, insulation, flooring and soundproofing. It can replace gypsum board in water-resistant backing for shower stalls.
6. Starch composites, having partial hydrophobicity and water-absorption power, can be compatible with concrete.

Sour starch prepared by enzymatic action on cassava is widely used for production of bakery products in Latin American countries. Pandeyuca is the popular bread in Colombia (Fig. 9.2).

Starchy flour produced by inoculum-provided fermentation of cassava tubers also has been found to have desirable properties for the production of bakery items (Moorthy *et al.*, 1993).

9.2 Sweet Potato Starch

Interest in utilization of native and modified sweet potato starch for food and non-food industrial uses is on the rise during the last decade. The most common application of native and

Fig. 9.2. Pandeyuca from cassava starch (source: M.S. Sajeev).

modified starches has been as noodle ingredients (Chen *et al.*, 2003a, b).

Sweet potato starch has been used as a cheaper potential replacement for mung bean starch in the production of starch noodle, which could be obtained with or without freezing (Chen *et al.*, 2003a; Lee *et al.*, 2006). Tan *et al.* (2006) have studied the relation of structural sweet potato starch for noodle quality (made with freezing) compared to mung bean.

The quality of cooked starch noodles made from small granules was superior to that of noodles made from large-sized granules and also native whole starch (Chen *et al.*, 2003a). Manipulation of processing variables and cooking conditions, and use of curdlan, could improve the texture and eating quality of sweet potato starch noodles made without freezing (Lee *et al.*, 2005, 2006). Acetylated sweet potato starch could substitute, in part, wheat flour in composite flour for producing noodles (Chen *et al.*, 2003b).

Sweet potato starch has been degraded and converted to various food and industrial products by chemical and enzymatic methods (Alves-Prado *et al.*, 2008; Chou *et al.*, 2010; He *et al.*, 2009; Yen *et al.*, 2010). These include fermentative production of lactic acid by *Rhizopus oryzae* (Yen *et al.*, 2010); degradation by use of various debranching enzymes and amylases to maltose and subsequent conversion to trehalose by treatment with a recombinant *Picrophilus torridus* trehalose synthase (Chou *et al.*, 2010); hydrolysis by β-amylase to produce reducing sugars (Sawai *et al.*, 2004); and by cyclomaltodextrin glucanotransferase from *Bacillus clausii* to produce β- and γ-cyclodextrins (Alves-Prado *et al.*, 2008). Bioethanol could be obtained by using genetically engineered *Zymomonas mobilis* (He *et al.*, 2009; Srichuwong *et al.*, 2012). CTCRI has also developed technology for large-scale production ethanol based on sweet potato starch.

Sweet potato starch has been used for producing edible film. The kinetics of water vapour adsorption of the film have been enumerated (Kim and Rhim, 2004). Edible films with antimicrobial properties by incorporation of functional ingredients such as origanum oil (Ehivet *et al.*, 2011) and potassium sorbate (Shen *et al.*, 2010) have been developed. A thermoplastic material using dialdehyde starch has also been reported (Zhang *et al.*, 2011).

Sweet potato starch has found multiple uses in pharmaceutical products. It has been formulated as film and pharmaceutical hard capsule (Bae *et al.*, 2008); solid dosage (Riley and Adebayo, 2010); and a binding agent for tablets (Odeku, 2013; Subhadhirasakul *et al.*, 2001). The micro-particles (10.3–13.1 mm) from starch were used as controlled drug release carriers (Liu *et al.*, 2007). Another novel medical use for sweet potato starch is as an ingredient in the production of porous structural materials for tissue-engineering scaffolds (Nakamatsu *et al.*, 2006; Torres *et al.*, 2007). There are still a great many new avenues to explore for the application of sweet potato starch, such as in biodegradable plastics and nanocomposites.

Starch is used in the food industry as a thickening, pasting or gelling agent in a variety of formulations. In Japan, sweet potato starch is sometimes used as a setting agent for a fish jelly paste called *kamaboko*. A starchy sheet jelly, eaten with vegetables, is made in China by spreading a thin layer of starch paste in the sun or oven to dry. Sweet potato starch has also been suggested as a replacement for other flours or starches in various traditional Korean products such as *mook* (a traditional starch gel made from acorn or buckwheat flour) or *kochujang* (a red pepper paste seasoning prepared with rice flour) (Bae and Lee, 1984). Sweet potato 'sago', prepared by moisturizing starch, shaking it in a vibrating screen and dry-heating it in a pan over fire can be converted into porridge by boiling in water.

9.3 Aroid Starches

Starches of *C. esculenta* and *X. sagittifolium* have found diverse applications (Antonio-Estrada *et al.*, 2009; Gonzalez-Soto *et al.*, 2011; Nasipuri, 1979; Owusu-Ansah *et al.*, 2005; Sit *et al.*, 2014c). In application as a tablet binder and disintegrant for some drugs, native starch of *C. esculenta* has properties comparable to potato starch (Nasipuri, 1979). The composition and properties of starches of *C. esculenta* and *X. sagittifolium* fulfilled the specifications of UK, US and Ghana authorities for tablet formulations (Owusu-Ansah *et al.*, 2005). Maltodextrin produced by the enzymatic hydrolysis of *C. esculenta* starch could be a potential substitute for the commercial maize maltodextrins, as these two possessed similar water- and oil-absorption properties (Antonio-Estrada *et al.*, 2009). Spherical aggregates of *C. esculenta* starch were produced by spray-drying for potential applications as an encapsulation matrix for controlled-release products (Gonzalez-Soto *et al.*, 2011). Incorporation of *C. esculenta* starch (2%) into tomato ketchup enhances the firmness and consistency of ketchup without affecting the colour and thus can be a replacement for maize starch in terms of sensory quality (Sit *et al.*, 2014c). Starch nanoparticles produced by recrystallization of pullulanase hydrolysed starch of *C. esculenta* were incorporated into maize starch films resulting in decreased water vapour permeability, and increased tensile strength and thermal stability of the resulting films (Dai *et al.*, 2015).

A potential area of application of taro starch is in cosmetics, where rice has been conventionally employed in face powder formulations, in view of its small granular size of 1–10 μm, which is similar to rice. Cosmetic dusting preparations using aerosol dispensing systems would also clearly benefit from the availability of starches that have small granule-size in view of the mechanical problems posed by aerosol valve devices and their small apertures. Dusting powders for surgical use can also be produced by using this starch. At one point in the development of the plastic industry, before petrochemical-derived packaging film materials became available, much effort was put into perfecting the viscose process for producing regenerated cellulose transparent packaging film. Attempts to create a similar product based on starch solutions to avoid the unpleasant and costly xanthate process were not successful because of the brittleness of the cast films. However, it was found that a taro-derived film was more successful (Simonds and Ellis, 1943) and presumably has been abandoned with the advent of low-cost synthetics. Since this early success suggests some unique property of certain taro starches, it clearly deserves reinvestigation against the current background of rising oil prices and projected scarcity of petrochemicals in the future.

It has been published that plastics based on such polymers as polystyrene, polyethylene and polyvinyl chloride can incorporate up to 40% of starch without affecting most of the original physical properties of the plastics. When taro starch is used in the plastics in an appropriate formulation, the result can be useful in

Table 9.3. Advantages of using taro starch in plastic production (Griffin, 1979).

No.	Taro starch	Minerals
1	Density comparable with plastics 1.49	High density ranging from 2.6 for silica to 4.6 for barites
2	Narrow particle size range. Low porosity. Very small particle size; can therefore be used in surface coatings or very thin films	Broad particle size range with often high fines content. Very low porosity, except certain chalks, dolomites and clays
3	Simple particle geometry approaching spherical and regular; minimum disturbance to melt rheology	Irregular particles, mostly fracture fragments from grinding operations. High surface area
4	Very low abrasive properties	Often extremely abrasive
5	Colourless and almost transparent, can yield translucent or transparent products	Usually colourless but of high refractive index
6	No significant metallic content; starch itself is an accepted food product	Transition metals may be released; possible interference with antioxidant function; possible toxicity questions
7	Thermally stable to 250°C	Thermally stable to very high temperatures
8	Not water-soluble, but hygroscopic. Also hygroscopic *in situ*	Some minerals retain traces of water tenaciously, but not normally hygroscopic *in situ*
9	Low fixed-gas permeability	Very low fixed gas permeability
10	Biodegradable formulations possible	Permanent in a biologically active environment

acceleration of the biodegradability of the parent polymer (Griffin, 1979).

The significance of the particle size of filler in plastics resides in both its influence on the processing properties and in the resulting solid composite. The former is evident especially in the viscosity increase produced by the suspended particles, which, being an undesirable effect, can be minimized using particles of spherical form, and taro starch can fill this need. The advantage of using taro starch in this application is shown in Table 9.3.

The lack of compatibility of native starch with plastics can be overcome by chemically modifying the starch so as to suppress the influence of the hydroxyl groups; but the modification may be restricted to the surface layer only, since this is the only part that is in contact with the polymer matrix.

Though the fine powder fillers can give rise to moisture adsorption problems due to higher-equilibrium moisture content compared to mineral fillers, a small dimensional change was associated with the uptake of moisture by starches, which was measured directly in an impressive experimental study by (Hellman and Melvin, 1950). One positive aspect of the hygroscopic nature of starch is the lower tendency of the surface of starch-extended plastics to accumulate electrostatic charges once they have come closer to the equilibrium moisture content. This is relevant in products that are susceptible to the attraction of dust.

9.4 Yam Starches

Yam starches in native and modified forms have been used in various food and non-food applications, such as tablet formulation, film formation and composite flour, and the yam species *D. alata* has been the most widely used starch.

9.4.1 Food uses

Native yam starches have been incorporated as ingredients in the production of pasta and bread (Nindjin *et al.*, 2011; Sobukola *et al.*,

2013). *D. rotundata* starch was used in the production of cookies (CTCRI). The starch of *D. rotundata* was formulated with brewers' spent grain to produce dietry pasta with a single-screw extruder (Sobukola *et al.*, 2013). Starch of *D. cayenensis-rotundata*, used to substitute up to 30% wheat, resulted in bread with a similar loaf volume as the control and acceptable sensory quality attributes (Nindjin *et al.*, 2011). Wheat flour could be partially substituted by yam starches in some wheat-based food products (Zaidul *et al.*, 2007, 2008). Otegbayo *et al.*, 2014 have examined different yams for food and non-food applications and the study showed that yam starches show large variability in their physicochemical and functional properties; hence they can have diverse applications in both food and non-food industries. The small granule size of *D. dumetorum* compared with other species indicated that it could find application as fat replacers (carbohydrate-based) in foods, since they can exhibit fat-mimetric properties, which produces textures desirable in frozen desserts, cookies and other low-fat and fat-free formulations. Yam starches with high swelling capacity (*D. rotundata*, *D. bulbifera*) can be used as thickeners, binders and gelling agents in foods. Though retrogradation and syneresis tendency of the starches could be a disadvantage in frozen products, they can be utilized in products were retrogradation is a desired quality, e.g. noodles, soups and sauces. Diverse gelation properties of the starches could be an advantage in different food applications: those with high-quality gelling ability could be used in foods such as jams and jellies, where elasticity is desired, while those with low gelling ability will be very useful in processing complementary diet.

9.4.2 Non-food uses

The native and cross-linked starches from *D. alata* and *D. trifida* were used to produce biodegradable starch films. The drying conditions and ingredient composition could be related to various physical properties (water barrier, mechanical and thermal) of the films (Mali *et al.*, 2002, 2004, 2005; Mali and Grossmann, 2003; Durango *et al.*, 2006; Ferreira *et al.*, 2009; Pérez *et al.*, 2012; Reis *et al.*, 2013). Incorporation of monoglyceride in the starch film prevented phase separation, while decreasing water permeability and increasing stability of the film by forming a V-type inclusion complex with the amylose component, fixed the starch components in the film matrix (Ferreira *et al.*, 2009). Cross-linked starch was used to make native film and the resulting sample had advantages like greater width, opalescence and water permeability compared to native starch. Cross-linking altered the steric hindrance of amylose resulting in less condensed internal structure of the film (Pérez *et al.*, 2012). Films with the addition of chitosan, which has antimicrobial properties against *Salmonella enteritidis*, have been reported for food packages and had good flexibility comparable to low-density polyethylene film (Durango *et al.*, 2006). Strawberries stored in starch film exhibited enhanced microbial stability from 14 to 21 days (Mali and Grossmann, 2003). Apart from the above-mentioned starch-based films, low-density polyethylene-based films with starch of *D. alata* as an ingredient were produced using a hot-melt compounding technique. The starch incorporation level up to 20% could give superior biodegradability (Ogbobe *et al.*, 1997). Yam starches can also be used as stabilizers in baking powder, aerosols, face powder or dusting powder in the cosmetic and textile industries (Otegbayo *et al.*, 2014).

9.4.3 Pharma applications

Yam starches from diverse species (*D. alata*, *D. dumetorum*, *D. oppositifolia*, *D. cayenensis*, *D. esculenta*, *D. polygonoides* and *D. rotundata*) have been used for tablet formulation (Riley *et al.*, 2006a, 2008; Zuluaga *et al.*, 2007; Odeku and Picker-Freyer, 2007, 2009a, 2009b; Odeku *et al.*, 2008; Nattapulwat *et al.*, 2009; Okunlola and Odeku, 2011; Odeku and Akinwande, 2012; Okunlola and Akingbala, 2013). Starches of *D. dumetorum* with smaller granules and higher specific surface area, being more compressible, provided tablets with acceptable crushing force, while *D. rotundata* and *D. alata* with larger granules and higher amylose content imparted lower compressibility, thereby preventing easy compaction (Liu *et al.*, 2007; Odeku and Picker-Freyer, 2007;

Torres *et al.*, 2007; Zaidul *et al.*, 2007). The dissolution speed of paracetamol was positively related to average powder diameter, mucilage surface tension and intrinsic viscosity of starch as the matrix (Riley *et al.*, 2008). Modified starches in various forms (cross-linked, hydroxypropylated, acid-thinned, carboxymethylated, and pre-gelatinized) have been used to provide the tablets with a range of properties. For instance, acid-thinned starch was found to be useful to improve the disintegrant efficiency (Odeku and Akinwande, 2012); as binders when high bond strength and fast dissolution are desired (Okunlola and Akingbala, 2013); and as a directly compressible excipient to improve the crushing force of tablets (Odeku and Picker-Freyer, 2009a). Acid-thinned starch has a lesser quantity of amorphous regions and smaller particle size, resulting in stronger and denser compacting and thus higher crushing force and tablet strength (Odeku and Picker-Freyer, 2009a). Incorporation of carboxymethyl starch (1.0–4.0%) increases the rate of disintegration of tablets (Nattapulwat *et al.*, 2009). Hydroxypropylation and cross-linking of starch raised the compaction and tablet crushing force (Odeku and Picker-Freyer, 2009b), and freeze-dried pre-gelatinized starch gave high crushing force and fast disintegration of the tablet (Odeku *et al.*, 2008). These properties of modified starches can be linked to starch's properties. For example, compared to native starches, pre-gelatinized and carboxymethylated starches have enhanced water solubility, thus giving tablets propensity to disintegrate fast. Therefore, careful selection of native and modified starches with desired physicochemical properties and structures can achieve required tableting characteristics (Riley *et al.*, 2006a; Odeku and Picker-Freyer, 2009b). Shanavas (2011) has studied the different yam starches for tablet preparations and identified native and modified yam starches having potential as excipient in tablets.

Composites that have been produced using yam starches incorporated with hectorite, and halloysite nanotubes with less than 30% clay content had a stronger gel network than those with higher clay contents as revealed by diverse rheological studies (Wilhelm *et al.*, 2005; Xie *et al.*, 2011). Incorporation of halloysite nanotubes into starch of *D. opposita* enhanced thermal stability, mechanical strength, and water-barrier ability of the composite (Xie *et al.*, 2011).

Yam starch has been used as a substrate in *Penicillium* sp. S-22 fermentation to yield enzymes that can efficiently degrade starch (Sun *et al.*, 2006). Yam starches can be used as a good substrate to produce bioethanol and other chemicals such as lactic acid, due to the high starch content of the tubers; however, there are only very few reports on large-scale utilization of yam starches.

9.5 Other Starches

9.5.1 Arrowroot starch

Starch from arrowroot is highly in demand for production of biscuits (Fig. 9.3). It is used because of its digestibility and flavour. The starch also finds use as infant food and food for convalescents due to its easy digestibility.

Fig. 9.3. Arrowroot starch biscuit (source: M.S. Sajeev).

Fig. 9.4. Biscuits using canna starch.

9.5.2 Curcuma starch

Starch is extracted from tubers of edible curcuma species at homestead level for use as infant food. In fact, in Kerala state in India, small transportable starch extractors are carried from home to home to extract the starch in the production sites.

9.5.3 *Canna edulis* starch

Resistant starch (RS) type III, which has several health effects, could be produced from starch with high amylose such as *Canna edulis* Kerr. by an autoclaving-cooling method (Astuti and Naufalin, 2011), and this starch was used in the preparation of functional biscuits, suitable for diabetics and those suffering from a gluten allergy (Fig. 9.4).

Canna is gaining wide popularity in Colombia, where there is growing demand for biscuits made from roots. In Vietnam, the crop is grown over 30,000 ha to make high-value noodles.

9.5.4 *Arracacha* starch

The small size of *Arracacha* starch grains makes it easy to digest, so it is good puréed or in soup for babies, the elderly or people with disabilities. The processed roots are used as a thickener for baby food and instant soups.

References

Alves-Prado, H.F., Carneiro, A.A.J., Pavezzi, F.C., Gomes, E., Boscolo, M. *et al*. (2008) Production of cyclodextrins by CGTase from *Bacillus clausii* using different starches as substrates. *Applied Biochemistry and Biotechnology* 146(1–3), 3–13. DOI: 10.1007/s12010-007-8093-z.

Antonio-Estrada, C., Bello-Perez, L.A., Martínez-Sanchez, C.E., Montañez-Soto, J.L., Jimenez-Hernandez, J. *et al*. (2009) Enzymatic production of maltodextrins from taro (*Colocasia esculenta*) starch. *CyTAe Journal of Food* 7, 233–241.

Astuti, S.D. and Naufalin, R. (2011) Formulation and characterization of functional biscuit consisting of *Canna edulis*, Kerr resistant starch type III, granulated palm sugar, and soy protein concentrate. *International Food Conference, Surabaya*, 28–29.

Bae, H.J., Cha, D.S., Whiteside, W.S. and Park, H.J. (2008) Film and pharmaceutical hard capsule formation properties of mungbean, water chestnut, and sweet potato starches. *Food Chemistry* 106(1), 96–105. DOI: 10.1016/j.foodchem.2007.05.070.

Bae, M.H. and Lee, S.W. (1984) Study on history and quality evaluation of various hot peppers. *The Korean Journal of Community Living Science* 1, 187–202.

Chen, Z., Schols, H.A. and Voragen, A.G.J. (2003a) Starch granule size strongly determines starch noodle processing and noodle quality. *Journal of Food Science* 68(5), 1584–1589. DOI: 10.1111/j.1365-2621.2003.tb12295.x.

Chen, Z., Schols, H.A. and Voragen, A.G.J. (2003b) The use of potato and sweet potato starches affects white salted noodle quality. *Journal of Food Science* 68(9), 2630–2637. DOI: 10.1111/j.1365-2621.2003.tb05781.x.

Chou, H.-H., Chang, S.-W., Lee, G.-C., Chen, Y.-S., Yeh, T. *et al.* (2010) Site-directed mutagenesis improves the thermostability of a recombinant *Picrophilus torridus* trehalose synthase and efficiency for the production of trehalose from sweet potato starch. *Food Chemistry* 119(3), 1017–1022. DOI: 10.1016/j.foodchem.2009.08.010.

Dai, L., Qiu, C., Xiong, L. and Sun, Q. (2015) Characterisation of corn starch-based films reinforced with taro starch nanoparticles. *Food Chemistry* 174, 82–88. DOI: 10.1016/j.foodchem.2014.11.005.

Durango, A.M., Soares, N.F.F., Benevides, S., Teixeira, J., Carvalho, M. *et al.* (2006) Development and evaluation of an edible antimicrobial film based on yam starch and chitosan. *Packaging Technology and Science* 19(1), 55–59. DOI: 10.1002/pts.713.

Ehivet, F.E., Min, B., Park, M.-K. and Oh, J.-H. (2011) Characterization and antimicrobial activity of sweet-potato starch-based edible film containing origanum (*Thymus capitatus*) oil. *Journal of Food Science* 76(1), C178–C184. DOI: 10.1111/j.1750-3841.2010.01961.x.

Ferreira, F.A.B., Grossmann, M.V.E., Mali, S., Yamashita, F. and Cardoso, L.P. (2009) Effect of relative humidities on microstructural, barrier and mechanical properties of yam starch-monoglyceride films. *Brazilian Archives of Biology and Technology* 52(6), 1505–1512. DOI: 10.1590/S1516-89132009000600022.

Gonzalez-Soto, R.A., de la Vega, B., García-Suarez, F.J., Agama-Acevedo, E. and Bello-Pérez, L.A. (2011) Preparation of spherical aggregates of taro starch granules. *LWT - Food Science and Technology* 44(10), 2064–2069. DOI: 10.1016/j.lwt.2011.06.018.

Griffin, G.J.L. (1979) Nonfood application of starch with special reference to potential uses of taro. In: Plucknett, D.L. (ed.) *Small-scale Processing and Storage of Tropical Root Crops*. Westview Press, Boulder, Colorado, pp. 275–299.

He, M.X., Feng, H., Bai, F., Li, Y., Liu, X. *et al.* (2009) Direct production of ethanol from raw sweet potato starch using genetically engineered *Zymomonas mobilis*. *African Journal of Microbiology Research* 3, 721–726.

Hellman, N.N. and Melvin, E.H. (1950) Surface area of starch and its role in water sorption. *Journal of the American Chemical Society* 72(11), 5186–5188. DOI: 10.1021/ja01167a110.

Kim, J.N. and Rhim, J.W. (2004) Kinetics of water vapor adsorption by sweet potato starch-based edible films. *Food Science and Biotechnology* 13, 616–621.

Lee, S.-Y., Woo, K.-S., Lim, J.-K., Kim, H.I. and Lim, S.-T. (2005) Effect of processing variables on texture of sweet potato starch noodles prepared in a nonfreezing process. *Cereal Chemistry Journal* 82(4), 475–478. DOI: 10.1094/CC-82-0475.

Lee, S.Y., Kim, J.Y., Lee, S.J. and Lim, S.T. (2006) Textural improvement of sweet potato starch noodles prepared without freezing using gums and other starches. *Food Science and Biotechnology* 15, 986–989.

Liu, C.-S., Desai, K.G.H., Meng, X.-H. and Chen, X.-G. (2007) Sweet potato starch microparticles as controlled drug release carriers: Preparation and in vitro drug release. *Drying Technology* 25(4), 689–693. DOI: 10.1080/07373930701290977.

Mali, S. and Grossmann, M.V.E. (2003) Effects of yam starch films on storability and quality of fresh strawberries (*Fragaria ananassa*). *Journal of Agricultural and Food Chemistry* 51(24), 7005–7011. DOI: 10.1021/jf034241c.

Mali, S., Grossmann, M.V.E., Garcia, M.A., Martino, M.N. and Zaritzky, N.E. (2002) Microstructural characterization of yam starch films. *Carbohydrate Polymers* 50(4), 379–386. DOI: 10.1016/S0144-8617(02)00058-9.

Mali, S., Grossmann, M.V.E., García, M.A., Martino, M.N. and Zaritzky, N.E. (2004) Barrier, mechanical and optical properties of plasticized yam starch films. *Carbohydrate Polymers* 56(2), 129–135. DOI: 10.1016/j.carbpol.2004.01.004.

Mali, S., Grossmann, M.V.E., García, M.A., Martino, M.N. and Zaritzky, N.E. (2005) Mechanical and thermal properties of yam starch films. *Food Hydrocolloids* 19(1), 157–164. DOI: 10.1016/j.foodhyd.2004.05.002.

Moorthy, S.N., George, M. and Padmaja, G. (1993) Functional properties of the starchy flour extracted from cassava on fermentation with a mixed culture inoculum. *Journal of the Science of Food and Agriculture* 61(4), 443–447.

Nakamatsu, J., Torres, F.G., Troncoso, O.P., Min-Lin, Y. and Boccaccini, A.R. (2006) Processing and characterization of porous structures from chitosan and starch for tissue engineering scaffolds. *Biomacromolecules* 7(12), 3345–3355. DOI: 10.1021/bm0605311.

Nasipuri, R.N. (1979) Evaluation of cocoyam starch as tablet binder and dis-integrant. *Pharmaceutica Acta Helvetiae* 54, 48–53.

Nattapulwat, N., Purkkao, N. and Suwithayapan, O. (2009) Preparation and application of carboxymethyl yam (*Dioscorea esculenta*) starch. *AAPS PharmSciTech* 10(1), 193–198. DOI: 10.1208/s12249-009-9194-5.

Nindjin, C., Amani, G.N. and Sindic, M. (2011) Effect of blend levels on composite wheat doughs performance made from yam and cassava native starches and bread quality. *Carbohydrate Polymers* 86(4), 1637–1645. DOI: 10.1016/j.carbpol.2011.06.076.

Odeku, O.A. (2013) Potentials of tropical starches as pharmaceutical excipients: a review. *Starch/Stärke* 65(1–2), 89–106. DOI: 10.1002/star.201200076.

Odeku, O.A. and Akinwande, B.L. (2012) Effect of the mode of incorporation on the disintegrant properties of acid modified water and white yam starches. *Saudi Pharmaceutical Journal* 20(2), 171–175. DOI: 10.1016/j.jsps.2011.09.001.

Odeku, O.A. and Picker-Freyer, K.M. (2007) Analysis of the material and tablet formation properties of four *Dioscorea* starches. *Starch/Stärke* 59(9), 430–444. DOI: 10.1002/star.200700619.

Odeku, O.A. and Picker-Freyer, K.M. (2009a) Characterization of acid modified *Dioscorea* starches as direct compression excipient. *Pharmaceutical Development and Technology* 14(3), 259–270. DOI: 10.1080/10837450802572367.

Odeku, O.A. and Picker-Freyer, K.M. (2009b) Evaluation of the material and tablet formation properties of modified forms of *Dioscorea* starches. *Drug Development and Industrial Pharmacy* 35(11), 1389–1406. DOI: 10.3109/03639040902960185.

Odeku, O., Schmid, W. and Picker-Freyer, K. (2008) Material and tablet properties of pregelatinized (thermally modified) *Dioscorea* starches. *European Journal of Pharmaceutics and Biopharmaceutics* 70(1), 357–371. DOI: 10.1016/j.ejpb.2008.04.011.

Ogbobe, O., Okeke, C.N. and Otashu, M. (1997) Coco-yam and water-yam starches: as alternative dual purpose polymer ingredients. *Indian Journal of Engineering and Materials Sciences* 4, 134–138.

Okunlola, A. and Akingbala, O. (2013) Characterization and evaluation of acid-modified starch of *Dioscorea oppositifolia* (Chinese yam) as a binder in chloroquine phosphate tablets. *Brazilian Journal of Pharmaceutical Sciences* 49(4), 699–708. DOI: 10.1590/S1984-82502013000400009.

Okunlola, A. and Odeku, O.A. (2011) Evaluation of starches obtained from four *Dioscorea* species as binding agent in chloroquine phosphate tablet formulations. *Saudi Pharmaceutical Journal* 19(2), 95–105. DOI: 10.1016/j.jsps.2011.01.002.

Otegbayo, B., Oguniyan, D. and Akinwumi, O. (2014) Physicochemical and functional characterization of yam starch for potential industrial applications. *Starch/Stärke* 66(3–4), 235–250. DOI: 10.1002/star.201300056.

Owusu-Ansah, E., Oppong-Boachie, F.K.. and Awuah, R.T. (2005) Chemical constituents of essential oil from the leaves of Ghanaian *Ocimum gratissimum* L. *Asian Journal of Chemistry* 17(3), 1446–1448.

Pérez, E., Segovia, X., Tapia, M.S. and Schroeder, M. (2012) Native and cross-linked modified *Dioscorea trifida* (cush-cush yam) starches as bio-matrices for edible films. *Journal of Cellular Plastics* 48(6), 545–556. DOI: 10.1177/0021955X12445603.

Reis, R.C., Côrrea, P.C., Devilla, I.A., Santos, E.S., Ascheri, D.P.R. *et al.* (2013) Drying of yam starch (*Discorea* spp.) and glycerol filmogenic solutions at different temperatures. *LWT—Food Science and Technology* 50, 651–656.

Riley, C.K. and Adebayo, S.A. (2010) A comparative investigation of the packing and flow properties of sweet potato (*Ipomoea batatas*) starches and their potential use in solid dosage formulations. *Starch/Stärke* 62(6), 285–293. DOI: 10.1002/star.200900201.

Riley, C.K., Adebayo, S.A., Wheatley, A.O. and Asemota, H.N. (2006a) Fundamental and derived properties of yam (*Dioscorea* spp.) starch powders and implications in tablet and capsule formulation. *Starch/Stärke* 58(8), 418–424. DOI: 10.1002/star.200600491.

Riley, C.K., Adebayo, S.A., Wheatley, A.O. and Asemota, H.N. (2008) The interplay between yam (*Dioscorea* spp.) starch botanical source, micromeritics and functionality in paracetamol granules for reconstitution. *European Journal of Pharmaceutics and Biopharmaceutics* 70, 326–334.

Sawai, J., Nakai, T., Hashimoto, A. and Shimizu, M. (2004) A comparison of the hydrolysis of sweet potato starch with beta-amylase and infrared radiation allows prediction of reducing sugar production. *International Journal of Food Science and Technology* 39(9), 967–974. DOI: 10.1111/j.1365-2621.2004.00865.x.

Shanavas, S. (2011) *Studies on modified starches useful for tablets and capsules*. Kannur University, Kerala.

Shen, X.L., Wu, J.M., Chen, Y. and Zhao, G. (2010) Antimicrobial and physical properties of sweet potato starch films incorporated with potassium sorbate or chitosan. *Food Hydrocolloids* 24(4), 285–290. DOI: 10.1016/j.foodhyd.2009.10.003.

Simonds, H.R. and Ellis, C. (1943) *Handbook of Plastics*. Chapman and Hall, New York.

Sit, N., Misra, S., Baruah, D., Badwaik, L.S. and Deka, S.C. (2014c) Physicochemical properties of taro and maize starch and their effect on texture, colour and sensory quality of tomato ketchup. *Starch/Stärke* 66(3–4), 294–302. DOI: 10.1002/star.201300120.

Sobukola, O.P., Babajide, J.M. and Ogunsade, O. (2013) Effect of brewers spent grain addition and extrusion parameters on some properties of extruded yam starch-based pasta. *Journal of Food Processing and Preservation* 37(5), 734–743. DOI: 10.1111/j.1745-4549.2012.00711.x.

Srichuwong, S., Orikasa, T., Matsuki, J., Shiina, T., Kobayashi, T. *et al.* (2012) Sweet potato having a low temperature-gelatinizing starch as a promising feedstock for bioethanol production. *Biomass and Bioenergy* 39, 120–127. DOI: 10.1016/j.biombioe.2011.12.023.

Subhadhirasakul, S., Yuenyoungsawad, S., Ketjinda, W., Phadoongsombut, N. and Faroongsarng, D. (2001) Study on tablet binding and disintegrating properties of alternative starches prepared from taro and sweet potato tubers. *Drug Development and Industrial Pharmacy* 27(1), 81–87. DOI: 10.1081/DDC-100000131.

Sun, H.Y., Ge, X.Y. and Zhang, W.G. (2006) Improved production of an enzyme that hydrolyses raw yam starch by *Penicillium* spp. S-22 using fed-batch fermentation. *Biotechnology Letters* 28, 1719–1723.

Tan, H.-Z., Gu, W.-Y., Zhou, J.-P., Wu, W.-G. and Xie, Y.-L. (2006) Comparative study on the starch noodle structure of sweet potato and mung bean. *Journal of Food Science* 71(8), C447–C455.

Torres, F.G., Boccaccini, A.R. and Troncoso, O.P. (2007) Microwave processing of starch-based porous structures for tissue engineering scaffolds. *Journal of Applied Polymer Science* 103(2), 1332–1339. DOI: 10.1002/app.25345.

Wilhelm, H.-M., Sierakowski, M.-R., Reicher, F., Wypych, F. and Souza, G.P. (2005) Dynamic rheological properties of yam starch/hectorite composite gels. *Polymer International* 54(5), 814–822. DOI: 10.1002/pi.1775.

Xie, Y., Chang, P.R., Wang, S., Yu, J. and Ma, X. (2011) Preparation and properties of halloysite nanotubes/plasticized *Dioscorea opposita* Thunb. starch composites. *Carbohydrate Polymers* 83(1), 186–191. DOI: 10.1016/j.carbpol.2010.07.039.

Yen, H.-W., Chen, T.-J., Pan, W.-C. and Wu, H.-J. (2010) Effects of neutralizing agents on lactic acid production by *Rhizopus oryzae* using sweet potato starch. *World Journal of Microbiology and Biotechnology* 26(3), 437–441. DOI: 10.1007/s11274-009-0186-0.

Zaidul, I.S.M., Norulaini, N.A.N., Omar, A.K.M., Yamauchi, H. and Noda, T. (2007) RVA analysis of mixtures of wheat flour and potato, sweet potato, yam, and cassava starches. *Carbohydrate Polymers* 69(4), 784–791. DOI: 10.1016/j.carbpol.2007.02.021.

Zaidul, I.S.M., Absar, N., Kim, S.-J., Suzuki, T., Karim, A.A. *et al.* (2008) DSC study of mixtures of wheat flour and potato, sweet potato, cassava, and yam starches. *Journal of Food Engineering* 86(1), 68–73. DOI: 10.1016/j.jfoodeng.2007.09.011.

Zhang, L., Zheng, Y., Zhang, Y. and Lin, T. (2013) Effect of starch and lignin on physico-chemical properties of phenol-starch resin and its resin core sand. *Starch - Stärke* 65(7-8), 666–678.

Zhang, S.D., Wang, X.L. and Wang, Y.Z. (2011) Properties of thermoplastic dialdehyde sweet potato starch affected by carbonyl contents. *Advanced Materials Research* 233–235, 2541–2544. DOI: 10.4028/www.scientific.net/AMR.233-235.2541.

Zuluaga, M.-F., Baena, Y., Mora, C.-E. and D'León, L.-F.P. (2007) Physicochemical characterization and application of yam (*Dioscorea cayenensis-rotundata*) starch as a pharmaceutical excipient. *Starch/Stärke* 59(7), 307–317. DOI: 10.1002/star.200600516.

10

Characterization and Analysis of Starches: Methodology for Starch Analysis

10.1 General Method for Starch Isolation

Tuber crops are rich in starch and the starch is easily extracted compared to cereals. In some cases, enzyme activity must be inhibited in order to prevent degradation during storage or processing. Mucilaginous substances can present a difficulty (yams and aroids), which as a rule can be overcome by dilution or some chemical and enzymatic treatment.

Procedure

It is necessary to use fresh material, since biochemical changes take place during storage. All non-starchy tissues are removed as much as possible. Depending on the quantity of material available, the tissue is homogenized in a mortar or a Waring Blendor with the possible addition of an amylase inhibitor, such as a 0.01M mercuric chloride solution. A cold 1% ammonium oxalate solution has been used with success for the extraction of starch from widely varying materials.

The suspension is filtered through muslin or cheesecloth and the homogenate is centrifuged immediately at 200–300 × *g*. Crude impurities can be removed from the starch layer by repeated gentle washing, followed by resuspension of the starch in water and centrifugation at 1500 × *g*. If the impurities settle down below the starch layer, it is recommended that the starch be resuspended in a small amount of water and centrifuged for a few seconds; this has to be repeated several times and each time the suspension is poured into a clean tube.

In the case of cassava, the starch settles quite fast and centrifugation may not be needed.

Repeated suspension in 0.1M sodium chloride, followed by shaking the saline suspension for several hours with toluene (0.1 vol), will denature and remove most of the protein without causing degradation of the starch. After sedimentation of the granules, the protein-toluene layer is discarded. The product may now be defatted by refluxing with 80% aqueous methanol, washed with acetone and ether and dried under reduced pressure at 50 pascal.

10.2 Separation of Amylose and Amylopectin

The method involves leaching of amylose from starch carefully dispersed in dilute alkali; it can be performed on a large scale at room temperature in the presence of air. It produces pure amylose which dissolves readily in cold water to form a comparatively stable solution, and leaves pure amylopectin in gel form insoluble in cold water. The yield can be nearly half of theoretical for both components.

10.2.1 Dispersion of the starch

To 13.2 litres of 0.157N sodium hydroxide in a glass-walled tank is added 160 g of drained freshly prepared starch, weighed wet and suspended by gentle shaking in 1050 ml of water. The temperature of the dispersion must be at least 22°C, but preferably near 25°C. The mixture should be stirred gently until it clears, and the stirring must not be too vigorous. The method of stirring is crucial; motor-driven stirrers or bubbling gas through the mixture must be avoided lest the gel structure of the amylopectin be damaged, making subsequent separation difficult. Gentle stirring with a plate glass strip (10 × 50 cm) as a paddle is ideal. The alkaline solution should be set aside for a total of 5 minutes; then 3660 ml of 5% sodium chloride is added; next the dispersion is neutralized with N-hydrochloric acid to pH of 6.5–7.5, with the aid of an external indicator or long-lead glass electrode assembly.

The amylose solution should be filtered through a 15 cm fritted glass filter (No. 3) used as a filter stick if possible and collected in 5L flasks. The solution must next be filtered through a No. 4 filter.

10.2.2 Isolation of amylose

The amylose retrogrades if stored as its saline solution for more than a few days, and so it should be separated from this solution as soon as possible. 1-butanol is used to precipitate the amylose selectively from the saline solution by the following procedure. The filtrate is saturated with redistilled 1-butanol and stirred gently with a motor-driven stirrer for about one hour at room temperature. The precipitated amylose-butanol complex is allowed to settle for two to three hours and the clear supernatant is siphoned off. The partially sedimented solid is then centrifuged at 3000 rpm for 15 minutes in polypropylene buckets. The complex is stirred in 1-butanol saturated water (500 ml) and reconcentrated by centrifuging as before. With some samples of starch, reprecipitations may be necessary. The complex is stable if stored in a stoppered container. It is readily soluble in water at room temperature. Most of the butanol can be removed from this solution by bubbling oxygen-free nitrogen through it for 10 minutes in a vessel heated in a boiling-water bath; 160 g of wet starch yields 6–7 g of amylose. The blue values for various preparations have been found to be between 1.35 and 1.41.

10.2.3 Purification of amylopectin

The amylopectin gel is centrifuged at 8000 rpm for 20 minutes at 20°C. The supernatant is discarded and 1% sodium chloride solution is added to the gel (100 ml/4 g of original wet starch) with stirring. The mixture is allowed to stand 20 hours at 18°C, and the gel is collected by centrifuging at 8000 rpm. A second washing is carried out under the same conditions. After centrifugation, the gel can be either suspended in water for immediate use (it is insoluble in cold water) or can be freeze-dried and stored; 160 g of wet starch yields approximately 24 g of amylopectin of blue value 0.13 at 680 μm.

10.3 Determination of Starch Content

10.3.1 Estimation of starch content

Though extracted starch looks pure, it is not 100% pure, and since many applications of starch depend on the starch purity, the determination of starch content is very important.

The principle behind the determination is hydrolysis of starch to glucose and determination of the glucose produced by any standard method.

Procedure

500 mg to 1 g of dry starch is taken in a conical flask and 20 ml 2N HCl is added and the suspension is heated on a heating mantle or hot plate. The starch will go into solution and heating is

continued with slow boiling for about 10 minutes. The flask is cooled and the solution is made up to 100 ml with distilled water. An aliquot of the dilute solution is used for determination of glucose content. Standard methods include glucose oxidase, Fehling's solution or anthrone reagent.

Calculation: % glucose in aliquot $\times 100 \times 0.9/$ Wt of starch used

10.3.2 Determination of glucose by glucose oxidase method

Glucose is a widely distributed simple sugar with an active aldehyde group. Estimation of glucose by glucose oxidase gives the true glucose concentration eliminating the interference by other reducing sugars.

Principle

Glucose oxidase catalyses the oxidation of α-D-glucose to D-glucono-1,5 lactone (gluconic acid) with the formation of hydrogen peroxide. The oxygen liberated from hydrogen peroxide by Peroxidase reacts with the O-dianisidine and oxidizes it to a red chromophore product.

$$\text{Glucose} + \text{oxygen} \xrightarrow{\text{glucose oxidase}} H_2O_2 + \text{Gluconic acid}$$

$$H_2O_2 + \text{O-dianisidine} \xrightarrow{\text{Peroxidase}} \text{Red coloured product}$$

Materials

1. Glucose oxidase peroxidize reagent: Dissolve 25 mg O-dianisidine completely in 1 ml of methanol. Add 49 ml of 0.1M phosphate buffer pH 6.5. Then add 5 mg of peroxidize and 5 mg of glucose oxidase to the above prepared O-dianisidine solution.
2. Standard: Dissolve 100 mg glucose in 100 ml water. Dilute 10 ml of this stock to 100 ml to obtain the working standard.

Procedure

1. To 0.5 ml of deproteinized plant extract (deproteinization is not necessary in samples with very low protein content) add 0.5 ml of distilled water and 1 ml glucose oxidase-peroxidase reagent.
2. Into a series of test tubes, pipette out 0 ml (blank), 0.2 ml, 0.4 ml, 0.6 ml, 0.8 ml and 1 ml of working standard glucose solution and make up to a volume of 1.0 ml with distilled water. Then add 1 ml of glucose oxidase-peroxidase reagent.
3. Incubate all the tubes at 35°C for 40 mins.
4. Terminate the reaction by the addition of 2 ml of 6N HCl.
5. Read the colour intensity at 540 nm.

Calculation

From the standard graph, the amount of glucose in the sample can be calculated (Malik and Singh, 1980).

10.3.3 Rapid method for determination of glucose

Procedure

To 10 ml of 1% ferricyanide solution, 5 ml of 2.5N NaOH solution is added in a conical flask. The solution is heated to boiling point and a few drops of methylene red indicator is added to provide a dark blue colour. While slow-boiling, the dilute hydrolysed solution is added using a pipette and the end-point is when the colour just disappears.

Calculation % starch = 10 × volume to which made up × 0.9/ wt of starch used (mg)

10.4 Physical Methods

10.4.1 Colour

Starch colour is often assessed by visual comparison with a standard product or by matching against certified colour standards such as the Munsel colour disks. However, instability of standards and the subjective nature of visual methods, except in the hands of colour experts, have led to the introduction of spectrophotometric procedures employing diffuse reflectance measurements. The colour methods recommended are similar to that employed for starch hydrolysis products except for the use of reflectance rather than transmittance values. When applied to maize starch and other products exhibiting similar colour characteristics, the method gives colour and brightness values which correlate with visual appearance through the use of the Commission Internationale de l'Eclairage (C.I.E) colour specification system. Colour values so obtained are related to excitation purity or colour saturation and brightness is related to luminous reflectance or lightness, when the sample hue is equivalent to that of maize starch or, more specifically, when the dominant wavelength is approximately 575 µm.

A Lovibond tintometer may also be used to compare the colour values.

10.4.2 Density

Commercial starches are obtained in powder, crystal, flake and lump forms by various commercial methods resulting in a variety of particle sizes and shapes. These physical differences, together with differences in origin and type or degree of modification or derivatization, affect absolute density of starch products only to a minor degree, but can influence bulk density to a large degree. Bulk density is important, particularly to starch producers, because of its relation to package design.

10.4.3 Bulk density

A 250 ml graduated cylinder, not less than 24 cm and not more than 26 cm in length, is tared carefully to the 250 ml graduation mark with loose sample. Filling is accomplished with the aid of a 7 cm, 60° powder funnel with a 1.27 cm ID stem, and one inch in length. The funnel is mounted with the stem inside the graduated cylinder so that the end of the square-cut stem is 6 cm above the 250 ml graduation mark. The sample is carefully added to the funnel with a spoon until the cylinder is loosely filled to the 250 ml graduation mark and the cylinder and contents are weighed.

The loose sample is then packed by vibration until the sample volume reaches a minimum value. This is usually accomplished by placing the cylinder, loosely supported, on the wooden deck of a paper jogger and vibrating for about five minutes at a frequency and amplitude which cause the cylinder to vibrate rhythmically with the wooden deck.

The packed sample volume is measured and the loose and packed bulk densities are calculated as follows:

$$\text{Bulk density, grams / millilitre (Loose)} = \frac{\text{Loose filled sample weight in grams}}{250\,\text{ml}}$$

$$\text{Bulk density, grams / millilitre (Packed)} = \frac{\text{Loose filled sample weight in grams}}{\text{Packed sample volume in millilitres}}$$

10.4.4 Determination of absolute density

The density of starch is of importance primarily for technological and engineering purposes and has only minor significance as a characterization in fundamental studies. Density can be determined readily by the pycnometer method using an inert non-solvent liquid as a suspending medium for the starch. Water and xylene are the most commonly used media.

A Hubbard Cormick 25 ml specific gravity bottle is employed; the dry specific gravity bottle (conditioned at room temperature, not oven-dried) is weighed and then filled with distilled water. The stopper is firmly inserted and the bottle is immersed to its neck in the constant temperature bath for 15 minutes. During this time, excess water will overflow through the capillary outlet. The bottle is then removed from the bath, carefully wiped dry and weighed to the nearest milligram. Volume of the bottle is computed from the weight of the contained water, divided by the density of water at 30°C. After drying, the bottle is similarly filled with xylene at 30°C and its density calculated from the weight of the contained xylene divided by the volume of the bottle.

To determine the density of a starch product, a 5–10 g sample of the pre-dried material is quickly transferred to a clean, dry specific gravity bottle, and the latter is immediately stoppered and weighed. Sufficient xylene is added to wash down and overlay the sample and the bottle is then placed in the constant temperature bath. The starch is allowed to settle; the bottle is filled with xylene and the stopper is carefully inserted. After equilibrating the bottle for 15 minutes in the constant-temperature bath, the meniscus is adjusted as previously.

Absolute density is calculated as follows:

$$\text{Density} = \frac{(a \times d)}{(a + b - c)}$$

a = weight of pre-dried starch sample in grams; b = weight of bottle filled with xylene; c = weight of bottle plus sample plus xylene; d = density of xylene. The weight of the empty bottles cancels out in the denominator expression (b – c).

10.5 Granular Methods

10.5.1 Microscopic techniques for determining starch granule properties

Starch occurs in plants as granules which are characteristic in size and shape for each plant species. The diameter of starch granules of various sizes is from 1 to 2 μm to about 200 μm. The shape may be spherical, egg-like or irregular. Microscopic observation is, therefore, the quickest and easiest way to identify the source of a starch.

Preparation of microscopic mounts

A drop of distilled water is placed near the centre of a microscope slide. The point of a dissecting needle is dipped into the starch sample and then into the water so that some of the starch is transferred to the drop of water. The amount of starch transferred should be enough to provide several granules in the microscopic field, but not enough to cause the granules to spread and jostle one another. The drop of starch suspension is covered with a cover slip. Care must be taken to exclude air bubbles.

The tip of a dissecting needle is used to gently 'stir' ungelatinized starch granules in the drop of water, to disperse them before the cover slip is put on the mount. Bits of starch gel or droplets of starch paste should not be stirred because fragile granules are broken by stirring, leading to false impressions.

If granules move in the field during observation, either the slide should be sealed or glycerol–water (1:1) should be used as a mounting medium.

To seal a mount, paraffin is melted in a broad vessel, such as a small evaporating dish. Do not overheat. A warmed metal rod a little longer than one side of the square slip is dipped into the molten paraffin, and then quickly laid on the edge of the cover slip and in contact with the slide. The rod is removed before the paraffin hardens, so that molten paraffin is left as a sealing layer on both cover slip and slide. The process is repeated until all four sides are sealed. Seals must be complete at all corners. If a circular cover slip must be sealed, a small, pointed brush is used to distribute the paraffin; the slide is placed on a turntable and, with the hand resting firmly on the table top, the brush is held against the edge of the slide as it is rotated on the turntable. Mounts that have been sealed can be kept for days, sometimes weeks, stored horizontally.

Identification of starch

A mount of the starch in water or in glycerol–water is prepared, and the slide is placed on the stage of a compound microscope and observed at a magnification of 300– 600× – the smaller the starch granules, the higher the magnification. The unknown granules are compared with known starches similarly prepared for observation or with photo-micrographs of known starches.

Microscopic determination of gelatinization temperature

The gelatinization temperature of starch is defined by microscopists as the temperature at which all birefringence is lost by all granules. The gelatinization temperature range is the temperature range between first evidence of loss of birefringence and final, total loss. Damaged granules or parts of granules that gelatinize in cold water are not permitted to influence the gelatinization temperature range. Note that care must be taken to define, 'gelatinization temperature' in presentations and publications, because the term is sometimes used to indicate a stage of swelling attained by the granules at a higher temperature than that at which birefringence is lost.

A 0.5% (by wt) suspension of the starch in distilled water is made at room temperature in a test tube. The tube is placed in a water bath at room temperature. A thermometer is hung in the starch suspension so that it touches neither wall nor bottom of the test tube. The temperature of the water bath is increased about 0.5°C each three minutes. The thermometer is used to stir the suspension very gently now and then to keep the temperature uniform throughout. Be sure that no starch sticks to the test tube wall above the level of the liquid. To remove a drop for microscopic observation, a clean pipette made of glass tubing drawn to a medium tip is used. Use a clean, dry pipette for each removal to avoid contamination or change of temperature of the suspension.

Determination of starch gelatinization temperature

Starch gelatinization is usually described as the rapid change from dispersion to paste that is observed when aqueous starch slurry is heated. Each starch type gelatinizes over a definite temperature range; therefore, measurement of gelatinization temperature is an important parameter in starch characterization. The first and most easily observed change in starch granules during heating is the rapid and complete loss of birefringence (polarization cross) of individual granules when viewed microscopically. Physical changes such as increase in viscosity and translucency have been used to measure gelatinization temperature of starch. However, since loss of birefringence precedes these other changes, it is generally the preferred criterion of gelatinization temperature.

A Kofler electrically heated microscopic hot stage and polarizing microscope or compound microscope equipped with polarizing attachments is required. The Kofler stage is a circular metal block heated by internal resistance coils. To prepare the stage for use, the thermometer is calibrated in the heat chamber by checking the melting point of known compounds. A uniform rate of temperature increase of about 2°C/min in the gelatinization range must be obtained by operating the hot stage through a variable transformer. A transformer setting of 23V will approximate the desired temperature rise, but calibration runs must be made to determine the correct setting.

The starch sample is slurried in water at a 0.1–0.2% concentration. A small drop of the suspension is spotted on a microscope slide and surrounded by a continuous and contiguous ring of high-viscosity mineral oil. A cover slip is then dropped on in such a way that the aqueous drop is completely surrounded by an oil barrier with no bubbles under the cover glass. The oil barrier prevents the escape of steam which fogs the glass buffle and prevents air channels from penetrating under the cover glass and disturbing the field. With the slide mounted on the hot stage and the hot stage in place on the microscope stage, heating is started about 20°C below the lowest temperature of the gelatinization range of the starch.

A uniformly distributed field of about 100–200 granules is observed using a magnification sufficient to clearly distinguish the polarization of the smallest granules. The field is watched continuously under normal lighting to determine when the first few granules change physical form. Loss of polarization process in theses granules can be immediately confirmed by inserting the analyser prism. On microscopes equipped with a rotating analyser, it may be found more convenient to make observation with the analyser rotated far enough to show distinct polarization crosses and still allow visualization of granule outlines.

Heating is continued, and the field is watched to record the temperatures corresponding with loss of birefringence by 2, 10, 25, 50, 75, 90 and 98% of the granules in the field. The 98% point is taken as the gelatinization temperature end-point. For most purposes one may need to record only the 2, 50 and 98% points, and, in some cases, only the temperature of the gelatinization end-point will be needed. Duplicate or triplicate runs are averaged for the best precision. Although the visual estimate of granule percentages seems crude at first, a little practice will produce a consistent agreement of about 1⁰ between duplicate runs.

10.5.2 AFM

Principle

Atomic force microscopy images are obtained by the measurement of the force on a sharp tip created by proximity to the sample surface. Starch samples are fixed on a freshly cut surface of mica. During scanning, the movement/deflection of the tip, which is fixed on a cantilever spring, can be captured by a laser-based device. Recording the change of the deflection by a photon detector in the course of scanning, a topographic image of the sample is plotted.

Based on the nature of the interactive force between the tip and the sample, major scanning modes of AFM include contact DC mode, deflection mode, and non-contact AC and tapping AC modes.

There are three major types of images obtained by AFM, including topography, frictional force (lateral force) and phase imaging. Topography imaging is mostly used for starch analysis to obtain the height information. The frictional force imaging is about the lateral deflections of the cantilever due to the forces on the cantilever parallel to the plane of sample surface, and it records the variations in surface friction. Phase imaging records the phase lag between the cantilever oscillation output signal and signal driving the cantilever to oscillate (Fig. 10.1).

The starch granules are embedded in order to immobilize them, and reduce the overall height variation of the sample surface. Physical destruction using a glass homogenizer is more suitable for observation of the bare inner surface of the starch than using digestion or cutting.

Methodology

AFM is performed with silicon nitride probes mounted on cantilevers with nominal spring constants of 0.064 and 0.032 N/m. Imaging is performed in the constant contact force mode with nominal imaging forces of 1.0–0.25 nN. Scan areas vary between 125 nm^2 and 5 μm, with scan frequencies between 1 and 30 Hz.

Starch granules are imaged after partial embedding at room temperature into a film of low-viscosity cyanoacrylate polymer. 20 μl Permabond C1 glue (viscosity = 2 mPa.s, quoted setting time of 2–10 s) is spun cast (1500 rpm) on to a~1.0 cm^2 of Biotrace r HP membrane attached to an AFM sample stub. Excess glue is eliminated due to the spinning, leaving a thin layer of glue. The spincaster is then stopped and starch granules sprinkled sparingly onto the wet glue surface through a 63 μm sieve (in order to break up large clusters of granules). The preparation rapidly polymerizes (30 s), with

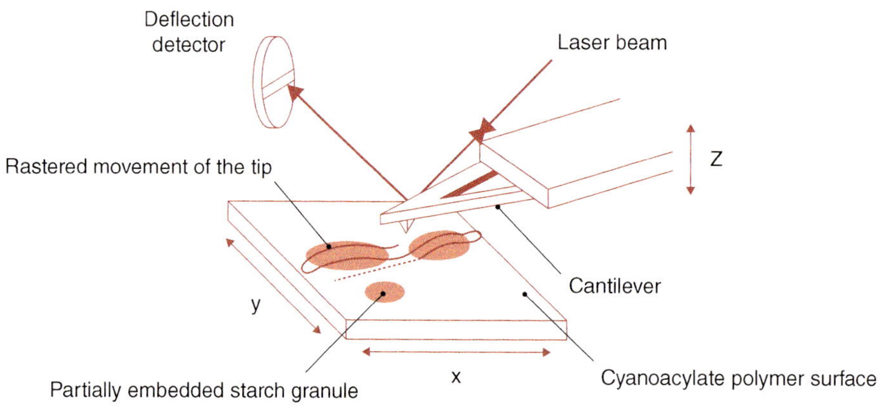

Fig. 10.1. Diagram of AFM imaging of partially embedded starch granules.

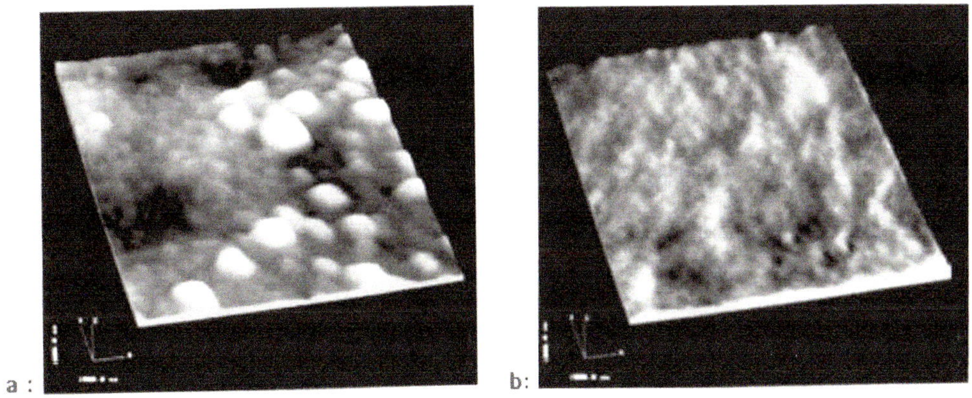

Fig. 10.2. **a.** AFM image of a typical surface region of potato starch granule (Scan size 1000 nm^2, z height difference of 73.8 nm). **b.** AFM image of a typical region of the Riband wheat starch granule surface (Scan size 1000 nm^2, z height difference of 52.6 nm).

granules partially embedded in the glue surface (Fig. 10.2a, b) (BeMiller and Whistler, 2009; Ohtani *et al*., 2000).

10.6 Chemical Methods

10.6.1 Acidity

Some modified starches and starch derivatives are processed in the presence of acid or acidic reagents. Since these affect physical properties of resultant starch pastes, they are neutralized in process and removed by washing. Neverthless, residues sometimes remain and these are determined by titration of a pasted sample or the filtrate from sample slurry.

Samples containing hard pellets or large lumps should be ground before analysis. When the sample contains a high concentration of solubles, which would make recovery of a filtrate difficult, it is first pasted and the paste is titrated. A 10 g sample is mixed thoroughly in 300 ml of distilled water at room temperature. The water used for sample preparation should be of such quality that 200 ml will require not more than 0.05 ml of 0.1N acid or base to obtain the methyl red or phenolphthalein end-points, respectively. The slurry is placed on a hot plate and heated to boiling point in about 15 min. The pasted sample is removed from the heat source; about 1 ml of phenolphthalein indicator is added and the sample is titrated with 0.1N sodium hydroxide.

Since different acids may contribute to the acidity of the sample, the value usually reported is the milliequivalents of acid per unit sample weight:

$$\text{Acidity (milliequivalents / gram)} = \frac{\text{Sample titre in millilitres} \times \text{NaOH normality}}{\text{Sample weight in grams}}$$

If the sample contains a negligible amount of soluble material, a slurry technique is sometimes used. A 10 g sample is mixed with 100 ml of good-quality distilled water, and the mixture is agitated continuously at a moderate rate for 30 minutes in a closed container. The mixture is gravity-filtered through a good-quality filter paper; the filtrate is collected in a clean and dry receiver and the first few millilitres of filtrate are discarded. Then, 50 ml of filtrate is pipetted into a clean Erlenmeyer flask, followed by dilution with 50 ml of good-quality distilled water and addition of phenolphthalein indicator. The diluted filtrate is titrated with standard 0.1N sodium hydroxide to a permanent pink end-point, and the acidity value is calculated as follows:

$$\text{Acidity (milliequivalents / gram)} = \frac{\text{Sample titre in millilitres} \times \text{NaOH normality} \times 100\text{ ml}}{\text{Sample weight in grams} \times 50\text{ ml}}$$

10.6.2 pH

Since the pH of a starch product affects the physical properties of pastes, most unmodified starches are adjusted to a pH level of about 5.0 during the final stages of processing. Some modified and derivatized starches, on the other hand, may be adjusted to a somewhat higher pH value.

A 20 g sample is dispersed in 100 ml of good-quality distilled water and stirred at a moderate rate until an equilibrium pH has been attained. A stirring time of 5 minutes is sufficient for most products, but some containing additives must be stirred for as long as 30 minutes to assure pH equilibrium. The agitation is stopped, and the pH is measured immediately at room temperature with a calibrated instrument.

10.6.3 Solubles

To measure the soluble content of an unmodified or slightly modified starch, an accurately weighed sample, about 20 g, is placed in a 250 ml bottle and slurried in 198 ml of distilled water at room temperature. The container is closed and the slurry is agitated continuously at a moderate rate for 30 mins. The suspension is gravity-filtered through a Whatman no. 12 or equivalent filter paper; filtrate is collected in a dry receiver, and the first 25 ml of filtrate is returned to the suspension (see equation at bottom of page).

10.6.4 Moisture

The moisture or volatile contents of unmodified and modified starches are most commonly determined by drying to constant weight in a vacuum oven at 100–120°C. Air-oven techniques are sometimes employed, but results are usually low. The azeotropic distillation method is the referee procedure of the maize wet-milling industry, but it is rarely used except for referee or calibration work because of its time-consuming nature.

The chemical method of Karl Fischer is frequently applied to starch products containing other volatiles which would be included in the weight loss obtained by oven drying, and it is similarly applied to those products showing evidence of decomposition at customary drying temperatures. Karl Fischer moisture values on starch products containing no other volatiles are in good agreement with oven results. Recent improvements in reagent stability and automatic titration equipment provide for rapid and accurate moisture determination.

Oven-drying method

About 5 g of sample is weighed accurately in a pre-dried, cooled and tared moisture dish (with cover). Aluminum dishes, about 5 cm in diameter and 2.5 cm tall, are commonly used, but dishes constructed of glass or other metals serve equally well. The dish and cover are placed in a vacuum oven operating at 120°C and the oven is maintained for four hours at a pressure below 100 mm (Hg). While the sample is drying, a small stream of air is blown through the drying train and oven.

The vacuum supply is then shut off, and the oven is slowly filled with air drawn through the drying train. The dish with the sample is removed, quickly covered, cooled in desiccators and weighed. Moisture or volatile content is calculated as follows:

$$\text{Per cent volatile} = \frac{\text{Weight loss in grams} \times 100}{\text{Sample weight in grams}}$$

Karl Fischer method

A quantity of sample containing 80–100 mg of water is weighed accurately in a dried, cooled and tared weighing tube, and the reagent buret is refilled. The glass stopper is removed from the beaker diaphragm, and the small end of the

$$\text{Per cent solubles} = \frac{\text{Residue weight in grams} \times 200\ \text{ml} \times 100}{\text{Aliquot volume in millilitres} \times \text{sample weight in grams}}$$

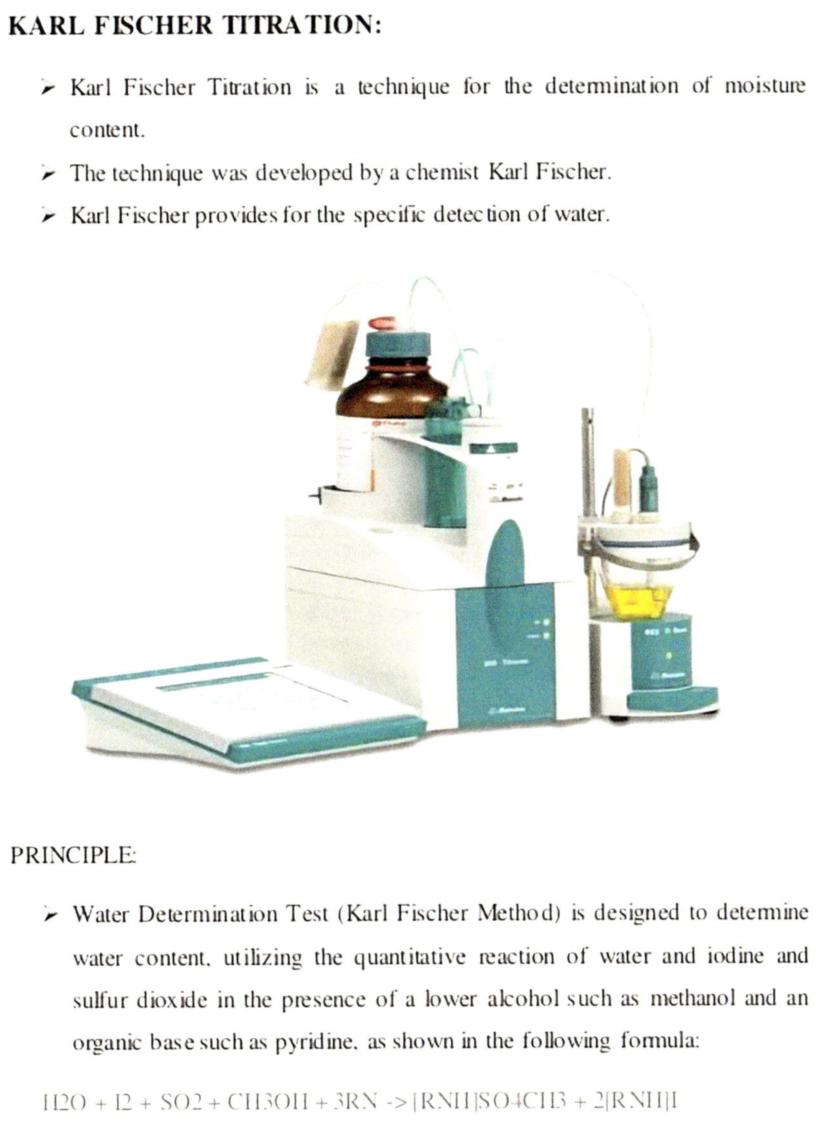

KARL FISCHER TITRATION:

- Karl Fischer Titration is a technique for the determination of moisture content.
- The technique was developed by a chemist Karl Fischer.
- Karl Fischer provides for the specific detection of water.

PRINCIPLE:

- Water Determination Test (Karl Fischer Method) is designed to determine water content, utilizing the quantitative reaction of water and iodine and sulfur dioxide in the presence of a lower alcohol such as methanol and an organic base such as pyridine, as shown in the following formula:

$H_2O + I_2 + SO_2 + CH_3OH + 3RN \rightarrow [RNH]SO_4CH_3 + 2[RNH]I$

Fig. 10.3. Karl Fischer titration apparatus.

weighing tube is inserted in the opening. The sample is allowed to flow into the prepared solvent, and the transfer is completed by rinsing the weighing tube with 10 ml of methanol delivered from a pipette. As before, the glass stopper is replaced; the 'titrate' switch is actuated, and the titration is allowed to proceed to completion. Another blank is run on 10 ml of methanol pipetted into prepared solvent through a dried weighing tube (Fig. 10.3).

Calculations

K = water equivalent of Karl Fischer reagent (milligrams of water per millilitre of reagent)

$$K = \frac{\text{Tartrate wt in mg} \times \text{Tartrate moisture \%}}{(\text{Tartrate titre} - \text{Blank titre}) \text{ in millilitres} \times 100}$$

Per cent water (sample)

$$= \frac{(\text{Sample titre} - \text{Blank titre}) \text{ in millilitres} \times K \times 100}{\text{Sample weight in milligrams}}$$

10.6.5 Ash

All commercial starches contain minor or trace quantities of uncombined inorganic materials. The bulk of these originate in the crop from which the starch is isolated but a small fraction may be contributed by water used in processing. The approximate concentration of these salts is customarily determined as the residue after ignition at a specified temperature.

About 5 g of the pre-ground and well-blended starch product is weighed in a pre-heated, cooled and tared platinum or silica dish. Platinum ware is preferred when the product contains only trace amounts of ash because silica dishes tend to lose weight during ignition. The dish and contents are heated carefully over an open flame or on a hot plate until the sample is thoroughly carbonized. Excessive foaming during this stage can usually be controlled by igniting the sample with an open flame or by heating the surface of the sample with an infrared heat lamp during the charring process.

The partially decomposed sample is then heated in a muffle furnace at 525°C for two to six hours until the residue is free from carbon. If complete ignition of carbon is slow or difficult, as is sometimes the case with high ash samples, the process may be hastened by cooling the dish, moistening the residue with a few drops of water, and then reheating.

When the residue is carbon-free, the dish is cooled to room temperature in a desiccator and weighed.

$$\text{Per cent ash} = \frac{\text{Residue weight in grams} \times 100}{\text{Sample weight in grams}}$$

10.6.6 Calcium

Commercial starches isolated from cereal and tuber crops contain minor or trace amounts of inorganic materials, and calcium salts are usually a minor fraction thereof. The bulk of this calcium probably pre-exists in the crop from which the starch is isolated, although some may arise from the processing waters. On the other hand, some commercial starches contain added calcium salts, particularly those which may be converted subsequently by enzymes.

To analyse the starch sample, not more than 10g is weighed in a platinum or Vycor dish of 100–150 ml capacity (larger samples are difficult to ignite) and ashed as above.

The dish and ash residue are cooled, and 10 ml of 0.1N hydrochloric acid is added; more acid may be required if the ash residue is large and highly alkaline; for example, if it contains sodium or calcium carbonate. The dish is covered with a watch glass and heated for 15 minutes on a steam bath to dissolve all calcium. The solution and any residue are quantitatively transferred to a beaker or flask for titration and diluted to about 50 ml with distilled water. If the calcium content is known to be high, the solution is diluted to a known volume and an aliquot containing not more than 1–2 mg of calcium is selected for titration.

To the sample solution or aliquot, 10 ml of 0.002M magnesium chloride is added for end-point improvement. Then, 4 ml of 8N potassium hydroxide is added, and the solution is allowed to stand about 5 minutes. Next, about 25 mg each of potassium cyanide and hydroxylamine hydrochloride are added and dissolved by mixing. About 0.2–0.3 g of hydroxynaphthol blue indicator is added and the solution is titrated with standard EDTA solution until the pink indicator colour disappears and a blue colour persists.

Knowing that the equivalent weights of EDTA and calcium are equal to the molecular weight and gram atomic weight, respectively, the calcium content of the sample is calculated from the titre as follows:

$$\text{Calcium in ppm} = \frac{\text{EDTA titre} \times \text{EDTA molarity} \times 0.040 \times 10^6}{\text{Sample weight in grams}}$$

10.6.7 Sulphur dioxide

Starches are frequently recovered in a processing system that incorporates small amounts of sulphurous acid as a preservative or decolourizing agent. The latter is commonly determined by the well-known 'Monier–Williams' procedure, which determines free as well as combined sulphur dioxide or sulphites frequently present in complex food products, particularly those containing reducing sugars. Sulphur dioxide is released from the sample by heating in dilute acid, and is removed by sweeping with an inert gas such as carbon dioxide or nitrogen. Free sulphur dioxide can be recovered from some products by sweeping an acidified solution of the sample with air under reduced pressure at room temperature. The liberated sulphur dioxide is usually absorbed in a dilute, neutral hydrogen peroxide solution in which it is oxidized to sulphur trioxide forming sulphuric acid; the latter is determined by titration or by gravimetric analysis.

The starch sample, up to 100 g depending on the sulphur dioxide content, is dispersed in about 200 ml of freshly boiled, cooled water. The slurry is transferred immediately to the reaction flask by means of the dropping funnel while regulating sample addition rate and gas flow rate through the apparatus so as to prevent drawback of hydrogen peroxide, inclusion of air or burning the sample. The mixture is brought to the boil and simmered for one hour, continuing the flow of carbon dioxide through the apparatus at a moderate rate.

Water flow through the condenser is discontinued just before the end of the boiling period when the delivery tube directly above the receiving flask becomes hot and the tube joining the upper end of the condenser is disconnected. The delivery tube and contents of the Peligot tube are washed into the receiving flask, and the solution is titrated with standard 0.1N sodium hydroxide solution to a bromophenol blue indicator end-point. A blank determination is made on all reagents, substituting freshly boiled water for the sample.

A gravimetric determination may be made following the titration if the neutral peroxide reagent is free of sulphate and other materials precipitated with barium in acid solution. The solution is acidified with hydrochloric acid and sulphate is precipitated by adding excess barium chloride solution. The precipitate is settled, recovered by filtration, dried, ignited, cooled and weighed as barium sulphate (see equation at bottom of page).

10.6.8 Phosphorus

A representative portion of the prepared sample of not more than 20 g and containing not more than 40 mg of phosphorus is weighed accurately into a 100 ml Vycor dish. Ten millilitres of 10% zinc acetate solution is added, distributing the solution uniformly through the sample by adding water if necessary. The sample is evaporated to dryness on a steam bath, heated on a hot plate until thoroughly charred, and then ignited for two hours in a muffle furnace at 550°C. The dish is cooled to room temperature and the residue is wetted by cautious addition of 3 ml of 29% nitric acid. The sample is again evaporated to dryness on a steam bath and dehydrated completely by brief heating on a hot plate, and the dish is returned to the muffle furnace at 550°C for 30 minutes. This process assures complete decomposition of organic material in the shortest time.

After cooling to room temperature, the sides of the dish are washed down with 10 ml of 29% nitric acid followed by 15 ml of distilled water. The dish is covered with a watch glass; the residue solution is heated to incipient boiling and held for ten minutes at that temperature to ensure complete hydrolysis of pyrophosphate to orthophosphate. After cooling to room temperature, the solution is quantitatively filtered through Whatman no. 1 paper into a volumetric flask of a size selected to provide a final concentration of 1–4 mg of phosphorus per 25 ml. The transfer is completed by washing the dish and filter paper with small portions of distilled water and the solution is diluted to volume and mixed thoroughly.

An aliquot selected to contain not more than 2.5 mg of phosphorus is pipetted into a 100 ml volumetric flask and 25 ml of water is added to another 100 ml volumetric flask serving for the reagent blank. The following reagents are added to both flasks, in the order stated, mixing after each addition: 10 ml of 29% nitric acid, 10 ml of 0.25% ammonium vanadate solution (prepared by dissolving 2.5 g of ammonium metavanadate in 600 ml of boiling water,

$$\text{Per cent sulphur dioxide} = \frac{(\text{Sample titre} - \text{Blank titre})\ \text{millilitres} \times \text{NaOH normality} \times 0.032 \times 100}{\text{Sample weight in grams}}$$

cooling to 60–70°C, adding 20 ml of con. nitric acid, cooling to room temperature and diluting to 1L with water), and 10 ml of 5% ammonium molybdate solution. Both solutions are diluted to volume with water, mixed thoroughly and allowed to stand for 10 minutes. Using the reagent blank in a 1 cm cuvette as the reference at zero absorbance, the absorbance of the sample solution at 460 µm is determined in a 1 cm matching cuvette. The phosphorus concentration (milligrams per 100 ml) of the sample solution is read from the calibration curve.

The calibration curve is prepared with the aid of a standard solution of potassium dihydrogen phosphate. Aliquots containing 0.5, 1.0, 1.5, 2.0 and 2.5 mg of phosphorus are pippeted into separate 100 ml volumetric flasks and 25 ml of water are added to another flask serving for the reagent blank. The following reagents are added to each flask, in the order stated, mixing after each addition: 10 ml of 29% nitric acid solution 10 ml of 0.25% ammonium vanadate solution, and 10 ml of 5% ammonium molybdate solution. All solutions are diluted to volume with water, mixed thoroughly and allowed to stand for 10 minutes.

Using the reagent blank in a 1 cm cuvette as the reference at zero absorbance, the absorbances of the standard solutions are determined at 460 µm in matching 1 cm cuvettes. A calibration curve, which is reproducible but should be checked when fresh reagents are used, is obtained by plotting standard solution absorbances (at 1 cm) versus phosphorus contents (mg per 100 ml) on linear coordinates. Phosphorus contents of sample solutions are ascertained by reference thereto (see equation at bottom of page).

10.6.9 Sulphur

A representative portion of the finely ground or powdered sample of known moisture or volatiles content, not more than 1.2 g, is weighed accurately and wrapped in a half-circle of 9 cm Whatman no.1 filter paper. The wrapped sample is placed in the ignition cup of a Parr oxygen bomb; the fuse wire is attached and positioned so that it touches the wrapped sample. Twenty millilitres of 6% hydrogen peroxide solution is pipetted into the base of the bomb. The cover is placed in position and tightened, and oxygen is added carefully to a pressure of 29 Atm. The oxygen inlet should be modified if necessary so that the oxygen stream is directed horizontally towards the interior wall of the bomb rather than towards the base. The charge is then ignited.

One minute or more after ignition, the bomb is fixed horizontally in a shaker, and the contents are agitated vigorously for 15 minutes to complete the oxidation of the lower oxides of sulphur to sulphate. The bomb is then placed in an upright position and allowed to stand a minute or two; residual pressure is carefully released, and the cover is removed and rinsed with a fine stream of water delivered from a wash bottle. Contents of the bomb are transferred quantitatively to a 250 ml beaker; the transfer is completed by washing the bomb with about 30 ml of distilled water in small portions.

Reagent-grade silver oxide powder (0.4 g) is slowly added to the sample solution which is then heated to boiling point and boiled for 5 minutes with continuous stirring. This treatment may be omitted if the sample is free of chlorides and phosphates which interfere in the subsequent titration with lead nitrate. The solution is cooled to room temperature, transferred quantitatively to a 100 ml volumetric flask and diluted to volume with water. The solution is mixed thoroughly, and the precipitate is allowed to settle; if settling does not yield a clear supernate, the solution should be gravity-filtered.

Fifty millilitres of the supernatant sample solution is pipetted into a small glass ion-exchange column (15 × 150 mm) with a glass reservoir (25 × 70 mm) at the top, having a

$$\text{Per cent phosphorous} = \frac{\text{P} \times \text{Dilution volume in millilitres} \times 100}{\text{Aliquot volume in millilitres} \times \text{Sample weight in grams} \times 1000}$$

In which P = Phosphorus content (milligrams/100ml) from calibration curve.
Per cent phosphate (PO_4) = per cent phosphorus × 3.065

capacity somewhat in excess of 50 ml. The lower end is constricted and a short length of rubber tubing and screw clamp are attached thereto for adjustment of flow rate. The column contains 10 ml (wet settled volume) of Nalcite HCR (H^+) cation-exchange resin, freshly regenerated with 60 ml of 10% nitric acid and washed with water until the effluent is essentially neutral. The sample solution is allowed to flow as rapidly as the column will permit and the effluent is collected in a 100 ml volumetric flask. Flow is continued until the solution meniscus reaches the upper surface of the resin bed; the column reservoir is rinsed with a fine stream of water delivered from a wash bottle, and the washings are allowed to drain to the upper surface of the resin bed. The sample solution in the column is displaced by adding 10 ml portions of water, allowing each portion to drain to the upper surface of the resin bed prior to the addition of the next portion. Washing is continued until 100 ml of effluent has been collected, and the effluent is mixed thoroughly.

An aliquot of the effluent, preferably not more than 20 ml, containing 0.1–1.5 mg of sulphur is pipetted into a 250 ml beaker and water is added to give a minimum volume of 10 ml. Two drops of Bromophenol Blue indicator solution (0.1% in ethanol) is added, and N ammonium hydroxide solution is added, drop-wise, until the solution colour is distinctly blue. Sufficient 0.16N nitric acid is added to discharge the blue colour and 1 ml of 20% acetic acid solution is added. The sample solution is then diluted with 65 volumes of ACS reagent-grade acetone, and four drops of freshly prepared dithizone indicator solution (0.1% in acetone) is added for each 60 ml of total volume. The green solution is then titrated, while stirring continuously with standard 0.01N lead nitrate solutions, until a pale pink endpoint persists. The standard lead nitrate solution should be prepared from reagent-grade material, and it can be standardized by titrating a 10 ml aliquot of standard 0.01N sulphuric acid solution in the manner described above (see equation at bottom of page).

Degree of substitution values for sulphate and p-toluene sulphonate derivatives of starch are calculated as follows:

$$\text{Degree of substitution (DS) sulphate} = 162\,S / 3200 - 102\,S$$

$$\text{Degree of substitution (DS) (p-toluene sulphonate)} = 162S/3200 - 154\,S$$

In which s = per cent sulphur (dry basis)

10.6.10 Nitrogen (protein)

Protein contents of starches are estimated from the determined nitrogen contents, using conversion factors appropriate for the particular crop. Direct calorimetric techniques for protein determination have been successfully applied to some cereal products but modifications of the Kjeldahl procedure are most commonly practised. Starch derivatives containing substituent groups with nitrogen – for example quarternary ammonium, amine, carbanilate or others – are analysed also by the Kjeldahl procedure. In the examination of these derivatives, the original starch is analysed simultaneously, and the increased nitrogen content is the basis for calculating the degree of substitution. The Dumas technique utilizing automated analysers can be applied conveniently to those products containing appreciable amounts of nitrogen.

Standard Kjeldahl digestion and distillation equipment is recommended. A starch sample weighing about 10 g is placed in an 800 ml Kjeldahl flask. Most commercial unmodified starches contain less than 0.1% nitrogen. If the sample is known to contain significantly more nitrogen (for example, a nitrogen containing derivative), a smaller sample may be more appropriate.

$$\text{Per cent nitrogen} = \frac{(B - A) \times \text{NaOH normality} \times 0.014 \times 100}{\text{Sample weight in grams}}$$

$$\text{Per cent sulphur} = \frac{(\text{Pb}(\text{NO}_3)_2 \text{ millilitres}) \times (\text{Pb}(\text{NO}_3)_2 \text{ normality}) \times 0.016 \times 100\,\text{ml} \times 100\,\text{ml} \times 100}{\text{Dry sample weight in grams} \times 50\,\text{ml} \times \text{ml of aliquot}}$$

Where A = sample titre (ml of standard sodium hydroxide solution); B = blank titre (ml of standard sodium hydroxide solution)

$$\text{Per cent protein} = \text{Per cent nitrogen} \times 6.25$$

10.6.11 Fatty materials

Starches contain trace amounts of fatty acid glycerides, usually less than 0.1%, which can be removed by extraction with hydrophobic solvents such as ether, hexane or carbon tetrachloride, using conventional Soxhlet or Butt-type extraction equipment. Most starches also contain about 0.5–0.6% of free fatty acids, which appear to be complexed with the linear fraction (amylose) of the native starch. Since these complexed fatty acids alter the physical and chemical properties of the starch, it is often of interest to know their concentration, and sometimes their removal is a prerequisite for accurate analysis, such as, for example, the determination of iodine affinity and amylose content. The complexed fatty acids can usually be removed by extended extraction with hot, hydrophilic solvents such as methanol, ethanol, 80% dioxane and 2-methoxyethanol (methyl cellosolve), so as to provide essentially fat-free starches for analysis. Total concentration of fatty substances (fats and fatty acids) is usually determined by a procedure involving partial hydrolysis of the starch, precipitation of the free fatty substances and recovery by solvent extraction.

The folded filter paper containing the dried residue is placed in a paper extraction shell (22 mm diameter by 80 mm height), and the top of the shell is closed with a cotton ball extracted previously with carbon tetrachloride or petroleum ether so as to be fat-free. Alternatively, glass extraction shells with fritted glass bottoms or Alundum thimbles may be used. The sample containing the shell is placed in a Butt-type extractor and attached to a standard extraction assembly consisting of a flat-bottomed flask, an extractor and a water-cooled condenser. A Soxhlet extractor may be used, alternatively, and complete assemblies with standard tapered glass joints are preferred (Fig. 10.4).

After attaching a previous dried, cooled and tared extraction flask containing about 50 ml of carbon tetrachloride (petroleum ether may be substituted), the assembly is placed on a heater. Heating is adjusted to produce 150–200 drops of condensed solvent per minute and the residue is extracted for three hours under these conditions. The extraction flask is disconnected from the assembly and carbon tetrachloride is evaporated on a steam bath until no odour of solvent remains. The flask and residue are dried for one hour in a vacuum oven at 100°C; prolonged drying at higher temperatures is avoided to prevent fat oxidation. After cooling the flask and residue in desiccators, the residue and flask are weighed and the total fat content is calculated.

$$\text{Per cent total fat} = \frac{\text{Residue weight in grams} \times 100}{\text{Sample weight in grams}}$$

10.6.12 Amylose

Amylose content is an important index for appraising food quality. The content of amylose was expressed as a percentage by mass on a dry basis and could be obtained by checking the absorbance from the calibration curve (standard curves).

Iodimetric determination of amylose

The absorbance of the blue colour produced in aqueous solutions of amylase upon the addition of tri-iodine ion has long been used as a convenient indicator of the amount of linear fraction present in solution. The method was put on a quantitative basis by McCready and Hassid (1943).

To a 50 ml flask is added 1 ml of an aqueous solution containing approximately 0.5 mg of linear fraction; 0.5 ml of N sodium hydroxide is added, and the mixture is warmed for three minutes in a boiling-water bath. After cooling, an exactly equivalent amount of approximately 1N hydrochloric acid is added, followed by 0.07–0.1 g of potassium hydrogen tartarate. Water is added to a volume of about 45 ml and then 0.5 ml of iodine solution (2 mg of iodine/ml; 20 mg of potassium iodide/ml). The solution is made up to 50 ml, mixed and allowed to stand for 20 minutes at room temperature. The absorbance is measured at 680 μm in a spectrophotometer

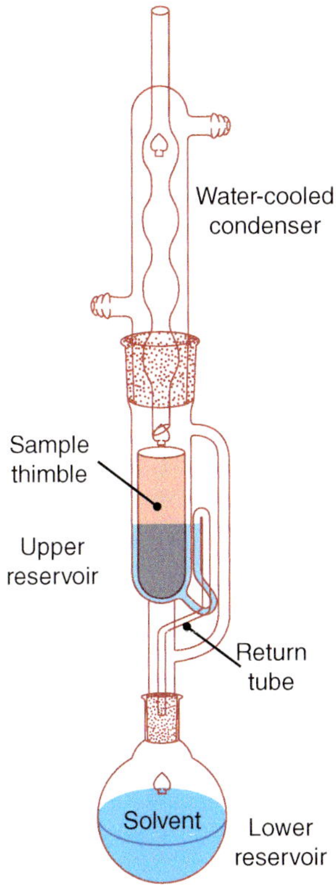

Fig. 10.4. Soxhlet apparatus.

using a 1 cm cell. It is not recommended to use an instrument which isolates the 680 μm region by means of filters. For the reference solution, an iodine solution of equal concentration is used.

Based on Beer-Lambert law the Blue value (B.V.) can be calculated from the equation:

$$\text{Blue value} = \frac{\text{Absorbance} \times 4}{C\,(\text{mg}/\text{dl})}$$

Amylose content calculated using Blue values of standard amylose solution.

Modified methods for amylose determination

Forty milligrams of starch is weighed into a conical flask (50 ml), 10 ml urea-DMSO (6M urea 10 ml + 90 ml DMSO) is added and heated for 15 minutes in boiling water with proper shaking. Keep in oven for one hour at 100°C. After cooling, pippette 1 ml aliquot into 100 ml standard flask then add 2 ml of a solution of potassium iodide with iodine in water (I_2KI) and absorption measured at 635 nm. Amylose content calculated using absorbance of standard amylose solution (20–80 mg/ml).

Micro method

To 20 ml amylose in a 100 ml conical flask, add 0.4 ml DMSO and dissolve by heating at 85°C for 15 minutes. Dilute to 100 ml with deionized water. Take 1 ml aliquot, 3 ml water and 1 ml of iodine solution; then allow the mixture to stand for 10 minutes. Measure absorbance at 600 nm. Calculate the amylose content using standard amylose solution (McGrance *et al.*, 1998).

Amylose by DSC

A 10–25 mg starch (dry) sample is weighed into a DSC pan, then add 50–70 μl of L-α-lysophosphatidyl chloride (LPC) (3% wt by volume) and the pan is thermatically sealed. The pan is heated from 20°C to 140°C at a rate of 5°C/min, cooled to 4°C and reheated to 140°C, the ΔH is obtained from the endotherm of the second heating. A standard graph was prepared by using potato amylose of different concentrations (Sievert and Holm, 1993). The amylose content is calculated by the formula:

$$\text{Amylose content} = \Delta H\ \text{sample}/\Delta H\ \text{standard amylose}$$

Determination of soluble amylose

100 mg of starch is weighed into a 100 ml conical flask, 40 ml distilled water is added and the starch is gelatinized by shaking in boiling water. The solution is transferred to a 100 ml standard flask using a fluted filter paper. Five millilitres of aliquot is transferred to a 50 ml standard flask, 2 ml of iodine solution is added and made up to 50 ml. Absorbance is measured at 635 nm and soluble amylose is determined by using absorbance of standard amylose solution (Shanthy *et al.*, 1980).

10.7 Structural Methods

10.7.1 XRD

Methodology

An X-ray diffractometer equipped with copper K α-radiation is commonly used. The moisture content of starch is determined to make sure the moisture content of all samples is constant. The sample of starch is exposed to the X-ray beam (30 kV, 15 mA). The scanning regions of the diffraction angle (2θ) were 4–35°C for starch.

Percentage content of relative crystallinity of starch can be calculated based on Herman's theory. The upper region ac over the middle dash line is the area of crystalline fraction and the lower region is the amorphous region. The crystallinity is calculated as ac/Ac, where Ac is the diffraction for 100% crystalline substance (Figs. 10.5 and 10.6) (Nara *et al.*, 1978; Ratnayake and Jackson, 2008; Zobel *et al.*, 1998).

10.7.2 GPC (gel permeation chromatography)

Gel permeation chromatography is a procedure to determine the amylose content and details about the molecular structure of amylose and amylopectin fractions. It involves debranching starch with an enzyme such as isoamylase and subsequent fractionation in a suitable column. The molecular weights of the separated fractions are determined by using standard dextrins or by chemical methods.

Procedure

Starch samples are debranched with isoamylase. Debranched starch is injected onto Sepharose CL 6B column using 0.25 M KOH as eluent. The molecular weight of fractions is studied using the phenol sulphuric acid method or using dextrin standards of known molecular weight.

A typical graph is given below (Fig. 10.7). The amylose peak comes first and the area under the peak is calculated to obtain amylose content.

10.7.3 MALLS

Molecular weight distribution of amylopectin is determined by an HPSEC-MALLS-RI system.

Principle

The HPSEC system with optimized operating conditions can bring out good seperaion of amylopectin from amylose. Molecular weight distributions of amylopectins, eluted by distilled-deionized water as the mobile phase and traced by an RI detector are obtained in the procedure (Fig. 10.8).

Methodology

The analysis described below is for preparation of starch aqueous dispersions for HPSEC.

Starch (120 mg) is evenly wetted with 1.2 ml of water and then dispersed in 10.8 ml of dimethyl sulphoxide (DMSO). The suspension is mechanically stirred while heating in a boiling-water bath for one hour and then stirred for 24 hours at 25°C. An aliquot (0.4 ml) of starch dispersion (1.0%, w/v) is mixed with five volumes of ethanol (2 ml) to precipitate starch. Ethanol-precipitated starch is separated by centrifugation at 6750 × *g* for 20 minutes. The starch pellet is then redissolved in boiling water (10 ml) and stirred for 30 minutes in a boiling-water bath. The hot sample solution is filtered through a nylon membrane filter (5.0 μm) and then injected into an HPSEC system. The final concentration of the starch solution filtrate injected (100 μl) is 0.4 mg/ml.

1. Molecular weight distribution of amylopectin determined by an HPSEC-MALLS-RI system

An HPSEC system consists of an injection valve (100 μl sample loop), a multi-angle laser-light-scattering detector and an RI detector. To separate amylopectin from amylose, a Shodex OH pak KB-G guard column and KB-806 and KB-804 analytical columns have been used. The temperature of the injector and columns are maintained at 55.0°C using a CH-460 column heater and a TC-50 controller. The temperature of the RI detector is set at 30.0°C. The mobile phase is distilled–deionized water (18.2 MΩ cm)

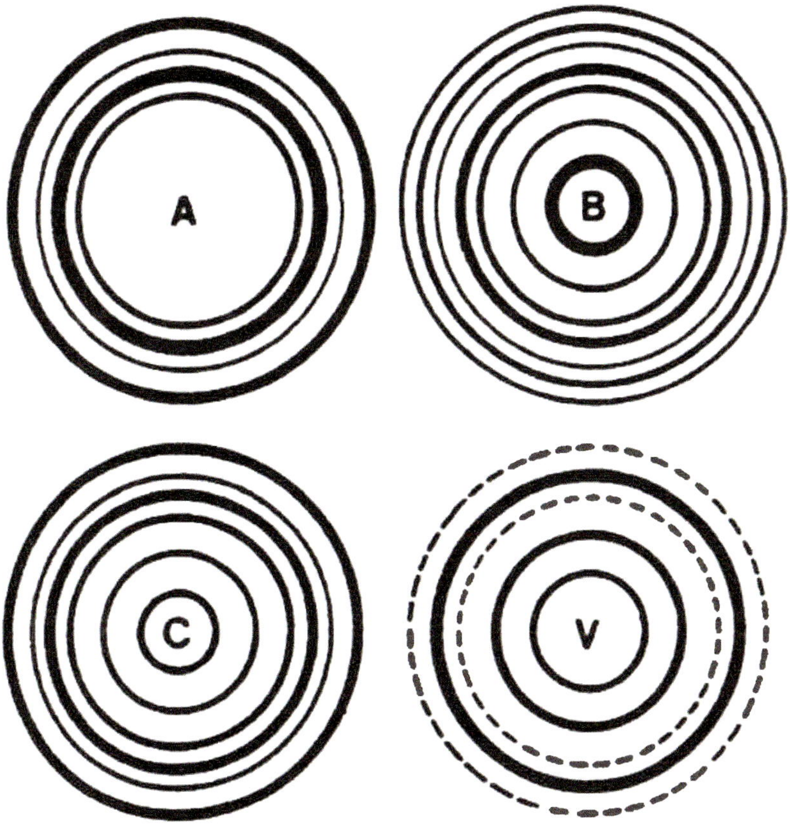

Fig. 10.5. Starch X-ray pattern.

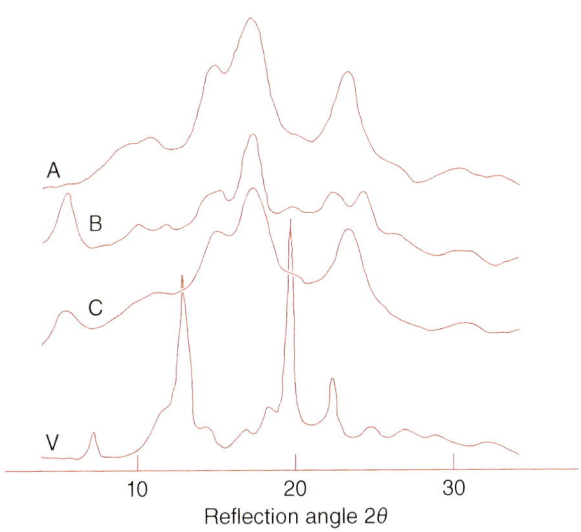

Fig. 10.6. X-ray diffractograms showing A, B, C patterns.

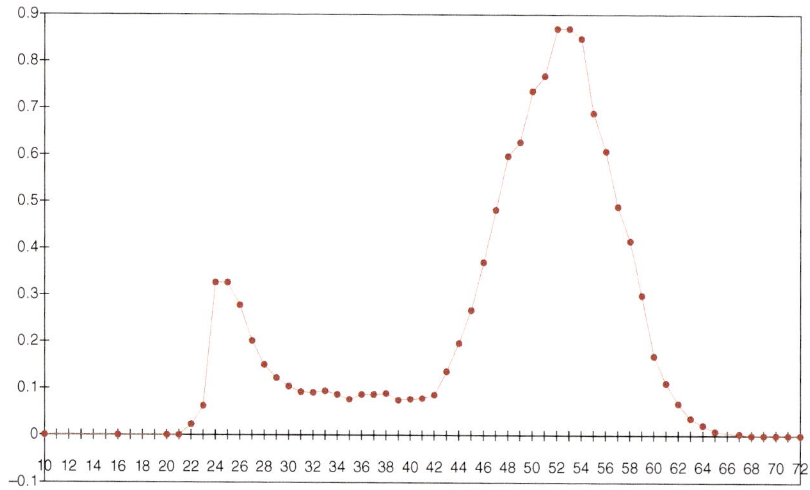

Fig. 10.7. GPC pattern of debranched starch.

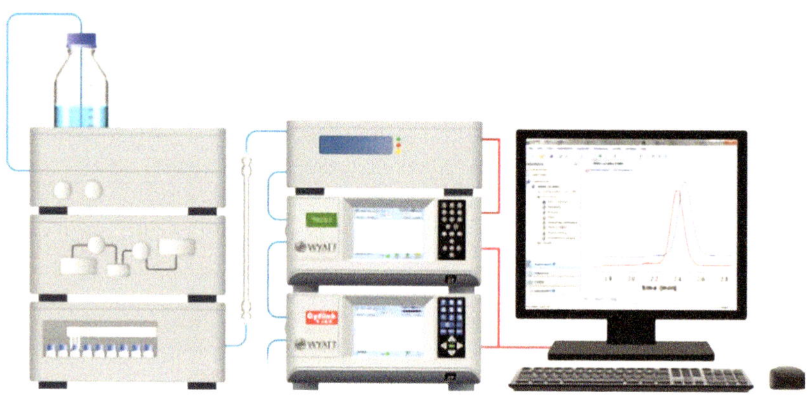

Fig. 10.8. SEC-MALLS equipment.

passed through in-line membrane filters (0.2 and 0.1 μm, Millipore, Bedford, MA) at a flow rate of 0.7 ml/min.

2. Data analysis

A pullulan standard, P-20 ($M_w = 2.28 \times 10^4$, 5 mg/ml), is used for normalization of multiangle photodiode detectors and for determination of delay volume (0.222 ml) between MALLS and RI detections. Data obtained from MALLS and RI detectors are analysed using Astra software (Version 4.7.07, Wyatt Technology, Santa Barbara, CA). M_w is calculated using the following equation:

$$K^* c / R_\theta = 1 / [M_w P(\theta)] - 2A_2 c \qquad (1)$$

where R_θ is the excess intensity of scattered light at angle θ; c is the sample concentration; M_w is the weight-average molecular weight; A_2 is a second viral coefficient; K^* is an optical parameter equal to $4\pi^2 n_0^2 (dn/dc)^2 / (\lambda_0^4 N_A)$ where n_0 is the solvent RI and dn/dc is the RI increment; N_A is Avogadro's number; and λ_0 is the wavelength of the scattered light in vacuum. Function $P(\theta)$ describes the angular dependence of scattered light. Expansion of $1/P(\theta)$ to first order gives:

Curve-fitting method in this study is based on a second-order Berry method ($\sqrt{(K*c/R_\theta)}$ vs $\sin^2(\theta/2)$), and a second virial coefficient, A_2, was set at zero (Wen *et al.*, 1996; Yokoyama *et al.*, 1998). M_w and R_z (equal to r_g in Eqn 2) values were calculated from intercept and slope of the Eqn (1) by extrapolating multiangle signals to zero angle, respectively. The d*n*/d*c* value of 0.146 ml/g was used in this calculation (Bello-Pérez *et al.*, 1998; Fishman *et al.*, 1996; Roger and Colonna, 1993).

Result

Molecular weight distributions of amylopectins, eluted by distilled–deionized water as the mobile phase and traced by an RI detector, are shown as examples in Figure 10.9. By using the HPSEC system with the experimental operating conditions, amylopectin can be separated from amylose, which leads to accurate measurements of amylopectin molecular weights (Fig. 10.9) (Yoo and Jane, 2002).

10.7.4 CP/MAS C13-NMR

Principle

To distinguish between double helices (short-range order) and single helices (amorphous chains), and give their relative estimate amounts. Amorphous starch shows broader C13-CP/MAS-NMR spectral features than ordered structures (Fig. 10.10).

Methodology

The high-resolution 13C CP/MAS NMR spectra are recorded at the resonance frequency of approximately 75 MHz carried out on the 300 MHz solid-state NMR spectrometer. A contact time of 1 ms is used for all the standard spectra with a recycle time of 2 s (recycle time is adjustable). The spectra are obtained at room temperature, averaging over at least 10,000 scans.

The C1 peak is a triplet for the A-type crystalline polymorph (indicated in Fig. 10.11), and a doublet for B-type crystalline starch (Flanagan *et al.*, 2015; Flores-Morales *et al.*, 2012; Fričová and Koval'aková, 2013; Gidley and Bociek, 1985).

10.7.5 Alkali number (reducing value)

Reducing values of native starches are extremely small because of the low incidence of reducing end groups in amylose and amylopectin. These polymers, theoretically, contain only one reducing end group per molecule, and a relative measure of the number of terminal reducing groups or an estimate of molecular size, can be obtained by determining the alkali number. This method involves degradation of the starch sample under controlled alkaline conditions; where in the terminal aldehydic groups are presumed to undergo enediol rearrangements, followed by decomposition to form simple organic acids. The concentration of such acids produced under rigorously controlled conditions is an indication of the concentration of reducing end groups.

Alkali number is inversely related to the molecular weight (see equation at bottom of page).

10.7.6 Ferricyanide number (reducing value)

Reducing value methods employing alkaline copper reagents are rarely used for the

$$1/P(\theta) = 1 + (16\pi^2/3\lambda^2)\langle rg^2\rangle \sin^2(\theta/2) + f_4 \sin^4(\theta/2) + \cdots \quad (2)$$

$$\text{Alkali number} = \frac{(\text{Net sample titre} \pm \text{Blank titre})\ \text{millilitres} \times \text{Acid normality} \times 10}{\text{Dry sample weight in grams}}$$

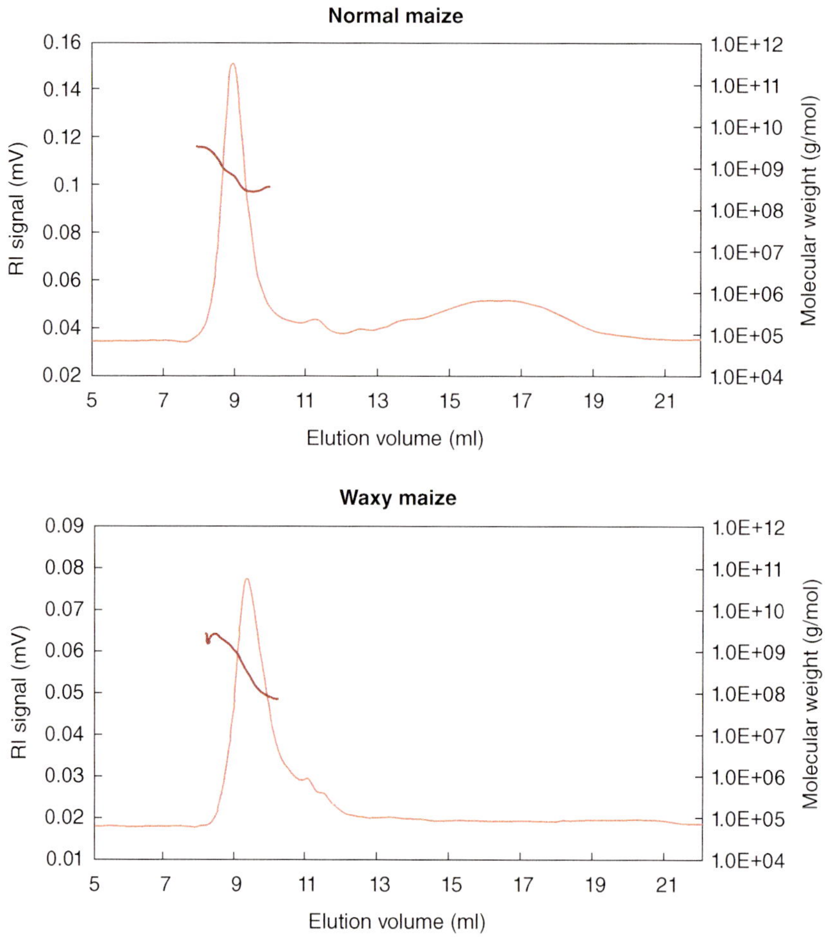

Fig. 10.9. Typical SEC-MALLS graph.

analysis of modified starches. Such products are analysed by alkaline reagents employing dinitrosalicylate or, more frequently, by alkaline ferricyanide reagents. The ferricyanide reducing method is preferred because it is the least influenced by sample size, amount of reagent or minor variations in other reaction parameters. The method described here is that recommended by Schosh.

The ferricyanide reagent is prepared by dissolving 16.5 g of potassium ferricyanide and 22 g of anhydrous sodium carbonate in distilled water and diluting to a volume of 1 litre. The solution should be allowed to stand three or four days and any precipitate which forms should be removed by filtration. The filtered reagent should be kept in a dark-brown glass bottle and stored in a dark cabinet when not in use since it deteriorates when exposed to light. Reagent prepared and stored in this manner is stable for several months.

The starch sample of known moisture or volatiles content is powdered or finely ground (60-mesh) prior to analysis. A sample weighing 200–250 mg, or less if the reducing value is high, is transferred to a clean and dry 8-ounce

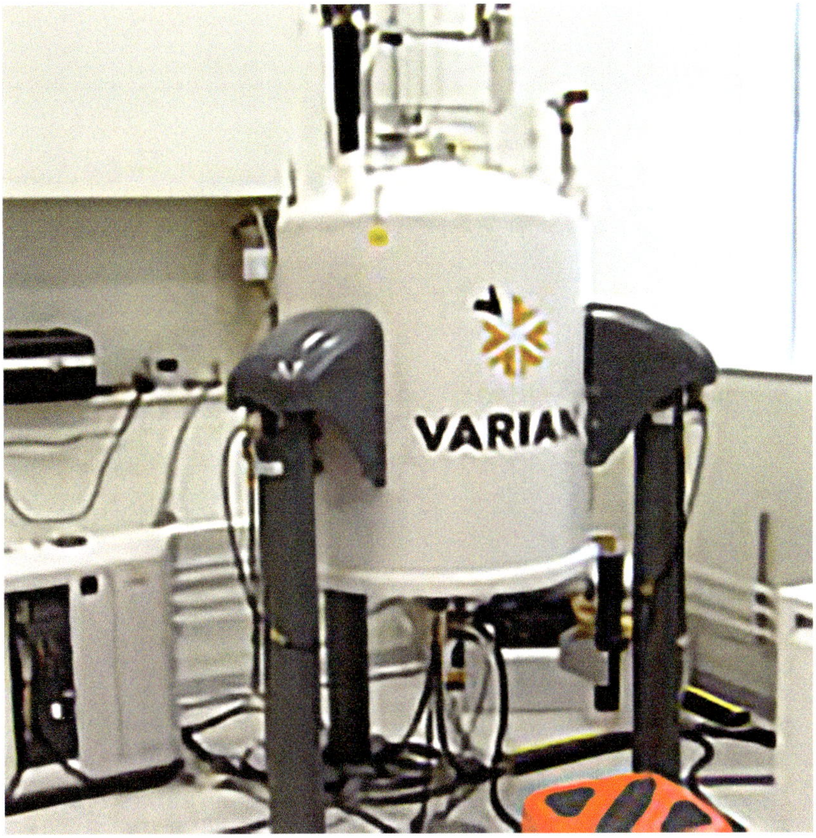

Fig. 10.10. CP/MAS C13-NMR equipment.

Pyrex bottle with a narrow mouth. If the material is pre-gelatinized or displays significant swelling tendencies in cold water, it is wetted completely by the addition of 2–3 ml of benzene. The sample is suspended in 25 ml of distilled water; the bottle is placed in a boiling-water bath and heated for several minutes while mixing the contents continuously so as to disperse and gelatinize the sample uniformly. If lumps appear in the sample dispersion at this point, the test should be repeated.

Next, 25 ml of alkaline ferricyanide reagent is added to the hot sample dispersion and the contents of the bottle are mixed. The bottle is closed with a one-hole rubber stopper containing a short length of capillary tubing which serves as a vent, and then placed immediately in a vigorously boiling water bath, so that the bottle is immersed to the neck in boiling water. The bottle is kept in the bath for exactly 15 minutes and the contents are swirled frequently to ensure rapid and uniform heat transfer.

After removal from the bath, the bottle and contents are cooled immediately in a stream of cold water or an ice-and-water bath; 60 ml of zinc sulphate–acetic acid solution (200 ml of glacial actic acid, 70 g of potassium chloride, and 20 g of zinc sulphate heptahydrate dissolved in water and diluted to 1 litre) is added carefully to avoid loss of sample through excessive foaming and the contents are mixed. Twenty millilitres of 20% potassium iodide solution is added, and the iodine liberated by reaction of iodide ion with excess ferricyanide ion is titrated with standard 0.05N sodium thiosulphate solution. The starch sample usually serves as the indicator for end-point detection, but if the product does not provide the characteristic intense blue or

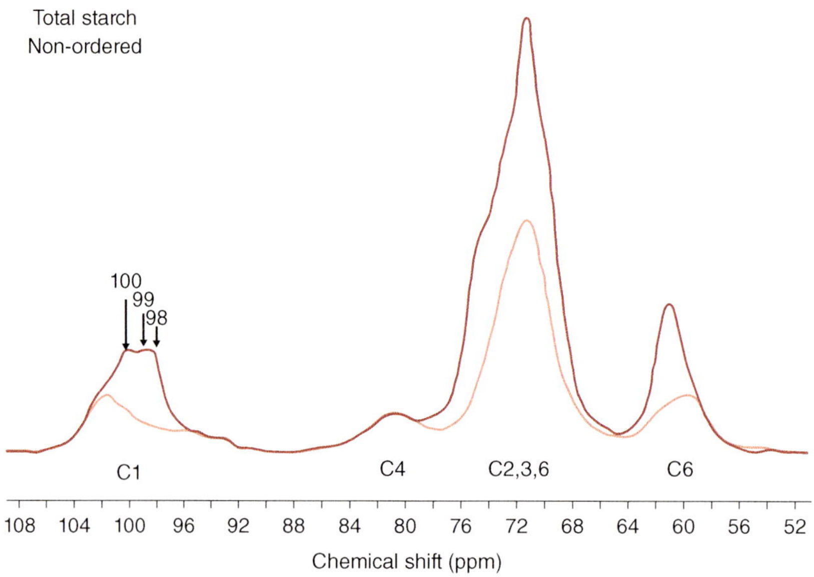

Fig. 10.11. Typical CP/MAS spectrum.

purple colour, starch indicator solution should be added. Near the end of the titration, the presence of insoluble, blue-black starch–iodine complex may obscure the end-point. In this case, the bottle should be stoppered and the contents shaken vigorously; the titration is then completed by drop-wise addition of standard thiosulphate solution and shaken after each addition.

A blank determination is performed on 25 ml of the ferricyanide reagent; 25 ml of water are added without heating, 60 ml of zinc sulphate–acetic acid reagent and 20 ml of 20% potassium iodide solution are added as described previously. The reaction mixture is titrated with standard 0.05N sodium thiosulphate solution, starch indicator solution is added as the end-point is approached, which can be judged by perceptible fading of the yellow ferricyanide colour. The ferricyanide number or reducing value expressed as the number of millilitres of 0.1N ferricyanide solution reacting with 1 g of dry sample can be calculated using the equation at the bottom of the page.

Determination of reducing value

Starch has one reducing end and the reducing value therefore provides information about the molecular weight. When starch is broken down, the reducing value increases, corresponding to the extent of breakdown.

Reagents and apparatus

1. Alkaline ferricyanide: 16.5 g of potassium ferricyanide and 22 g of anhydrous sodium carbonate are dissolved in water and the solution is made up to 1L.
2. Zinc sulphate–acetic acid: 200 ml of glacial acetic acid, 70 g of potassium chloride and 20 g of zinc sulphate heptahydrate are

$$\text{Ferricyanide number} = \frac{(\text{Blank titre} - \text{Sample titre})\text{ in millilitres} \times \text{thiosulphate normality} \times 10}{\text{Dry sample weight in grams}}$$

dissolved in water, and the solution is made up to 1L.

3. 20% Potassium iodide: the solution of 100 g of potassium iodide in 500 ml of distilled water should be stored in a dark-glass bottle away from light and discarded if it shows any tinge of yellow iodine colour. The addition of a pellet of sodium hydroxide is claimed to stabilize the solution.
4. Starch indicator: most conveniently, 2 g of potato starch is cooked with stirring in 400 ml of distilled water for 15 minutes and the paste is then 'homogenized' in a Waring Blendor for five minutes.
5. 0.05N standard sodium thiosulphate solution

Method

Sample: granular starch weighing 200–250 mg is transferred to the Pyrex bottle; 2–3 ml benzene is added to keep the sample wet; 25 ml of water is then added. The bottle is heated in the boiling-water bath with frequent agitation until uniform dispersion of sample is achieved. To the hot solution, 25 ml of alkaline ferricyanide reagent is added; the bottle is closed with a vented rubber stopper and heated in the boiling-water bath for exactly 15 minutes with occasional agitation. The bottle is then removed from the bath and cooled rapidly in cold water. Sixty millilitres of zinc sulphate–acetic acid solution is added with care to avoid loss of sample through excessive frothing and the content is mixed properly. Twenty millilitres of 20% potassium iodide solution is added and the liberated iodine is titrated with 0.05N sodium thiosulphate. Blanks are run (without heating) on 25 ml of ferricyanide reagent, adding 25 ml of water and 60 ml of zinc sulphate–acetic acid reagent, thoroughly mixing and then adding 20 ml of 20% potassium iodide.

10.8 Rheological Methods

10.8.1 Swelling power and solubility

When starch is pasted in hot water, the individual granules swell and a portion of the starch substance dissolves in the surrounding aqueous medium. The degree of swelling and the amount of soluble material will depend on the starch species and on the type and extent of modifications; for example, chemical cross-bonding, introduction of derivative groups and so on. For the full characterization of the starch, the swelling power and solubility may be determined at 5°C intervals over the entire pasting range (60–95°C) and these values plotted against temperature of pasting. For practical purposes, it is usually adequate to determine swelling power and solubility at a temperature of 85°C, which has been found to be optimum for single-point characterization.

Procedure

Moisture content of the starch should first be determined in conventional fashion. An appropriate amount of the starch is accurately weighed and quantitatively washed with distilled water into a dry, tared, 250 ml centrifuge bottle, pre-weighed on a sensitive torsion balance. Sufficient distilled water is added to give a total volume of water equivalent to 180 g. The bottle and a 200 rpm stirring motor are arranged in proper position on a vertical metal rod, which in turn is attached to a support

	Ferricyanide number	
Digestion time	With 5 ml of ferricyanide	With 25 ml of ferricyanide
0.25	1.12	1.19
0.5	1.16	1.30
1.0	1.32	1.57
2.0	1.48	2.07
3.0	1.59	2.47
Extrapolated to zero time	1.07	1.07

stand by a single clamp. The starch suspension is stirred at approximately 200 rpm or just sufficient to keep the starch completely and uniformly suspended. Excessive speed should be avoided, since it may cause fragmentation of swollen granules.

The bottle is closed with a rubber stopper, inverted several times to mix the paste, then centrifuged 15 minutes at 2200 rpm. By means of a glass tube attached to a clean dry suction flask, the clear supernatant is drawn off to within 0.6 cm of the top surface of the sedimented paste. A 50 ml aliquot of this solution is transferred by pipette to a tared, 100 ml nickel evaporating dish and evaporated to dryness on a steam bath. The dish is dried for four hours in a vacuum oven at 120°C, cooled in a desiccator and weighed.

To obtain the swelling power, the last 0.6 cm of supernatant above the sedimented starch is drawn off by suction as carefully and quantitatively as possible and discarded. The bottle and contents are then weighed on the torsion balance to give the weight of the swollen starch granules. A correction must be made for the soluble starch, in order to provide a measure of swelling of the swollen but undissolved granules. The following calculation and the equation at the bottom of the page can be used:

$$\%\,\text{Soluble (dry weight)} = \frac{\text{Weight of soluble starch} \times 100}{\text{Weight of sample on dry basis}}$$

The method is applicable to all unmodified granular starches, though high-amylose starches should be pasted at 100°C. Both swelling and solubility are affected if the pH is outside the range 4.5–7.0.

10.8.2 Water-binding capacity

A suspension of 1 g starch in 15 ml distilled water is agitated in a mechanical stirer for one hour and centrifuged at 3000 × *g* for 10 mins. The water layer is decanted and the residue is weighed. WBC is calculated as percentage weight of the water bound by the starch (Medcalf and Gilles, 1965).

10.8.3 Hot-plate viscosity

Substantially, all starch applications involve the use of cooked pastes. In some instances, the user receives the product in granular form and prepares cooked pastes to suit their particular needs, while in other instances the starch supplier precooks and dries the material so as to supply the market with pre-gelatinized products.

Methods for the determination of starch paste viscosities have been the subject of considerable research and much controversy. Most methods in common use measure consistency, of which viscosity is the principal component. The dynamic method developed by Myers and Knauss (Myers *et al.*, 1962) gives both viscosity and co-efficient of rigidity of the cooked paste, the latter being a function of elasticity.

Viscosity of starch pastes

When aqueous suspensions of granular starches are heated above the gelatinization temperature, the granules become highly hydrated and swell to many times their original volume. Continued heating, particularly in the presence of shear, produces cooked pastes that are mixtures of swollen granules, granule fragments and molecularly dispersed starch molecules leached from the granules.

Reviews by Fetzer, Mason and Fetzer reveal a variety of procedures based on rotational devices and flow-through orifices. Most methods in common use measure consistency of which viscosity is the principal component. The dynamic method developed by Myers and Knauss gives both viscosity and coefficient of rigidity of a cooked starch paste, the latter being a function of elasticity.

$$\text{Swelling power} = \frac{\text{Weight of sedimented paste} \times 100}{\text{Weight of sample on dry basis} \times (100 - \%\,\text{soluble, dry basis})}$$

Procedure with the Brabender amylograph

The Brabender amylograph (Fig. 10.12) is a rotational instrument which permits continuous determination of viscosity or consistency while cooking and cooling the starch paste. Similar information can be obtained with the visco-amylograph. The rotating sample cup or bowl contains a number of fixed vertical pins and is driven at constant speed by a synchronous motor. A circular metal disk with several metal pins projecting vertically downward into the sample serves as the sensing element. Rotation of the viscous sample exerts a force on the sensing element which is dynamically balanced by a calibrated torsion spring. Application of this restoring force is accompanied by an angular deflection of the sensing element shaft, and this is recorded continuously on a strip chart recorder.

The sample cup is positioned in an electrically heated air bath and a cooling coil projects into the cup. Heating and cooling are controlled by a mechanically operated thermoregulator which maintains a constant temperature or which increases or decreases the temperature at a constant rate of 1.5°C/min. Meanwhile, the viscosity or consistency is recorded continuously. The instrument can also be calibrated with standard oils so that results can be presented optionally in absolute viscosity units.

A weighed sample containing an appropriate amount of starch dry substance is placed in a 500 ml beaker; the sample weight depends on the sample source and may range from as little as 2 g for an unmodified potato starch to 150 g for a highly modified product such as dextrin. It is desirable to select a sample weight or concentration that will produce a scale selection of 50% or more at the important points in the viscosity recording during the cooking and cooling cycle. About 250 ml of distilled water are added to the sample and the mixture is shaken to form a lump-free suspension. At this point, additives may be incorporated if their effects are being investigated or a buffer may be added, or the pH adjusted so as to control pH of the paste during the analysis. The slurry is transferred quantitatively to a 500 ml volumetric flask, diluted to volume with water at 25°C and mixed. The sample slurry is then transferred to the sample cup in the instrument; the sensing element is placed in position, and the instrument head is lowered into the operating position, the thermoregulator having previously been adjusted to 25°C. A short-stem thermometer can be mounted adjacent to the thermoregulator so that it dips into the sample, permitting temperature determination independent of the thermoregulator.

Fig. 10.12. Brabender viscometer.

When the Brabender curves for several concentrations of a given sample are reported and superimposed on rectangular co-ordinates, a family of curves is obtained (Fig. 10.13). These contain five successive points of significant interest.

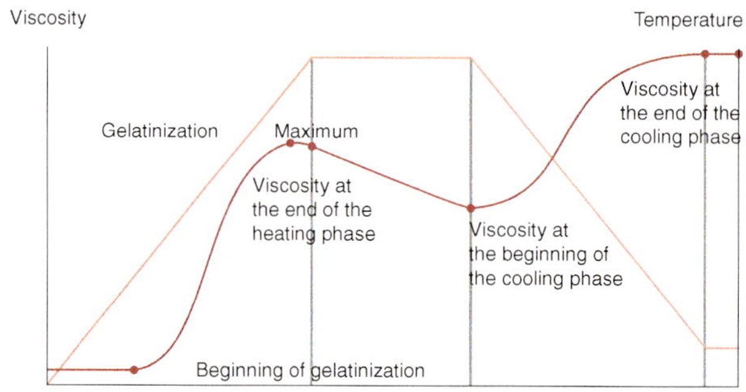

Fig. 10.13. Typical Brabender viscograph.

1. The peak viscosity, irrespective of the temperature, indicates the highest viscosity that the user might encounter during the preparation of a usable paste.
2. The viscosity of the paste when it reaches a temperature of 95°C, in relation to the peak viscosity, reflects the ease of cooking of the starch.
3. The viscosity after cooking 1 hour at 95°C indicates the stability or breakdown of the paste.
4. The viscosity of the cooked paste after cooling to 50°C is a measure of the setback produced by cooling.
5. The final viscosity after stirring 1 hour at 50°C indicates the stability of the cooked paste as it might be used.

Equations 1 and 2 at the bottom of the page relate to the Brabender amylograph.

Procedure with Brookfield viscometer

This instrument (Fig. 10.14) employs a rotating cylinder or disc, referred to as a spindle, driven by a constant-speed synchronous motor. The drive shaft is connected to the spindle shaft through a torsion spring which provides the restoring force to balance the viscous resistance to rotation of the spindle.

To evaluate a starch sample, an aqueous suspension containing an appropriate percentage of dry substance is stirred in a metal beaker or other suitable container until free of lumps. Actual concentration may be as low as 2–3% for an unmodified starch or lower for a cross-bonded product, and as high as 50–60% for a highly modified starch product such as dextrin. The beaker is then placed in a boiling-water or steam bath and the starch suspension is cooked while stirring in a prescribed manner. The time of heating will usually be about 30 minutes, but will depend on the volume of suspension and type of sample and it should be sufficiently long to bring the paste temperature to about 95°C. In routine sample evaluation, the type and vigour of stirring, the rate of temperature rise and the cooking time should be reproduced from sample to sample.

Rapid ViscoAnalyser (RVA)

RVA has advantages in that it requires fewer samples and is faster. New models, which can

$$\text{mg of iodine bound by 100 mg sample} = \frac{\text{Ml of } 8.33\times10^{-4}\,\text{M KIO}_3\times0.634}{\text{Sample wt in mg}\times0.25} \quad (1)$$

$$\%\text{ of amylose} = \left\{\frac{\text{mg of iodine bound by 100 mg of sample}}{\text{mg of iodine bound by 100 mg of pure sample}}\right\}\times100 \quad (2)$$

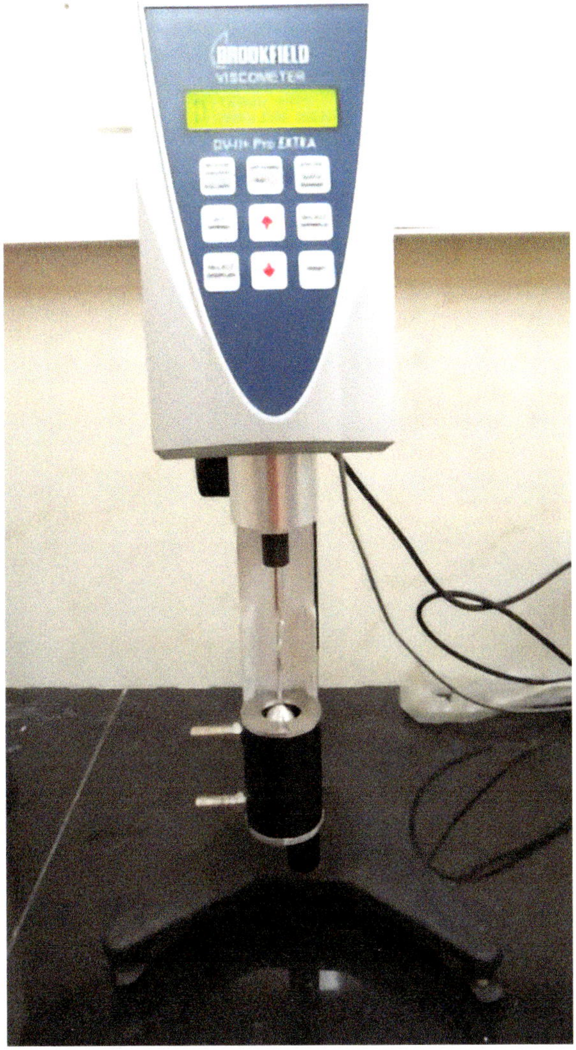

Fig. 10.14. Brookfield viscometer.

operate above 100°C have also been developed. The procedure is the same as with the Brabender viscometer. The required quantity of starch is weighed into aluminium cannisters and fitted into the equipment. The temperature sequence is programmed to obtain heating up to 95°C, holding and cooling at the same rate (Fig. 10.15).

10.8.4 Rheology

A rheometer is an instrument that measures both viscosity and viscoelasticity of fluids, semi-solids and solids. It provides information about the materials:

- Viscosity: functions of shear rate or stress, time and temperature dependence;

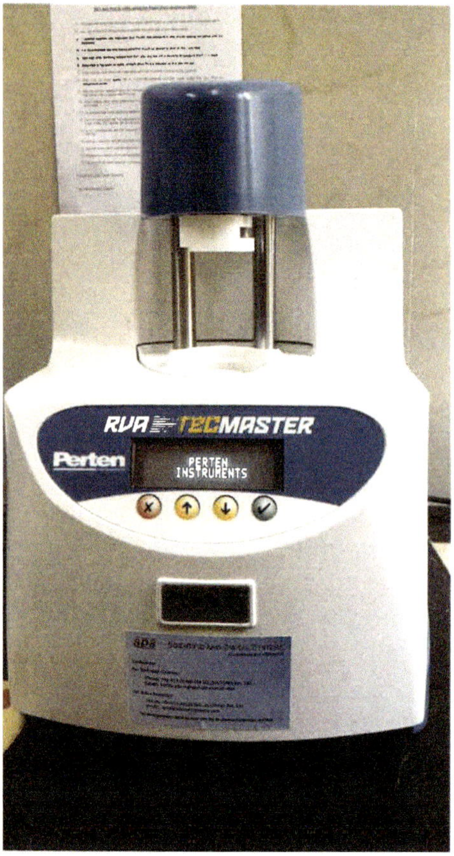

Fig. 10.15. Rapid ViscoAnalyser.

- Viscoelastic properties (G′, G″, tan δ) with respect to time, temperature, frequency and stress/strain.

Rotational test

a. Flow curves

Starch gels for flow curves were prepared in distilled water by placing 10% (w/v) of suspension in a boiling-water bath for 20 minutes, cooled and placed in the rheometer plate. Shear rate was increased from 0.1 to 100/s logerithmetically in a stepped ramp mode. Flow curve test was run at 30°C for samples and plots were made of shear rate versus viscosity and shear rate versus shear stress. The yield stress and infinite shear viscosity were determined using the Casson model (Hackley and Ferraris, 2001).

b. Thixotropy/shear/structural decomposition/recovery test

The thixotropy test was done in the 10% starch gel and the experiment was preceded by the loop test, which was carried out at three different intervals. Starch was first subjected to constant shear rate of 1/s for 10 s followed by increasing the shear rate to 100/s for 10 s. In the last interval the shear rate was released and the recovery of the starch sample was measured for 300 s. The recovery was measured using the 3 Interval Thixotropy Test (3ITT)

Dynamic oscillatory measurements

a. Frequency sweep measurements

The viscoelastic properties were measured using an oscillatory test in the same system as in the rotational test. Before starting the test, the linear viscoelastic region (LVR) of the starch gels was measured by the strain amplitude sweep test with strain ranging from 0.1 to 5% as, and a frequency of 0.1–20 Hz. Once the LVR was determined, the rheometer was programmed for running frequency sweep (0.1–10 Hz) and strain of 1%. The temperature was fixed to 30°C for the analysis. The three main parameters determined in the dynamic oscillation tests were storage modulus G', loss modulus G'' and loss tangent tan δ.

Geometries			
Concentric cylinders	Cone and Plate	Parallel plate	Torsion Rectangular
Very low to medium viscosity	Very low to high viscosity	Very low viscosity to soft solids	Very soft to very rigid solids

b. Temperature sweep test

For the temperature sweep measurements, 20% (w/v) of starch suspension was prepared by taking 20 g dry starch sample in 100 ml distilled water. During the temperature sweep measurements the strain was fixed to 1% and frequency to 10 Hz. The temperature was increased from 30°C to 90°C using the Peltier system of rheometer. The storage modulus G', loss modulus G" and loss tangent tan δ values of the starch samples were determined during the heating process.

10.9 Thermal Methods

10.9.1 DSC (differential scanning calorimetry)

Differential scanning calorimetry (Fig. 10.16) measures the difference between the heat flows from the sample and reference sides of a sensor as a function of temperature or time. It measures the heat flow associated with structure and changes in structure as a function of time and temperature in a controlled atmosphere. The two most frequent thermal signatures that are studied are starch gelatinization and retrogradation. Differential scanning calorimetry is particularly well suited to investigate the phase transitions of starch/water systems because it allows: (i) study of starch gelatinization over a wide range of starch:water ratio; (ii) determination of gelatinization temperatures above 100°C; and (iii) estimation of transition enthalpies. DSC measures the difference between the heat flows from the sample and reference sides of a sensor as a function of temperature or time.

Procedure

Five milligrams of dry starch is accurately weighed with a microbalance into an aluminium pan; 10 or 15 microlitres of water are added and the pan is sealed hermetically. The pan is transferred to the heating chamber and heated to the desired temperature using a programmed rate of heating. Sometimes, the sample is allowed to cool and reheated to gain more information on reversible and irreversible transitions.

The T_{onset}, T_{peak}, T_{end} and ΔH are obtained from the thermograph. A typical graph is given below (Fig. 10.17).

Modulated DSC is an important development which can be useful in giving information about reversible and non-reversible transformations in polymers. The method uses a sinusoidal heating mode. Non-reversible signals are the endothermic relaxation of amorphous materials, gelatinization, recrystallization and protein denaturation. Reversible events include glass transition and simultaneous crystallization. In modulated DSC, the temperature programme involves a continuous heating–cooling cycle, helpful in identifying reversible and non-reversible processes. Gelatinization of starch is a non-reversible process, whereas melting of the amylose-lipid complex is reversible. MDSC separates the two processes and provides more information on starch transformations (Fig. 10.18).

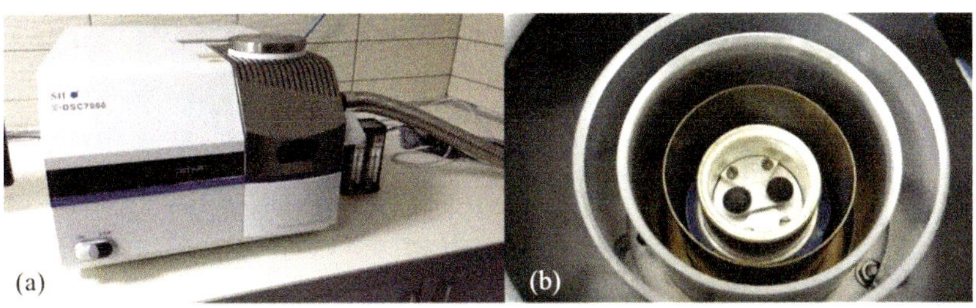

Fig. 10.16. DSC equipment.

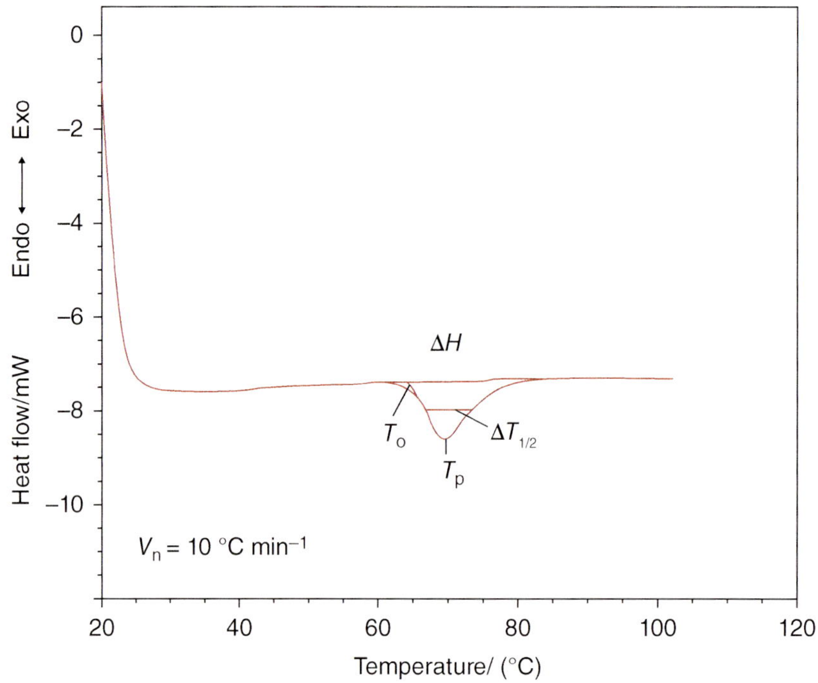

Fig. 10.17. Typical DSC thermogram.

10.10 Starch Digestibility

10.10.1 *In vitro* starch digestibility

In vitro starch digestibility was determined using the method of Englyst *et al.* (1992) with modification by Chung *et al.* (2009). Porcine pancreatic amylase was dispersed in sterile distilled water (4 ml) and centrifuged at 1500 × *g* for 12 minutes. The supernatant (2.7 ml) was transferred to a beaker glass and amyloglucosidase (0.3 ml) and invertase (0.2 ml) were added to the solution. This enzyme solution was freshly prepared for each determination of starch digestibility.

Starch (100 mg) and 4 ml of 0.5 M sodium acetate buffer (pH 5.2) were added to each test tube. Enzyme solution (1 ml) and 20 glass beads (4 mm diameter) were added to each tube and the tubes were incubated in a shaking water bath (37°C, 120 rpm). After 20 minutes, aliquots (0.1 ml) were collected and mixed with 1 ml of 80% ethanol. The solution was incubated again in a shaking water bath (37°C, 120 rpm) and an aliquot (1 ml) was collected after 100 minutes and mixed with 1 ml of 80% ethanol. The shaking of the water bath was not stopped during the sampling period. The aliquot was taken after 20 minutes and designated as G_{20} (rapidly available glucose, RAG) and that taken after 120 minutes was designated as G_{120} (slowly available glucose, SAG). The G_{20} and G_{120} were centrifuged at 1500 × *g* for 2 minutes to obtain a clear supernatant for glucose determination.

The remaining solution was removed from the shaking water bath, shaken vigorously to break up any large particles and incubated in a boiling-water bath for 30 minutes. The test tubes were shaken again and cooled in ice-water for 15–20 minutes. Thereafter, 10 ml of 7 M KOH were added to the aliquot, mixed well and incubated in a shaking water bath containing ice-water (5–10°C) for 30 minutes. An aliquot (1 ml) was collected, added to 10 ml of 0.5 M acetic acid and 0.2 ml of amyloglucidase, incubated at 70°C for 30 minutes, placed in the boiling-water bath for 10 minutes, cooled to

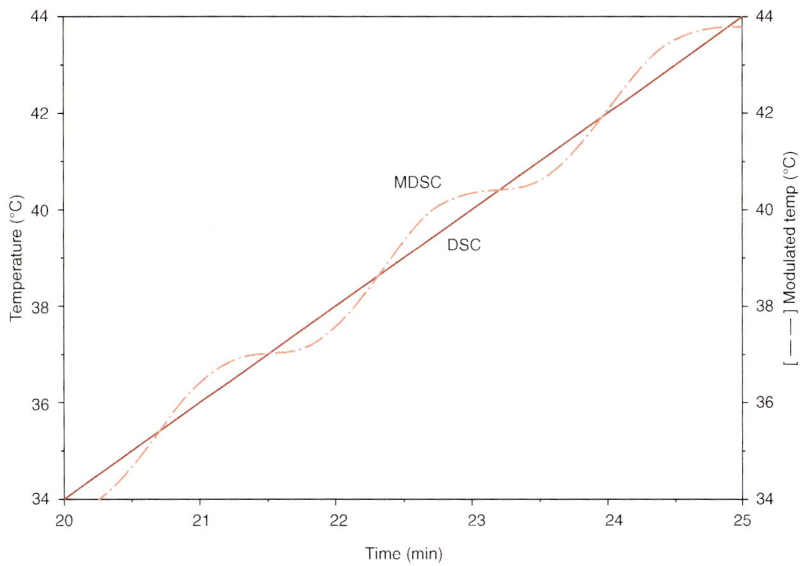

Fig. 10.18. Comparison of heating mode of normal and modulated DSC.

room temperature, diluted with 40 ml of distilled water and centrifuged at 1500 × *g* for 5 minutes. An additional aliquot (0.1 ml) was then collected for Total Glucose (TG) measurement. The determination of Free Glucose (FG) was conducted as follows: starch (400 mg) and 5 ml of 0.5 M sodium acetate buffer (pH 5.2) were added to screw-cap test tubes, shaken well, incubated in a boiling-water bath for 30 minutes and cooled to room temperature. An aliquot (1 ml) was collected and 2 ml of 80% ethanol was added, shaken well and centrifuged at 1500 × *g* for 5 minutes. The supernatant (1 ml) was taken and 5 ml of distilled water was added and shaken well for the determination of FG. The hydrolysed glucose content was measured using glucose oxidase-peroxidase reagent. Aliquots (0.1 ml) were collected and 3 ml of GOPOD reagent was added, incubated at 40–50°C for 20 minutes and cooled at room temperature. Then the absorbance was measured at 510 nm.

Starch classification based on the rate of hydrolysis included rapidly digestible starch (RDS, digested within 20 minutes), slowly digestible starch (SDS, digested between 20 and 120 minutes) and resistant starch (RS, undigested after 120 minutes). The digestible starch fractions and available starch fractions were calculated as follows:

$$RAG = G_{20}$$
$$SAG = G_{120} - G_{20}$$
$$TS = (TG - FG) \times 0.9$$
$$RDS = (G_{20} - FG) \times 0.9$$
$$SDS = (G_{120} - G_{20}) \times 0.9$$

$$RS = \frac{TS - RDS - SDS}{TS} \times 100\,(\%)$$

10.11 Derivative Methods

10.11.1 Acetyl

In the analysis of low degrees of substitution (DS) products, about 5 g of finely ground sample is weighed accurately and placed in a 250 ml Erlenmeyer flask and suspended in 50 ml of distilled water. After addition of a few drops of phenolphthalein indicator, the suspension is titrated with 0.1N sodium hydroxide to a permanent pink end-point. Then, 25 ml of 0.45N sodium hydroxide solution is added, taking care not to wet the neck of the flask. The flask is sealed tightly with a rubber stopper and shaken vigorously for 30 minutes (a mechanical shaker is recommended).

After shaking, the stopper is carefully removed and washed down, together with the walls of the flask, with a fine stream of distilled water. The saponified sample mixture, containing excess alkali, is then titrated with standard 0.2N hydrochloric acid to the disappearance of the phenolphthalein colour.

In the analysis of high DS starch acetates, about 1 g of finely ground sample, either predried or of known volatiles content, is weighed accurately and placed in a 250 ml Erlenmeyer flask, and 50 ml of 75% ethanol in distilled water is added. The flask is loosely stoppered and the contents are warmed to 50°C while stirring and held at this temperature for 30 minutes. After cooling to room temperature, 40 ml of 0.5N potassium hydroxide solution is added while swirling; the flask is stoppered and allowed to stand for 72 hours at room temperature with occasional swirling. The excess alkali is titrated with standard 0.3N hydrochloric acid using phenolphthalein as indicator. Additional alkali, which might leach from the sample, is titrated after the mixture is permitted to stand an additional 2 hours (see equations at the bottom of the page).

10.11.2 Alkoxyl (methoxyl and ethoxyl)

A portion of the prepared and dried sample containing about 2–4 mg of methoxyl is weighed accurately in a small glass vial. This may be constructed of a short length (15 mm) of glass tubing with one end closed and flattened to facilitate weighing. The sample need not be dry if the moisture or volatiles content is known, but it must be free of reagent residues and solvents, which react with hydroiodic acid giving volatile alkyl iodides.

The heating mantle is removed from the reaction vessel and the latter is allowed to cool a few minutes. The reaction vessel is disconnected from the apparatus; the glass vial with sample is introduced, and the vessel is quickly reconnected to the apparatus. The heating mantle is repositioned and heat is reapplied so that the solvent mixture reflexes for a period of 40 minutes while carbon dioxide flows through the system at a rate of 70–100 bubbles/min. The reaction is frequently rather vigorous immediately after sample introduction, which requires some adjustment of the carbon dioxide flow rate.

The receivers are removed from the apparatus without interrupting the carbon dioxide flow rate. With the aid of distilled water, the absorbing solutions in both receivers are transferred quantitatively to a 250 ml flask containing 5 ml of 25% sodium acetate solution, and the solution is diluted to about 100 ml. With water, concentrated formic acid is added, dropwise, until the yellow colour of the bromine is discharged, and 12 drops in excess are added. The solution is mixed and allowed to stand about two minutes.

Ten millilitres of 1:9 sulphuric acid solution is added followed by 10 ml of 15% potassium iodide solution. The liberated iodine is titrated immediately and rapidly with standard 0.1 N sodium thiosulphate solutions. When the end-point is approached, as evidenced by nearly complete elimination of the yellow colour of the iodine, 1 ml of starch indicator solution is added and titration is completed by drop-wise addition of the thiosulphate solution until the characteristic blue starch-iodine colour disappears.

$$\text{Per cent acetyl (dry basis)} = \frac{(\text{Blank titre} - \text{Sample titre})\ \text{millilitres} \times \text{Acid normality} \times 0.043 \times 100}{\text{Sample weight in grams (dry basis)}}$$

$$\text{Degree of Substitution (DS)} = 162A / 4300 - 42A$$

A = per cent acetyl (dry basis)

A blank determination is performed on all reagents in the same manner (see equation 1 at bottom of page).

Degree of Substitution (DS) = 162M / 3100 – 14M

M = per cent methoxyl (dry basis)

10.11.3 Hydroxyalkoxyl (hydroxyethyl and hydroxypropyl)

A portion of the prepared sample containing not more than 1 g of dry starch and about 25 mg of hydroxyethyl group (expressed as C_2H_4O) is weighed accurately in a small glass vial. This may be constructed of a short length (15 mm) of glass tubing with one end closed and flattened to facilitate weighing. Drying prior to analysis is not essential; small amounts of moisture contributed by the sample are not detrimental.

The receivers are removed from the apparatus in reverse order without interrupting the gas flow rate and before the heat source is removed. Contents of the last receiver (potassium iodide solution) are transferred competitively to a 250 ml Erlenmeyer flask with the aid of distilled water. Contents of the spiral receiver are drained into the same flask and transfer is made quantitative by rinsing with distilled water. Ten millilitres of 5% sulphuric acid solution are added, and liberated iodine is titrated immediately with standard 0.05 N sodium thiosulphate solutions. When the end-point is approached, as evidenced by nearly complete elimination of the yellow colour of the iodine, 1 ml of starch indicator solution is added and the titration is completed by drop-wise addition of the thiosulphate solution until the characteristic blue starch-iodine colour disappears (see equations 2 and 3 at bottom of page).

HE = total per cent C_2H_4O (dry basis)

Hydroxypropyl starch is analysed in the same way but milli-equivalent weights of 0.058 and 0.029 are substituted in the two equations for calculating % C_2H_4O (equivalent to the propyl iodide and propylene, respectively). The equation for calculating DS becomes:

Degree of substitution (DS) = 162HP / 5800 – 58HP

HP = per cent C_3H_6O (dry basis)

10.11.4 Carboxyl

Samples containing high carboxyl content, particularly derivatives that swell excessively in water at room temperature, are best freed of cations by leaching the sample with about 25 ml of 2N hydrochloric acid in 70% methanol (v/v). The sample is recovered by filtration through a medium-porosity fritted-glass crucible or funnel and washed with 70% methanol (v/v) until the washings are free of chloride.

The demineralized sample is transferred quantitatively to a beaker and dispersed in about 300 ml of distilled water. The dispersion is heated in a steam bath or boiling-water bath and stirred continuously until the starch gelatinizes. Heating is continued for about 15 minutes to ensure complete gelatinization. The pasted sample is titrated hot, or after cooling, with 0.1N sodium hydroxide solution to a phenolphthalein end-point. More dilute alkali solutions are recommended when titrating samples having low carboxyl contents.

Titers obtained by the preceding procedure represent the carboxyl group and other acidity native to the starch. Research has shown that

$$\text{Per cent methoxyl}(-OCH_3) = \frac{(\text{Sample titre} - \text{Blank titre})\ \text{millilitres} \times \text{Thiosulphate normality} \times 0.031 \times 100}{\text{Dry sample weight in grams} \times 6} \quad (1)$$

$$\text{Per cent } C_2H_4O(-C_2H_5I) = \frac{(\text{Blank titre} - \text{Sample titre})\ \text{millilitres} \times \text{Thiocyanate normality} \times 0.044 \times 100}{\text{Dry sample weight in grams}} \quad (2)$$

$$\text{Per cent } C_2H_4O(-C_2H_4) = \frac{(\text{Blank titre} - \text{Sample titre})\ \text{ml} \times \text{Thiocyanate normality} \times 0.044 \times 100}{\text{Dry sample weight in grams}} \quad (3)$$

Degree of Substitution (DS) = 162HE / 4400 – 44HE

the latter is due largely to fatty acids complexed with the amylase present in common starches. Since contributions from this source vary with starch origin and history, the unoxidized or underivatized original starch should be titrated to provide a blank value. The same weight of starch as that used for demineralization should be pasted in about 300 ml of distilled water and titrated with sodium hydroxide solution in the same manner as outlined above. Acid leaching and washing are not required prior to titration (see equations 1 and 2 on this page).

10.11.5 Carbonyl

Hydroxylamine method

The hydroxylamine reagent is prepared by dissolving 25 g of reagent-grade hydroxylamine hydrochloride in water and adding 100 ml of 0.5N NaOH. The mixture is diluted to 500 ml and mixed. The reagent is unstable and should be discarded when its age exceeds two days.

A portion of the powdered or finely ground (60-mesh) oxidized starch, containing not more than 10 g of dry starch and not more than 40 mg of carbonyl (expressed as >C=O) is weighed accurately and transferred to a beaker. The sample size is selected so that the blank sample titre difference does not exceed about 15 ml of 0.1N hydrochloric acid solution, so as to provide a substantial reagent excess during the reaction. The sample is gelatinized by the addition of 100 ml boiling distilled water; the resultant dispersion is cooled to 40°C, adjusted to pH3.2 and transferred quantitatively to a 500 ml glass stoppered flask with the aid of distilled water.

Exactly 60 ml of hydroxylamine reagent is added to the sample solution, and the flask is stoppered and placed in a water bath at 40°C. After four hours, the excess hydroxylamine is determined by rapid titration of the reaction mixture to pH3.2 with standard 0.1N hydrochloric acid solution. End-point detection as well as pH adjustment prior to reagent addition is accomplished with the aid of a pH meter. A blank determination on all reagents should be performed in the same manner, and an identical weight of the original unoxidized starch should be included whenever it is available.

$$\text{Per cent carboxyl} = \frac{(\text{Sample titre} - \text{Blank titre})\ \text{millilitres} \times \text{Alkali normality} \times 0.045 \times 100}{\text{Sample weight in grams}} \qquad (1)$$

$$\text{Per cent carboxyl} = \frac{\text{HCl millilitres} \times \text{HCl normality} \times 0.045 \times 100}{\text{Sample weight in grams}} \qquad (2)$$

$$\text{Per cent carboxyl} (> C = O) = \frac{(\text{Blank titre} - \text{Sample titre})\ \text{millilitres} \times \text{Acid normality} \times 0.028 \times 100}{\text{Dry sample weight in grams}}$$

References

Bello-Pérez, L., Roger, P., Baud, B. and Colonna, P. (1998) Macromolecular features of starches determined by aqueous high-performance size exclusion chromatography. *Journal of Cereal Science* 27(3), 267–278.

BeMiller, J.N. and Whistler, R.L. (eds) (2009) *Starch: Chemistry and Technology*. Academic Press, New York.

Chung, H.-J., Liu, Q. and Hoover, R. (2009) Impact of annealing and heat-moisture treatment on rapidly digestible, slowly digestible and resistant starch levels in native and gelatinized corn, pea and lentil starches. *Carbohydrate Polymers* 75(3), 436–447. DOI: 10.1016/j.carbpol.2008.08.006.

Englyst, H.N., Kingman, S.M. and Cummings, J.H. (1992) Classification and measurement of nutritionally important starch fractions. *European Journal of Clinical Nutrition* 46 Suppl 2, S33–S50.

Fishman, M.L., Rodriguez, L. and Chau, H.K. (1996) Molar masses and sizes of starches by high-performance size-exclusion chromatography with on-line multi-angle laser light scattering detection. *Journal of Agricultural and Food Chemistry* 44(10), 3182–3188.

Flanagan, B.M., Gidley, M.J. and Warren, F.J. (2015) Rapid quantification of starch molecular order through multivariate modelling of (13)C CP/MAS NMR spectra. *Chemical Communications* 51(80), 14856–14858. DOI: 10.1039/C5CC06144J.

Flores-Morales, A., Jiménez-Estrada, M. and Mora-Escobedo, R. (2012) Determination of the structural changes by FT-IR, Raman, and CP/MAS 13C NMR spectroscopy on retrograded starch of maize tortillas. *Carbohydrate Polymers* 87(1), 61–68. DOI: 10.1016/j.carbpol.2011.07.011.

Fričová, O. and Koval'aková, M. (2013) Solid-State ^{13}C CP MAS NMR spectroscopy as a tool for detection of (1→ 3, 1→ 6)-β-D-glucan in products prepared from *Pleurotus ostreatus*. *ISRN Analytical Chemistry* 4pp. DOI: 10.1155/2013/248164.

Gidley, M.J. and Bociek, S.M. (1985) Molecular organization in starches: A carbon 13 CP/MAS NMR study. *Journal of the American Chemical Society* 107(24), 7040–7044. DOI: 10.1021/ja00310a047.

Hackley, V.A. and Ferraris, C.F. (2001) Guide to rheological nomenclature: measurements in ceramic particulate systems. *National Institute of Standards and Technology*, Special Publication 946, 31.

Krishnaveni, S., Balasubramanian, T. and Sadasivam, S. (1984) Sugar distribution in sweet stalk sorghum. *Food Chemistry* 15(3), 229–232. DOI: 10.1016/0308-8146(84)90007-4.

Malik, C.P. and Singh, M.B. (1980) *Plant Enzymology and Histoenzymology*. Kalyani Publishers, New Delhi.

McCready, R.M. and Hassid, W.Z. (1943) The separation and quantitative estimation of amylose and amylopectin in potato starch. *Journal of the American Chemical Society* 65, 1154–1157.

McGrance, S.J., Cornell, H.J. and Rix, C.J. (1998) A simple and rapid colorimetric method for the determination of amylose in starch products. *Starch/Stärke* 50(4), 158–163. DOI: 10.1002/(SICI)1521-379X(199804)50:4<158::AID-STAR158>3.0.CO;2-7.

Medcalf, D.G. and Gilles, K.A. (1965) Wheat Starches. 1 Comparison of Physicochemical Properties. *Cereal Chemistry* 42, 558–568.

Myers, R.R., Knauss, C.J. and Hoffman, R.D. (1962) Dynamic rheology of modified starches. *Journal of Applied Polymer Science* 6(24), 659–666.

Nara, S., Mori, A. and Komiya, T. (1978) Study on relative crystallinity of moist potato starch. *Starch/Stärke* 30(4), 111–114. DOI: 10.1002/star.19780300403.

Ohtani, T., Yoshino, T., Hagiwara, S. and Maekawa, T. (2000) High-resolution imaging of starch granule structure using atomic force microscopy. *Starch/Stärke* 52(5), 150–153. DOI: 10.1002/1521-379X(200006)52:5<150::AID-STAR150>3.0.CO;2-F.

Ratnaningsih, N., Suparmo , Harmayani, E. and Marsono, Y. (2017) *In vitro* starch digestibility and estimated glycemic index of Indonesian cowpea starch (*Vigna unguiculata*). *Pakistan Journal of Nutrition* 16(1), 1–8. DOI: 10.3923/pjn.2017.1.8.

Ratnayake, W.S. and Jackson, D.S. (2008) Starch gelatinization. *Advances in Food and Nutrition Research* 55, 221–268.

Roger, P. and Colonna, P. (1993) Evidence of the presence of large aggregates contamination amylose solutions. *Carbohydrate Polymers* 21, 83–89.

Shanthy, A.P., Sowbhagya, C.M. and Bhattacharya, K.R. (1980) Simplified determination of water-insoluble amylose content of rice. *Starch/Stärke* 32(12), 409–411. DOI: 10.1002/star.19800321204.

Sievert, D. and Holm, J. (1993) Determination of amylose by differential scanning calorimetry. *Starch/Stärke* 45(4), 136–139. DOI: 10.1002/star.19930450405.

Wen, J., Arakawa, T. and Philo, J.S. (1996) Size-exclusion chromatography with on-line light-scattering, absorbance, and refractive index detectors for studying proteins and their interactions. *Analytical Biochemistry* 240(2), 155–166.

Yokoyama, W., Renner-Nantz, J.J. and Shoemaker, C.F. (1998) Starch molecular mass and size by size-exclusion chromatography in DMSO-LiBr coupled with multiple angle laser light scattering. *Cereal Chemistry Journal* 75(4), 530–535.

Yoo, S. and Jane, J.L. (2002) Molecular weights and gyration radii of amylopectins determined by high-performance size-exclusion chromatography equipped with multi-angle laser-light scattering and refractive index detectors. *Carbohydrate Polymers* 49(3), 307–314. DOI: 10.1016/S0144-8617(01)00339-3.

Zobel, H.F., Young, S.N. and Rocca, L.A. (1998) Starch gelatinization: An X-ray diffraction study. *Cereal Chemistry* 65(6), 443–446.

11

Conclusions and Future Prospects

11.1 General Introduction

It is evident from the rapid growth of research on tropical tuber crops that the tuber crop starches have huge potential. Such wide variability in the physicochemical and functional properties of tropical tuber starches is not observed in cereal starches; so these starches can find diverse applications in food and industry. Presently, tropical tuber crops are not given the same status as cereal crops and much more funding is needed to further develop these crops. All the tuber crops together presently contribute 5.4% of the total energy needs for humans, from 61.1 kg of tubers consumed per head annually. Of this, the contribution of the TTCs is 3.9% energy, which works out to be 72.2% of the total from all the tuber crops. It has been suggested that this contribution should be doubled to 8% by 2030.

This has become imperative in view of the emerging food crisis and the ongoing changes in climate, with the attendant problems of temperature rise, drought, sea-level rise, and the unpredictable spurt in extreme environmental variations. In this scenario, the innate advantages and climate resilience of TTCs to such extreme and unpredictable variations are becoming recognized. These factors include cassava and yams against water and temperature stress; and taro, tannia and swamp taro against flooding (Nayar, 2014).

Some of the areas that can really improve the status of the tropical tuber crops include:

1. Reduce the period of maturity of all the TTCs to 110–150 days.
2. Improve the tuber shape and make them more definitive in shape.
3. Improve plant architecture to attain a harvest index of 0.70.
4. Eliminate antinutritional factors and improve nutritive values.
5. Impart resistance to the three major biotic problems afflicting the TTCs: CMD (cassava mosaic disease), taro leaf blight and sweet potato weevil.
6. Introduce determinacy in cassava and yams.
7. Increase the post-harvest storage life.
8. Reduce the loss during post-harvest storage, transportation and processing.
9. Use biotechnological means to alter starch characteristics like amylose content, chain length and phosphorus content in starch.
10. It is also suggested that there should be a 'starch bank', which acts as a repository of different starches and provides information on all physicochemical and functional properties for various applications to researchers and industries.
11. An organization to carry out exclusive research on tuber starches may be

established in any of the TTC-growing countries.

Research should be done to address the following major issues to be able to better understand and utilize starch:

1. There appears to be some duplication of research efforts in various laboratories (e.g. starch modifications) to address 'issues' that have already been resolved years ago. A more thorough literature review is needed to avoid repetition.
2. Results from different reports are sometimes difficult to compare directly due to variation in methods used in specific studies. Quality parameters (e.g. amylose content) can be rather sensitive to the measuring method used. Global initiatives to standardize methodology with the same recommended experimental protocols are needed to maximize research efforts. In addition, the method used in each study should be clearly specified.
3. There is considerable confusion in uses of nomenclature, especially for the starch molecular structure (e.g. definitions of A-chain of amylopectin and cluster). This confusion in starch terminology further contributes to the difficulties of directly comparing data from different studies. The unification of nomenclature needs global scientific initiatives.
4. Exploring genetic means and environmental factors to diversify the variations in composition, structure and properties of starch will be useful. Starches with high amylose (e.g. amylose content >50%) and novel amylopectin structure for specific applications need to be developed.
5. A recent trend is to use 'green' technology to develop novel products in an environmentally friendly way. There are opportunities to use simple but novel techniques such as ultrasound and microwave treatments to modify starch by chemical modification processes, to efficiently achieve novel properties in a time- and energy-saving way.
6. Though the efforts in chemical modifications of starch are considerable, diverse nutritional properties of the modified starches affecting human health have not been thoroughly elucidated.

In addition, modified starch based on TTCs has very good potential in food and industry, as indicated below:

- Modified starches in different forms, such as grafted, acetylated and phosphate ester derivative in the form of microsphere or hydrogel are suitable for drug delivery. For example, high-amylose maize starch has been reported to have good sustained release properties due to its excellent gel-forming capacity.
- Polylactide polymers used in biomedical applications like resorbable sutures, prosthetic devices and drug delivery.
- Acetylated and hydroxyethyl starches are now mainly used as plasma volume expanders, for the treatment of patients suffering from trauma, heavy blood loss and cancer.
- Starch has also been used to produce novel and satisfactory artificial RBCs with good oxygen-carrying capacity; produced by encapsulating haemoglobin (Hb) with long-chain fatty-acids-grafted potato starch in a self-assembly way.
- Starch-based biodegradable bone cements (grafts of Starch-PVA or Starch-Polycaprolactone or Hydroxylapatite-reinforced starch composites) can provide immediate structural support and degrade from the site of application. Moreover, they can be combined with bioactive particles, which allow new bone growth to be induced in both the interface of cement–bone and the volume left by polymer degradation.
- Starch nanoparticles, nanospheres and nanogels in the construction of nanoscale sensors, tissues, mechanical devices and drug-delivery systems. The use of biodegradable microparticles as a dosage form for the administration of active substances, especially as a means of delivering proteins. Starch microparticles have been used for the nasal delivery of drugs and for the delivery of vaccines administered orally and intramuscularly. Bioadhesive systems based on polysaccharide microparticles enhance the systemic absorption of conventional drugs and polypeptides across the nasal mucosa; a major area of application of microparticles is as dry powder inhalation

formulation for asthma and for deep-lung delivery of various drugs.

- Porous starch formed by heating of starch with a limited amount of water, and also by treatment of gelatinized starch with ethanol, or by extrusion leading to disordered amylose and amylopectin chains, is useful in the production of extrusion-blown films from starch and poly(ethylene co-acrylic acid) graft polymers. Up to 50% starch incorporation is possible and uses can be found as filters for cigarette smoke and wound dressing. Such 'biopac' packages are becoming popular; they are light, good heat insulators and shock absorbers. Trays, cups, boxes etc. can be useful in storage of food.
- Oligosaccharides, beneficial to health, improve the intestinal microflora by increasing the number of bifidobacterium (fructo-OS, galacto-OS, xylo-OS, lactosucrose, lactulose, raffinose). They can prevent or resist formation of dental caries (palatinose, lactulose, trehalose); help uptake of minerals (fructo-OS, galacto-OS, lactosucrose, trehalose); control the serum lipid concentration (galacto-OS); activate the immune system (chito-OS,nigero-OS); and show antitumour activity (agaro OS).
- In view of the harmful effects of sucrose, glucose and other sweeteners used in food, confectionery and pharma, a new-generation product called 'sugar alcohols' have come into the limelight. Sugar alcohols, or polyols, are sugar replacers. Recent technical advances have added to the range of sugar alcohols available for food use and expanded the applications of these sugar replacers in diet and health-oriented foods. They have been found to be useful in sugar-free and reduced-sugar products, in foods intended for diabetics, and, most recently, in new low-carbohydrate diets. The main sugar alcohols having good potential include: (1) Mannitol, having a higher molecular symmetry used in certain food and non-food applications as a non-hygroscopic nutritive sweetener, starter for polyether-polyols and in the production of PU foams; (2) Erythritol is an interesting polyol with two primary and two secondary hydroxyl groups, used as non-toxic, non-cariogenic and non-caloric dietary sweetener, and its various derivatives are used in protein chemistry, hair curling, treatment of coronary disease, and as special solvents.
- Starch hydrolysates improve moisture stabilization of resin and prevent oozing of formaldehyde in hot press and show better burning behaviour in urea–formaldehyde resins.
- Maltodextrin as substitute for organic carrier liquids in ceramic moulding and maltodextrin cross-linked with glyoxal or gluteraldehyde prevents crack formation during heating.
- Dialdehyde starches with hydroxyl, amino or sulphydryl groups produce bioplastics of good tensile strength and elongation, act as scavengers for urea and ammonia as well as in enzyme immobilization, e.g. urease.
- Starch as xanthide or xanthate salts into natural rubber has a superior vulcanizing effect useful in shoe soles and tyre tread. Further cyanoethylation of xanthide of starch improved the life of the product.
- Ethylene when directly polymerized in the presence of starch, facilitated by alumoxan and a transition metal complex, gives high polythene–starch composites with up to 95% starch content that has biomedical applications.
- Starch delays the setting time of cement which may be advantageous when more time is required for casting. It adheres strongly to cement grains and hence leads to a higher viscosity and less slump compared with corn starch. This improves the durability of concrete. Starch is used in weight aggregate in roof tiles, insulation, flooring and sound proofing, water resistant backing for shower stalls and as a replacement for gypsum.

The suggested strategy for individual TTCs is outlined below.

11.2 Cassava

World cassava use is expected to reach up to around 290 million tonnes in the near future (Westby, 2002). The cheap, large, and sustainable supply of cassava starch provides a potential basis for diverse applications. The various structural and functional properties of this starch in the majority of reports include (Zhu, 2015b):

1. The amylose content ranges from 0% to 30.3%.
2. The starch granules are oval, truncated and rounded with a size of 2–32 mm; the size distribution is either unimodal or bimodal.
3. A- or C_a-type polymorph.
4. Clusters of amylopectin have a size of DP 58–78 and 8.2–11.1 chains and 7.3–8.9 blocks per cluster.
5. Great diversity in swelling and solubilization, gelatinization and rheological properties, retrogradation, and enzyme susceptibility of starch exists; this diversity is achieved through natural selection, genetic modifications of starch biosynthetic enzymes, manipulations of environments and agronomic practice, and varying processing conditions.

Research on cassava starch is being carried out in a number of laboratories all over the world. Further research on high- and low-amylose starches for various applications to make them available in large quantities and economically feasible is needed. The production of modified starches in bulk needs to be standardized. Modification by physical methods as a substitute for chemical modifications deserves further efforts. Bioenergy and applications in biopharma and bioengineering are potential arenas for cassava starch. The construction industry, especially concrete, can absorb large quantities of cassava starch. Standardization of quality parameters for food in industrial applications needs urgent attention.

11.3 Sweet Potato

The potential supply of sweet potato starch is large and cheap. The observed trend from literature during the recent years showed sweet potato starch is gaining increased research focus. The major advances in our knowledge of sweet potato starch include: (1) Genetic resources from various biogeographic regions were assessed and genetic modifications achieved, providing starches with novel structures and properties; (2) The structural basis of these novel physicochemical properties have been largely evaluated on the molecular and granular levels, especially towards a better understanding of internal amylopectin structure–physical properties relations; (3) The cluster structure of amylopectin has been modelled; (4) The physicochemical properties of native starch were further diversified by chemical/physical/enzymatic modifications; (5) Possible uses of starch in noodle making, film and thermoplastic, in medical and pharmarelated applications and as substrate for functional ingredients were explored in various laboratories all over the world (Zhu and Wang, 2014).

To further improve the possible utilization of sweet potato starch, efforts are needed in the following areas: (1) More genetic resources could be assessed to discover genotypes with novel properties; (2) Various means of genetic modifications and environmental alterations could be employed to generate genotypes with greater diversity in physicochemical properties (e.g. high amylose); (3) Nutritional characterization and production of slowly digestible and resistant starch fractions; (4) The genetic basis of the heterogeneity of the shape of starch granule; (5) Visualization of blocked unit and superhelical structure of the granule, and as a long-term goal, their functions in determining physicochemical properties of starch; (6) More types of chemical/physical/enzymatic modifications conducted to further vary the existing pool of functional properties; (7) Product development approach to develop foods with sweet potato starch as functional ingredients; (8) Potential applications of sweet potato starch in non-food industry, and a better definition of structure properties application relationship; (9) Starch from orange flesh and anthocyanin containing sweet potato varieties for use in noodle, infant food and extruded products.

11.4 Aroids

There is a great diversity in the structure and properties of aroid starches from different species, and between genotypes of the same species. There appears to be a lack of common features of starches from diverse species, and this may be due to limited genotypes tested and lack of comparative studies. Some of the results reported appear spurious or suspicious, which is likely due to the misidentification of correct species (for example, confusion with yams).

In order to better understand and utilize aroid starches, the following points should be considered:

1. Scientific names must be strictly and correctly included in any report. In order to avoid any confusion, the common names for *A. campanulata*, *C. esculenta*, and *X. sagittifolium* are suggested to be elephant foot yam, taro, and coco yam, respectively, for scientific reports. Local names should also be mentioned in scientific literature to help the ecotype variation research and sample collection in the field. Food scientists and chemists should work closely with plant biologists for correct identification of the species.
2. Genetic resources for starch quality should be extensively evaluated in a systematic way (Lebot *et al.*, 2004). Most of the studies were sporadic and focused on only one or a limited number of genotypes. Starches from some aroid species such as *C. merkusii* and *A. macrorrhiza* (Manner, 2011) are little known. Waxy and high-amylose (>40%) genotypes of the aroids can be developed through genetic means. How diverse agronomic practices influence the structures and properties of aroid starches can be studied as a way to diversify the variations in the properties and structures.

The diversity in the composition, properties and structures of aroid starches suggests a range of applications. For example, varieties with low amylose content could be desirable for their good storage and freeze–thaw stability. Varieties with high amylose content could be used for glycaemic profile management. Resistant and slowly digestible starches can be obtained from high amylose varieties for the regions where the high occurrence of chronic diseases such as diabetes and obesity exist and aroids are part of the staple food (e.g. in some Pacific Islands). High amylose varieties could also be used for the formation of biodegradable and edible films. Small granule starches (e.g. *C. esculenta*) are favoured for smooth and creamy texture in food applications, but may give technical problems in their production. Small-granule starches may also be suitable as fillers in biocomposites and as ingredients for cosmetics. The structural basis for physicochemical properties and applications of aroid starches should be better explored. More diverse modifications can be conducted to obtain a wider range of properties of starch, which can be further applied to food and industrial uses.

Taro maltodextrins could be a replacement for the commercial maltodextrins from maize starch with similar water- and oil-absorption properties (Antonio-Estrada *et al.*, 2009). Taro starch with comparable binding and disintegrating properties to potato starch specified by British, US and Ghana authorities can be used for some soluble and insoluble drug formulations (Nasipuri, 1979).

Starch spherical aggregates produced by spray-drying and taro starch suspension had a protein content of 4.5% and their pasting, thermal, and solubility suggested possible application for encapsulation (Gonzalez-Soto *et al.*, 2011). Up to 2% incorporation of starch from taro increased the firmness and consistency of ketchup without affecting the colour (Sit *et al.*, 2014). Incorporation of *C. esculenta* starch nanoparticles enhanced the mechanical properties and thermal stability of starch film (Dai *et al.*, 2015).

Waxy and high-amylose (>40%) genotypes of the aroids need to be developed through genetic means; diverse agronomic-environmental practices influence the structures and properties of aroid starches.

11.5 Yams

There is a great diversity in composition, morphology, structure and properties of starches within the same species and among diverse species. The major findings from various studies can be summarized as follows (Zhu, 2015a): (1) Starch isolation methods are very sensitive to species/variety as some of them can be rich in non-starch polysaccharides and proteins; (2) Amylose content of ranges from 0% to 40%;

D. dumetorum and *D. esculenta* starches have lower amylose content than other species; (3) Polymorphism of yam starches includes A-, B- or C-types, depending on specific species; (4) The granule size of yam starches ranges from 1 to 90 µm and *D. dumetorum* and *D. esculenta* have the smallest with a range of 1 to 10µm; (5) Starch diversity can be manipulated by agronomic practices such as harvesting time and post-harvest storage conditions.

There are numerous opportunities to further explore and better utilize yam starches, which include the following: (1) Standardization of methods and nomenclatures should be established on a global scale since methods, instrumentations and nomenclature can influence starch characterization; (2) Scaling up of starch-extraction methods should be conducted; (3) Genotypes with no and high amylose (>40%) content from various species can be developed through breeding and by genetic means. Starches from lesser-known yam species such as *D. nummularia*, *D. transversa* and *D. pentaphylla* need to be studied; (4) Novel types of modifications such as ultrasound treatment, γ-irradiation, high-pressure treatment, and their dual/triple combined modifications should be carried out on more starch genotypes with a greater diversity in structure. (5) Role of yam starch in applications such as bread- and pasta-making and film formation need to be better elucidated.

11.6 Minor Starches

There is a lot of scope for further research on the minor tuber starches. However the difficulty in extraction of starch from these crops and scant knowledge about their properties has restricted their applications. Use of biotechnological techniques to modify the starch properties appears to be absent. Modification of the starches by physical techniques is relatively unknown. Research should be focussed to improve the extraction efficiency of the starch from the minor tubers, to make them economically feasible. They have some special characteristics, which need to be exploited in food and industry; typical examples are high amylose content of *Canna edulis* starch, extremely elevated levels of amylopectin in Arracacha starch (*Arracacia xanthorrhiza*) and wide variation in granular size in many other starches. Thus these starches offer good scope for use in place of chemically modified starch and offer a green alternative. Arrowroot starches have not received the attention they deserve in production, processing and utilization.

References

Antonio-Estrada, C., Bello-Perez, L.A., Martínez-Sanchez, C.E., Montañez-Soto, J.L., Jimenez-Hernandez, J. *et al.* (2009) Enzymatic production of maltodextrins from taro (*Colocasia esculenta*) starch. *CyTA-Journal of Food* 7(3), 233–241.

Dai, L., Qiu, C., Xiong, L. and Sun, Q. (2015) Characterisation of corn starch-based films reinforced with taro starch nanoparticles. *Food Chemistry* 174, 82–88. DOI: 10.1016/j.foodchem.2014.11.005.

Gonzalez-Soto, R.A., de la Vega, B., García-Suarez, F.J., Agama-Acevedo, E. and Bello-Pérez, L.A. (2011) Preparation of spherical aggregates of taro starch granules. *LWT - Food Science and Technology* 44(10), 2064–2069. DOI: 10.1016/j.lwt.2011.06.018.

Lebot, V., Prana, M.S., Kreike, N., van Heck, H., Pardales, J. *et al.* (2004) Characterisation of taro (*Colocasia esculenta* (L.) Schott) genetic resources in Southeast Asia and Oceania. *Genetic Resources and Crop Evolution* 51(4), 381–392. DOI: 10.1023/B:GRES.0000023453.30948.4d.

Manner, H.I. (2011) Farm and forestry production and marketing profile for giant swamp taro (*Cyrtosperma merkusii*). In: Elevitch, C.R. (ed.) *Speciality Crops for Pacific Islands Agroforestry*. Permanent Agriculture Resources (PAR), Hawaii, Holualoa.

Nasipuri, R.N. (1979) Evaluation of cocoyam starch as tablet binder and disintegrant. *Pharmaceutica Acta Helvetiae* 54(2), 48–53.

Nayar, N.M. (2014) The contribution of tropical tuber crops towards food security. *Journal of Root Crops* 40, 1–12.

Sit, N., Misra, S., Baruah, D., Badwaik, L.S. and Deka, S.C. (2014) Physicochemical properties of taro and maize starch and their effect on texture, colour and sensory quality of tomato ketchup. *Starch/Stärke* 66(3–4), 294–302. DOI: 10.1002/star.201300120.

Westby, A. (2002) Cassava utilization, storage and small-scale processing. In: Hillocks, R.J., Thresh, J.M. and Bellotti, A. (eds) *Cassava: Biology, Production and Utilization*. CAB International, Wallingford, UK, pp. 281–300.

Zhu, F. and Wang, S. (2014) Physicochemical properties, molecular structure, and uses of sweet potato starch. *Trends in Food Science & Technology* 36(2), 68–78. DOI: 10.1016/j.tifs.2014.01.008.

Zhu, F. (2015a) Isolation, composition, structure, properties, modifications, and uses of yam starch. *Comprehensive Reviews in Food Science and Food Safety* 14(4), 357–386. DOI: 10.1111/1541-4337.12134.

Zhu, F. (2015b) Composition, structure, physicochemical properties, and modifications of cassava starch. *Carbohydrate Polymers* 122, 456–480. DOI: 10.1016/j.carbpol.2014.10.063.

Appendix

Appendix: Starch-based Patents

In this chapter, patents that deal with starch-based products are listed. Most of these have been generated during the last decade. Although they do not specify the source of the starch, tropical tuber starches can be assumed to be used in all these processes.

No.	Patents	Author	Patent number	Application
1	Pre-gelatinized starches	Bayer Crop Science GmbH, Frankfurt, Germany	WO 03/054024 A1 (2003)	food items, especially instant puddings, feedstuff, adhesives and paints
2	Depilatory paint thickener comprising an amylopectin starch	Cooperatieve Verkoop- en Productievereniging van Aardappelmeel en Derivaten Avebe B.A., Veendam, Netherlands	US 6,589,293 B1 (2003)	depilatory paint thickener comprising a starch containing essentially only amylopectin molecules
3	Uses of flaked cationic potato starch	Economy Mud Products Company, Houston, Texas	US 2003/0129210 A1 (2003)	personal care and agricultural drift control, or as an adhesive, a flocculent or a coagulant
4	Extrusion of starch	Cooperatieve Verkoop- en Productievereniging van Aardappelmeel en Derivaten Avebe B.A., Veendam, Netherlands	US 2003/0141637 A1 (2003)	Extrusion of starch
5	The Procter & Gamble Company, Cincinnati, Ohio		US 2003/0146537 A1 (2003)	Process for making non-thermoplastic starch fibres

No.	Patents	Author	Patent number	Application
6	Solid, orally dispersible pharmaceutical formulation	Roquette Freres, Lestrem, France	WO 03/061623 A1 (2003)	solid, orally dispersible pharmaceutical formulations.
7		Cerestar Holding B.V., Sas van Gent, Netherlands	WO 03/061404 A1 (2003)	Starch n-alkenyl succinate as texturizing agent for UHT-treated products.
8	Non-thermoplastic starch fibres and starch composition and process for making same	The Procter & Gamble Company, Cincinnati, Ohio	WO 03/066942 A1 (2003)	Non-thermoplastic starch fibres having no melting point and having apparent peak wet tensile stress greater than about 0.2 megapascals (MPa)
9	Novel starch derivatives and a method for their preparation	Valtion Teknillinen Tutkimuskeskus, Finland	WO 03/068823 A1 (2003)	used as adhesives, hot-melt adhesives, preparation of polymers
10	Novel starch-based adhesives	Valtion Teknillinen Tutkimuskeskus, Finland	WO 03/068884 A1 (2003)	The invention relates to an adhesive formulation that comprises a starch-based adhesive dissolved in water
11	Modified starch from plants	Bayer CropScience GmbH, Frankfurt, Germany	US 6,610,843 B1 (2003)	transgenic plant cells and plants which, due to a reduced activity of a disproportioning enzyme (D enzyme), synthesize a modified starch
12	Starch derivatives	Supramol Parenteral Colloids, Rosbach-Rodhein, Germany	WO 03/070772 A1 (2003)	starch active substance conjugates, pharmaceutical applications
13	Conjugates of hydroxyalkyl starch and a protein	Fresenius Kabi Deutschland GmbH, Bad Homburg, Germany	WO 2005/092390 A2 (2005)	method of producing these conjugates and the use of these conjugates
14	Conjugates of hydroxyethyl starch and erythropoietin	Fresenius Kabi Deutschland GmbH, Bad Homburg, Germany	WO 2005/092369A2 (2005)	conjugates of hydroxyethyl starch and erythropoietin, (conjugates comprise a linking compound)
15	Use of pre-cooked cereal and tubular starch in high-protein food products	Massoud Kazemzadeh, Bloomington, Minnesota	US 2005/0220980A1	Food applications. high-protein snacks
16	Method for the treatment of starch	Hannu Ketola, Hameenlinna, Finland	US 2005/0229925 A1	The modified starch used for surface sizing of paper

No.	Patents	Author	Patent number	Application
17	Encapsulation of oxygen-sensitive agents	National Starch and Chemical Company, Bridgewater, New Jersey	US 2005/0233002A1	encapsulated materials are useful in food product industry
18	Amylose starch products as sizing agents for textile yarns	BASF Plant Science GmbH, Ludwigshafen, Germany	WO 2005/098121 A1	sizing agent for sizing natural and synthetic yarns (textile industry)
19	Highly breathable, biodegradable film bag	Novamont SpA, Novara, Italy	WO 2005/097875 A1	thermoplastic polymer insoluble in water
20	Method of improving properties of starch containing food and property improving agent	Ajinomoto Co., Inc., Tokyo	WO 2005/096839 A1	starch containing food and property improving agent
21	Starch compositions and methods of making starch compositions	David J.Neivandt, Bangor, Maine Joseph M. Genco, Orono, Maine Mark A. Paradis, Old Town, Maine	US 2005/0236127 A1	forming an additive for a paper industry
22	A new generation of starchy products	Cerestar Holding B.V., Sas van Gent, Netherlands	WO 2005/100407A1	food products, feed products, cosmetics, pharma products, agrochemicals and industrial applications
23	Pharmaceutical compositions comprising an amphiphilic starch	Vectura Limited, Chippenham, UK	US 2005/099674 A1	pharmaceutical applications
24	Emulsifier	National Starch and Chemical Investment Holding Corporation, Delaware	EP 1 584 370 A2 (2005)	cosmetic applications
25	Nucleic acids from cassava encoding starch branching enzyme II (SBEII) and their use	National Starch and Chemical Investment Holding Corporation, New Castle, Delaware	US 6,956,148 B1 (2005)	methods of using the nucleic acids to alter gene expression in cassava to obtain starch
26	Hydroxyalkyl starch derivatives	Norbert Zander, Meine, Harald Conradt, Braunschweig, Wolfram Eichner, Butzbach, Germany	US 2005/0234230 A1	pharmaceutical composition derivative
27	pharmaceutical composition comprising said hydroxyalkyl starch derivative		US 2005/0233048 A1	Food and pharmaceutical applications
28	Starch hydrolysis products for size enlargement	National Starch and Chemical Company, Bridgewater, New Jersey	US 2007/0054027A1	pharmaceuticals, detergents and agricultural products

No.	Patents	Author	Patent number	Application
29	Translucent food	Cindy Semeijn, Groningen, Zhenghong Chen, Foxhol, Henk Arie Schols, Wageningen, Pieter Lykle Buwalda, Groningen, Netherlands	US 2007/0059403A1	translucent food products
30	Plants which synthesize a modified starch, process for the production thereof and modified starch	Bayer BioScience GmbH, Potsdam, Germany	US 2007/0067874A1	production of transgenic plant cells with modified starch and viscosity properties
31	A water-soluble biodegradable material	Li Xiaolu, Beijing	WO 2007/033595A1	has excellent simple beam impact strength, tensile yield strength, biodegradability
32	Biodegradable compositions based on nanoparticulate starch	Novamont SpA, Novara, Italy	WO 2008037749 (2008)	biodegradable multiphase compositions comprising a continuous phase composed of a matrix
33	Biodegradable multiphase compositions based on starch	Novamont SpA, Novara, Italy	WO 2008037744 (2008)	the present invention relates to biodegradable multiphase compositions characterized by comprising three phases
34	Branched starch derivative, process for production thereof, and moulded article comprising the branched starch derivative	Hayashibara Biochem Laboratory, Okayama, Japan	WO 2008044588	starch derivative having an anti-ageing property
35	Granules for use in formulations of personal care and/or homecare products and products containing said granules	Bentonit Uniao Nordeste SA, Sao Paulo, Brazil	WO 2008043163 (2008)	homecare products, visual agents, delivery systems, exfoliating and/or abrasive agents
36	Method of producing a wood-based product	Akzo Nobel Coating International B.V., Arnhem, Netherlands	US 2010043970 (2010)	wood-based product, water-based adhesive
37	Biodegradable material for injection moulding and articles obtained therewith	Novamont SpA, Novara, Italy	US 2010016470 (2010)	starch-based biodegradable material (healthcare)

No.	Patents	Author	Patent number	Application
38	Biodegradable material based on opened starch	Bioneedle Technologies Group B.V., Eindhoven, Netherlands	US 2010015185 (2010)	biodegradable material (healthcare)
39	Multicomponent starch binder compositions	S.D. Warren Company, Boston, Massachusetts	WO 2010008751 (2010)	plasticizer for the starch, material for use as binders for coatings for paper and other surfaces
40	Processes for producing fermentation products	Novozymes A/S, Bagsvaerd, Denmark (for all countries except USA)	WO 2010008841 (2010)	producing a fermentation product for food and bevarage industry
41	Polysaccharide and dimethicone copolyol as emulsifier for cosmetic compositions	National Starch & Chemical Co. Investment Holding Corporation, Wilmington, Delaware	US 6277893 B1 (2001)	cosmetic compositions
42	Biodegradable polymeric composition	Novamont SpA, Milan, Italy	US 6277899 B1 (2001)	synthetic biodegardeable thermoplastic polymers
43	Sheet and product based on foamed shaped starch	Novamont SpA, Novara, Italy	US 2001/0014388	materials in the form of foam sheet comprising destructured or complexed starch expanded as a continuous phase
44	Lubricant compositions and methods	Lee County Mosquito Control District, USA	US 2001/0014711	manufacturing a lubricant comprising a superabsorbent polymer
45	Method for manufacturing high-cationic starch solutions	Antti Likitalo, Turku, Finland; Jouko Kaki, Turku, Finland	US 2001/0014736	method for producing starch solutions of high cation equivalent value with a DS of 0.1–1.0 is provided
46	Method and apparatus for producing a pre-gelled starch product and normally sticky extrudates with minimal or no surfactant	Gordon R. Huber, Sabetha, Kansas; Bradley S. Strahm, Sabetha, Kansas; Brian S. Plattner, Sabetha, Kansas; Douglas D. Renyer, Wetmore, Kansas; Robert D. Sunderland, Sabetha, Kansas	US 2001/0015138	extrusion of various starting materials (e.g. starch-bearing grains or starch fractions thereof, proteinaceous materials and/or nutraceuticals) to yield improved products, especially cold-water swelling or soluble starch products
47	Electro-rheological fluid comprising dried water-soluble starch as a conductive particle	Agency for Defense Development, Daejon, Korea	US 2001/0015421	an automatic and aerospace industry
48	Method of manufacturing a maltose-rich syrup	Roquette Freres, Lestrem, France	US 6284498 B1 (2001)	pharmaceutical industry

No.	Patents	Author	Patent number	Application
49	Melt-processable starch compositions	Humatro Corporation, New York	EP 1132 427 A1 (2001)	the present invention relates to starch compositions that contain starch and additives
50	Starch-based core products for use in rolled-paper goods	National Starch and Chemical Company Investment Holding Corporation, Wilmington, Delaware	US 6284359 B1 (2001)	biodegradable product useful in winding or rolling and storage of paper goods
51	Cross-linked high amylose starch resistant to amylase as a matrix for the slow release of biologically active compounds	Vincent Lenaerts, Westmount, Canada; Francois Chouinard, Lorraine, Canada; Mircea Alexandru Mateescu, Pompilia Ispas-Szabo, Montreal, Canada	US 6284273 B1 (2001)	solid slow-release oral pharmaceutical products
52	Device for tissue engineering a bone equivalent	IsoTis N.V., Bilthoven, Netherlands	US 2001/0021530	device for tissue engineering a bone equivalent comprising a scaffold material
53	Paper-coating composition comprising gellan gum/starch blend	Penford Corporation, Bellevue, Washington; Pharmacia Corporation, St. Louis, Missouri	US 6290814 B1 (2001)	an improved method for sizing paper which comprises the step of coating paper with a composition comprising gellan gum and a derivatized starch
54	Rubber containing starch reinforcement and tyre having component thereof	The Goodyear Tire & Rubber Company, Akron, Ohio	US 2001/0025068 (2001)	rubber composition containing a combination of starch, modified starch and/or starch/plasticizer
55	Cold-water soluble starch aldehydes and the method of preparation thereof	National Starch and Chemical Company, Bridgewater, New Jersey	US 2001/0025102	stable, cold-water soluble, ready-for-use starch aldehyde compositions and the method of preparation of such compositions
56	Liquid laundry starch concentrate	Henkel Ecolab GmbH & Co, OHG, Düsseldorf, Germany	US 6,444,023 B1 (2002)	process for preserving a liquid laundry starch concentrate containing starch or starch derivatives
57	Processed blend useful as culinary thickeners comprising a modified starch and a flour	National Starch and Chemical Investment Holding Corporation, New Castle, Delaware	EP 1 241 216 A1 (2002)	cooking purposes; food industry

No.	Patents	Author	Patent number	Application
58	Protein/starch paper-coating compositions and method of use	Midwest Grain Products, Acheson, Kansas	US 2002/0121222	improved compositions for coating paper stock
59	Hydrophobically esterified starch products and process of making the same	Hercules Incorporated, Wilmington, Delaware	US 2002/0123624	The process can be enzymatic where an enzyme is used as a catalyst for the reaction. The process can also be chemical where no enzyme is used.
60	Time-saving method for preparing tapioca starch balls and the product thereof	Patrick Po-Young Ling, Aguora Hills, California; Janie Huoy-Jen Lai, Aguora Hills, California	US 2002/0127311	beverage industry, a method of preparing tapioca starch balls
61	Biodegradable packaging laminate, a method of producing the packaging laminate, and packaging containers produced from the packaging laminate	Mikael Berlin, Lund, Sweden; Karin Henningsson, Molndal, Sweden	US 2002/0127358	packaging biodegradable laminate for liquid foods having excellent liquid and oxygen gas barrier properties; food industry
62	Treatment composition based on pre-gelatinized starch and process for applying it to a metal especially steel, surface	Usinor, Puteaux, France	US 6,447,590 B1 (2002)	Treating a metal surface with the composition provides temporary corrosion protection and prelubrication
63	Dairy Products Supply, Emmen, Netherlands; methods for preparing and using B-starch	Pfeifer & Langen, Cologne, Germany		vegetable carbohydrates in animal feeds, food and dairy products
64	Enzymatically modified hydrophobic starch	Grain Processing Corporation, Muscatine, Indiana	US 2002/0132309	method for absorbing fluid comprises applying a fluid-absorbing effective amount of the dried, ground granules
65	Starch compositions and methods for use in papermaking	Cargill, Inc., Minneapolis, Minnesota	US 6,451,170 B1 (2002)	methods of using those starches in papermaking are disclosed
66	Oxidation of starch	Cooperative Verkoop-en Productievereniging van Aardappelmeel en Derivaten Avebe B.A., Veendam, Netherlands	US 6,777,548 B1 (2004)	invention relates to a process for the oxidation of starch; root or tuber starch comprising at least 95 wt% of amylopectin, based on dry substance of the starch

No.	Patents	Author	Patent number	Application
67	Curable starch composition, modified starch, preparation method and articles	Kansai Paint Co., Ltd, Amagasaki-shi, Hyogo-ken, Japan	EP 1447415 A1 (2004)	pharmaceutical industry
68	Starch-loaded polyvinyl alcohol copolymer film for packaging non-liquid product and method for making the same	Andrew P. Verall, Crown Point, Indiana; P. Scott Bening, Crown Point, Indiana; Karen A. Kugler, Porter, Indiana	US 2004/0161559A1	water-soluble film for packaging a non-liquid product; package employing the film
69	Method for preparing a gellable starch product	Coöperatieve Verkoop – en Productievereniging van Aardappelmeel en Derivaten Avebe B.A., Veendam, Netherlands	WO 2004/069877	food applications
70	Enzymatic treatment of starchy food products for shortening the tempering step	Novozymes North America, Inc., New York	US 2004/0166201A1	food applications
71	Additive for rubber elastomers	Don-B Corporation, Vermilion, Ohio	WO 2004/072165	invention directed to an additive/process increased tensile strength, abrasion resistance, elasticity, and tear strength
72	Bicomponent fibres comprising a thermoplastic polymer surrounding a starch-rich core	The Procter & Gable Company, Cincinnati, Ohio	US 6,783,854 B2 (2004)	bicomponent fibre comprising one thermoplastic polymer component and a plasticizer
73	Cationic starch graft copolymers and novel process for the preparation of cationic starch graft copolymers	Bernd Hauschel, Leverkusen, Germany; Bernd Thiele, Odenthal, Germany; Thomas Hubbe, Köln, Germany; Bernhard Jansen, Köln, Germany	US 2004/0170596	cationic starch graft copolymer
74	Melt-processable starch compositions	Bailey, Valerie Ann Florence, Kentucky; Mackey, Larry Neil, Fairfield, Ohio; Trokhan, Paul Dennis, Hamilton, Ohio	WO 2004/074367	thermoplastic equipment; the composition is especiallly suitable for uniaxial and biaxial extensional processes to make fibres, films, foams and the like
75	Emulsification of alkenyl succinic anhydride size	Georgia-Pacific Resins, Inc., Atlanta, Georgia	US 6,787,574 B1 (2004)	grafted cationic acrylamide starch co-polymer as an emulsification agent

No.	Patents	Author	Patent number	Application
76	Starch compositions and methods of making starch compositions	University of Maine Board of Trustees, Bangor, Maine	WO 2004/076551	combining the cooked starch with a polymer, the polymer containing anionic groups or potential anionic groups
77	Method for manufacturing a shape body containing a starch, a homogenized mass containing starch and a device for manufacturing a soft capsule	Peter Greither, Kirchberg, Switzerland	US 6,790,495 B1 (2004)	invention relates to a method for manufacturing a starch-containing shape body and a homogenized, starch-containing mass
78	Starch-based polymers	Dow Global Technologies, Inc., Midland, Michigan	WO 2004/078807	useful as a surface sizing agent for paper, paperboard or cardboard
79	Enhanced amylose production in plants	BASF Plant Science GmbH, Ludwigshafen, Germany	WO 2004/078983A2	methods for increasing the amylose content in plants, preferably in potato plants, by expressing a starch biosynthesis-enhancing protein
80	Method of producing starch having high less- digestible starch content	J-Oil Mills, Inc., Tokyo, Japan; Osaka Prefecture University Public Corp., Osaka, Japan	EP2138514 (A1)	It is intended to provide a method of easily and economically producing starch an increased content of less digestible starch.
81	Oral disintegrating tablet having masked bitter taste and method for production thereof	Kissei Parmaceutical Co., Matusmoto-shi, Japan	US 2009311321 (A1)	The present invention relates to a method of producing solubility-reduced, beta-glucan-containing products and uses thereof as raw materials in aqueous fractionation to produce starch and beta-glucan concentrates.
82	A method for liquefying and saccharifying starch quickly	Hunan Agricultural University, China	WO 2009146596 (A1)	method for liquefying and saccharifying starch quickly
83	Methods for producing super-absorbent polymers for use in agricultural applications	Absorbent Technologies, Inc., Beaverton, Oregon	US 2009304835 (A1)	Methods and systems for producing super-absorbent polymer particles for use in agricultural applications

No.	Patents	Author	Patent number	Application
84	Production of sustained release starch product	Tate and Lyle Ingredients Americas, Inc., Decatur, Illinois	WO 2009149136 (A2)	process for producing a sustained release starch product comprising (1) combining and mixing starch, hydrocolloid, and water to form a starch material; (2) drying the starch material to form a starch product
85	Use of high osmotic liquid composition in manufacturing medicament for enhancing wound healing	Chaoying Zhao, Shanghai, China	EP 2127659 (A1)	promoting wound healing and medicines
86	Desizing process	Novozymes North America, Inc., New York	US 2009317893 (A1)	manufacture of a fabric
87	Semi-moist oral delivery system	Ez-Med Company, Lake Charles, Louisiana	US RE41108 (E1)	The subject invention is a carrier or product formed of a matrix having starch, sugar, fat, polyhydric alcohol and water.
88	Microencapsulate and process for the manufacture	Feyecon Development & Implementation B.V., Weesp, Netherlands; Friesland Brands B.V., Meppel, Netherlands	WO 2010014011 (A1)	foodstuffs, beverages, nutritional compositions or pharmaceutical applications
89	Non-sticky, free-flowing comestible and a process for its preparation	Cerestar Holding, B.V., La Sans von Gent, Netherlands	US 2010021590 (A1)	product may be used as a food ingredient, for instance in bakery products, cereal products, ice cream and dairy products
90	Emulsifier	H.J. De Vries, Sellingen, Netherlands; C. Semeijn, Groningen, Netherlands; P. L. Buwalda, Mondriaanstraat, Netherlands	US 2010029928 (A1)	invention relates to an emulsifier, a method for preparing aid emulsifier, and to its use in various applications, primarily food and cosmetics
91	Pharmaceutical pellets comprising modified starch and therapeutic applications	Henkel Corp., Bay Point, California; J. P. Remon, Belgium; C. Vervaet, Belgium; P. Foreman, USA	WO 2010048389 (A2)	vaginal use compositions (pharmaceutical applications)
92	Chewable soft gel capsules	Viva Pharmaceuticals, Inc., Richmond, California	US 2010092548 (A1)	medicines, pharmaceutical compositions, nutraceuticals, vitamins, nutritional supplements

No.	Patents	Author	Patent number	Application
93	Immuno-stimulating polysaccharides isolated from *Curcuma xanthorrhiza* and manufacturing method thereof	J.-K. Hwang, Gyeonggi-do, Korea; A.-J. Kim, Seoul; J.-H. Sohn, Seoul; K.-L. Han, Seoul; S.-H. Lee, Gyeonggi-do, Korea; J.-H. Choo, Gyeonggi-do, Korea	US 2010048885 (A1)	drug and immunological applications
94	Hydroxyalkyl starch derivatives	Fresenius Kabi Deutschland GmbH, Bad Homburg, Germany	EP 2154160 (A1)	pharmaceutical applications
95	Injection moulding material containing starch and plant protein	Kimberly-Clark Worldwide, Inc., Neenah, Wisconsin	US 2010159170 (A1)	pharmaceutical applications
96	Antiperspirant or deodorant compositions	Unilever N.V., Rotterdam, Netherlands	WO 2010049235 (A1)	Cosmetics and pharmaceutical applications
97	Biodegradable and renewable film	Kimberly Clark Worldwide, Inc., Neenah, Wisconsin	US 2010159203 (A1)	biodegradable and renewable film formed from a thermoplastic composition that contains at least one starch and one plant protein
98	Thermoplastic starch formed from an enzymatically debranched starch	Kimberly Clark Worldwide Inc., Neenah, Wisconsin	US 2010159777 (A1)	thermoplastic starch for use in a melt-processed composition (e.g. fibre, non-woven web, etc.)
99	Hydroxypropylated starch as a processing aid to improve resistant starch total dietary fibr (tdf) retention in direct expansion extrusion applications	Tate & Lyle Ingredients Americas, Decatur, Illinois	WO 2010085630 (A1)	dietary applications
100	High-performance film-coating compositions	Ideal Cures Private Ltd, Mumbai, India	WO 2010052727 (A1)	coating of pharmaceutical tablets, nutritional supplements, food products, confectionery, agricultural seeds
101	Viscosity regulator and method for its production	B. Volkert, Berlin; A. Lehmann, Potsdam, Germany; H. Nerenz, Hoxter, Germany; A. Achrader, Bevern, Germany	US 2010158831 (A1)	cosmetic or pharmaceutical preparations

No.	Patents	Author	Patent number	Application
102	Colouring composition comprising starch derivatives as a hydrocolloid	C. Hansen, A/S, Hoersholm, Denmark	US 2010166875 (A1)	manufacture of food and pharmaceutical products
103	Milk replacer	Cargill, Inc., Wayzata, Minnesota	US 2010221386 (A1)	emulsifying starch has use as a milk replacer in the production of food products such as dairy products, bakery products, confectionery and convenience food products
104	Pet food	Kao Corp., Tokyo	US 2010209550	pet food industry
105	Resveratrol compositions	DSM Nutritional Products Ltd, Kaiseraugst, Switzerland	WO 2010089104 (A2)	food and beverage industry
106	Heat-moisture treated carbohydrates and uses	The UCAN Company, Woodbridge, Connecticut	WO 2010077635 (A2)	healthcare
107	Maltobionate as antioxidant in food products	Novozymes, A/S, Bagsvaerd, Denmark	US 2010173044 (A1)	food industry
108	Method for producing sturgeon and salmon caviar substitute	P.P. Pyvavarov, Ukraine; P. Y. Pyvavarov, Ukraine	WO 2010065007 (A1)	food industry
109	Cream substitute	P.L. Buwalda, Groningen, Netherlands; A. Sein, Leiden, Netherlands	US 2010143542 (A1)	food industry
110	High-performance film-coating compositions	Ideal Cures Private Ltd, Mumbai, India	WO 2010052727 (A1)	for use in coating of pharmaceutical tablets, nutritional supplements, food products, confectionery, agricultural seeds
111	Viscosity regulator, method for the production	B. Volkert, Berlin; A. Lehmann, Potsdam, Germany; H. Nerenz, Hoxter, Germany, A. Achrader, Bevern, Germany	US 2010158831 (A1)	cosmetic or pharmaceutical preparations
112	Colouring composition comprising starch derivatives as a hydrocolloid	C. Hansen, A/S, Hoersholm, Denmark	US 2010166875 (A1)	food and pharmaceutical products
113	Solid pharmaceutical composition comprising amlodipine and losartan	Hanmi Pharm. Co. Ltd, Gyeonggi-do, Korea	WO 2010085047 (A2)	pharmaceutical applications

No.	Patents	Author	Patent number	Application
114	Biodegradable and renewable film	Kimberly Clark Worldwide, Inc., Neenah, Wisconsin	US 2010159203 (A1)	biodegradable and renewable film that may be employed in a wide variety of applications
115	Thermoplastic starch formed from an enzymatically debranched starch	Kimberly Clark Worldwide, Inc., Neenah, Wisconsin	US 2010159777 (A1)	thermoplastic starch for use in a melt-processed composition (e.g. fibre, non-woven web etc.)
116	Hydroxypropylated starch as a processing aid to improve resistant starch total dietary fibr (tdf) retention in direct expansion extrusion applications	Tate & Lyle Ingredients Americas, Decatur, Illinois, USA	WO 2010085630 (A1)	retention in direct expansion extrusion applications
117	Use of starch dispersions as binder in coating compositions and process for preparing the starch dispersions	Ecosynthetix, Inc., Lansing, Michigan	EP2251484 (A1)	dispersion may be used as coating colour for paper
118	Fabrication method for porous carbon fibres	Korea Advanced Institute of Science and Technology, Daejeon, Korea	EP 2241658 (A1)	fabricated porous carbon fibre applicable in manufacturing electrodes for a super-capacitor, fuel cell
119	Method for producing a bio-based polymeric shoe component	Feng Chia University, Taichung, Taiwan; Pou Chien Technology Co., Ltd, Tachung City, Taiwan	US 2010237527	method for producing a bio-based polymeric shoe component includes: preparing a blend from a composition including 5–50 wt parts of a modified starch
120	Aqueous composition containing at least one soluble gelatinized anionic starch	Roquette Freres, Lestrem, France	US 2010243188 (A1)	water treatment or the production of cosmetic products
121	Granulated powder containing vegetable proteins and maltodextrins; method for producing same, and uses	Roquette Freres, Lestrem, France	WO 2010100369 (A1)	animal proteins in the preparation of food products

No.	Patents	Author	Patent number	Application
122	Artificial snow and method for making same	J. Quee, Vancouver; E. Quee, Vancouver	US 2011065841 (A1)	non-toxic, non-static, environmentally benign artificial snow product is made by extruding a mixture of starch, polyvinyl alcohol, hydrogenated soy flakes and talc to form an extruded starch
123	Method and device for producing electricity and conversion products such as ethanol	J.P.M. Sanders, Groningen, Netherlands; K.P.H. Meesters, Amersfoort, Netherlands; J.P.G. van der Broek, Tynaarlo, Netherlands	US 2011059498 (A1)	invention relates to a method for producing electricity and conversion products such as ethanol
124	Use of dispersions of cross-linked cationic starch in papermaking	Ecosynthetix, Inc., Lansing, Michigan	EP 2273008 (A2)	used in papermaking, can be used as a wet-end additive or in surface coating
125	Pulverous formulation of a fat-soluble active ingredient	DSM ASSETS B.V., Heerlen, Netherlands	US 2010260894 (A1)	food composition, especially beverage
126	Process for making a healthy snack food	Frito Lay Trading Company GmbH, Bern, Switzerland	US 2010266734 (A1)	used in the pharmaceutical and food industries
127	Coating composition and process for baked products to impart the attributcs of fried products	Board of Trustees of the University of Arkansas, Little Rock, Arkansas	US 2010272871 (A1)	food industry
128	System and method for preparing microwavable fried food products	Discovery Foods LLC., Hayward, California	US 2010229281 (A1)	food industry
129	Starchy foods containing anit-sticking pectin and/or high-acyl gellan gum	CP Kelco Aps, Lille Skensved, Denmark	WO 2010139646 (A1)	food product and anti-sticking pectin
130	Non-digestible hydroxypropyl starch hydrolysate, method for production thereof and food and beverage	Mathsutani Chemical Industry Co. Ltd, Titami-shi Hyogo, Japan	US 2010330257 (A1)	food products and beverage industry
131	Carotenoid compositions dispersed in a matrix comprising starch	DSMIP Assets B.V., Heerlen, Netherlands	EP 2272380 (A1)	additives for food, beverages, animal feeds, cosmetics or drugs
132	Microfoam and its manufacturing method	K.-N. Koo, Seoul; S. Kitamura, Kobe, Japan	US 2010227149 (A1)	Foamed materials and methods and systems of producing the same

No.	Patents	Author	Patent number	Application
133	Barley Cultivar BZ493-46e. Acidic gas- permeated carboxyalkyl starch particles; extrudates and process for making the same	Archer Daniels Midland Company, Decatur, Illinois	WO 2010096742 (A2)	The present disclosure relates to particles comprising carboxyalkyl starch that are permeated with an acidic gas, and their uses as absorbent materials.
134	Granulated seasoning salt and preparation method	CJ Cheiljedang Corp., Irvine, California	US 2010209573 (A1)	gelatinized tapioca starch solution as a binder and an anti-caking agent
135	Composition for producing starch foam resistant to moisture and freeze–thaw cycles	C.R. Bastos, Rio de Janeiro; P. Ponce, Sao Paolo, Brazil; L.G. Carr, Corning, Sao Paolo, Brazil; D.P. Fernandez, Sao Caetano do Sul, Brazil; L.A. Benevolo, Sao Paolo, Brazil	US 2010199884 (A1)	plastic materials for packaging have increased dramatically in the last two decades, replacing more traditional materials such as paper, glass and metals
136	Methods and materials for increasing starch biosynthesis in plants	University of Florida Research Foundation, Inc., Gainesville, Florida	US 2010199385 (A1)	The subject invention concerns materials and methods for providing plants or plant tissue with increased starch biosynthesis
137	Process for gelatinizing starch using a biodegradable polymer material bearing aldehyde groups	Nederlandse Organisatie voor Toegepastnatuur Wetenschappelijk Onderzoek TNO, Delft, Netherlands	US 2010285185 (A1)	invention further provides a granulate, a shaped starch product, a food product, or a blown starch film comprising such a carbohydrate polymer
138	Biodegradable packaging materials with enhanced oxygen barrier performance	Mead Westvaco Corp., Richmond, Virginia	EP 2263872 (A1)	biodegradable packaging material (100) havingexcellent oxygen barrier performance. The biodegradable structure may include homopolymer or co-polymer of lactic acid-based monomers; paper-based substrate
139	Biodegradable absorbent material and method of manufacture	W. Chambers, Lebanon, Oregon	WO 2010144575 (A1)	Biodegradable graft co-polymers derived from a carbohydrate. Product work-up typically involves optional neutralization and drying

No.	Patents	Author	Patent number	Application
140	Starch-based composites and process of manufacture	A. Netravali, Ithaca, New York	US 2010291822 (A1)	curing the lignocellulosic reinforcement material at a sufficient temperature and a sufficient pressure for a sufficient period of time to produce a thermoset biodegradable composition
141	Flexible thermoplastic films and articles	Kimberly-Clark Worldwide, Inc., Neenah, Wisconsin	WO 2010131134 (A2)	biodegradable, polyolefin-based material composition having incorporated therein thermoplastic starch particles
142	Pre-gelatinized starches as carrier materials for liquid components	Cargill, Inc., Wayzata, Minnesota	US 2010330369 (A1)	The present invention provides a liquid-loaded starch material comprising a solid carrier material consisting of pre-gelatinized, non-granular starch material which consists of flake-shaped starch particles
143	Starch-based fat replacer by crystallization of enzyme-modified starch and high-pressure shearing	Tate and Lyle Ingredients Americas Inc., Decatur, Illinois	WO 2010033475 (A2)	food industry
144	Dietary fibre of tapioca origin	Lyckeby Culinar AB, Fjalkinge, Sweden	WO 2010030221 (A1)	food industry
145	Food and beverage ingredient delivery systems	Novozymes, A/S, Bagsvaerd, Denmark	WO 2010029151 (A1)	foods and beverages industry
146	Wallboard and ceiling tiles and method of manufacturing same	M.A. Baig, Lindenhurst, Illinois	US 2009273113 (A1)	Method of producing a composite material suitable for the manufacture of gypsum wallboard and ceiling tile applications

Index

Page numbers in **bold** refer to figures and tables.